Synthese Library

Studies in Epistemology, Logic, Methodology, and Philosophy of Science

Volume 399

The aim of *Synthese Library* is to provide a forum for the best current work in the methodology and philosophy of science and in epistemology. A wide variety of different approaches have traditionally been represented in the Library, and every effort is made to maintain this variety, not for its own sake, but because we believe that there are many fruitful and illuminating approaches to the philosophy of science and related disciplines.

Special attention is paid to methodological studies which illustrate the interplay of empirical and philosophical viewpoints and to contributions to the formal (logical, set-theoretical, mathematical, information-theoretical, decision-theoretical, etc.) methodology of empirical sciences. Likewise, the applications of logical methods to epistemology as well as philosophically and methodologically relevant studies in logic are strongly encouraged. The emphasis on logic will be tempered by interest in the psychological, historical, and sociological aspects of science.

Besides monographs *Synthese Library* publishes thematically unified anthologies and edited volumes with a well-defined topical focus inside the aim and scope of the book series. The contributions in the volumes are expected to be focused and structurally organized in accordance with the central theme(s), and should be tied together by an extensive editorial introduction or set of introductions if the volume is divided into parts. An extensive bibliography and index are mandatory.

More information about this series at http://www.springer.com/series/6607

Theo A. F. Kuipers

Nomic Truth Approximation Revisited

Springer

Theo A. F. Kuipers
Department of Theoretical Philosophy
University of Groningen
Groningen, The Netherlands

Synthese Library
ISBN 978-3-319-98387-5 ISBN 978-3-319-98388-2 (eBook)
https://doi.org/10.1007/978-3-319-98388-2

Library of Congress Control Number: 2018952386

This Springer imprint is published by the registered company Springer Nature Switzerland AG.
The registered company address is: Gewerbestrasse 11, 6330 Cham, Switzerland

Contents

Chapter 1
Background and Overview

Abstract Nomic truth approximation and its relation to confirmation and empirical progress were the leading themes of my book *From instrumentalism to constructive realism* of 2000. The present book offers again a synthesis of articles. It concerns in the first place six chapters (2, 3, 4, 5, 6, and 7) dealing with a very broad generalization of the theory of nomic truth approximation, discovered in 2012. Moreover it presents adapted and otherwise revised versions of nine previously published articles (8–16) about extensions, related kinds of reasoning, including belief revision, and the resulting epistemological position. This introductory chapter starts with a global description of this background. Then it presents the main conclusions of *From instrumentalism to constructive realism*, which roughly remain valid. This is followed by an overview of all chapters and next a list with references to the origin of them. It concludes with some indications for new research in order to compare the presented approach with other approaches.

Keywords Background · Constructive realism · Extensions · *From Instrumentalism to Constructive Realism* · ICR · Generalization · Nomic truth approximation · Overview · Related kinds of reasoning

1.1 Background: Extensions, Related Kinds of Reasoning, and a Far-Reaching Generalization

Nomic truth approximation is the idea that one theory can be closer to, or more similar to, the truth about what is nomically, e.g. physically, chemically, biologically, possible than another theory. E.g. although Einstein's theory may well be false, there is much reason to assume that it is closer to the nomic truth than Newton's. Similarly, it may well be that Darwin's theory of the origin of species has false consequences, but there is much reason to assume that it is closer to the truth than the creationist theory. Another well-known example is the heliocentric cosmological theory of Copernicus replacing the geocentric theory of Ptolemy. It has always been my view that nomic truth approximation is the ultimate goal of theory oriented empirical science. Moreover, I assume a constructive realist point of view,

T. A. F. Kuipers, *Nomic Truth Approximation Revisited*, Synthese Library 399,
https://doi.org/10.1007/978-3-319-98388-2_1

i.e., the view that the actual and the nomic truth are unique products of the chosen language and the world.

Nomic truth approximation and its relation to confirmation and empirical progress were the leading themes of my book *From instrumentalism to constructive realism* (Kuipers 2000). The title summarized the conclusion that the instrumentalist methodology, rather than the falsificationist methodology, is functional for nomic truth approximation.

The book was in fact a synthesis of the most relevant papers that I had published on the topic from 1982 onwards. The next section provides a survey of the main conclusions of that book. After completing the 2000-book, I have published several new papers dealing with ideas that are more or less directly related to nomic truth approximation:

- the idea that nomic truth approximation is not only possible by revising theories about a given domain, but also by revising domains in view of a given theory,
- the idea that non-empirical, including aesthetic, factors, notably beauty, could be functional for nomic truth approximation,
- the idea that concluding to nomic truth approximation can be seen as a kind of abduction, leading to a challenge for computational modelling,
- the idea that the abductive nature also leads to a severe correction of the current notion of inference to the best explanation,
- the idea that nomic truth approximation in practice requires several pragmatic decisions,
- the idea that there must be some kind of probabilistic version of the HD-method, a 'Hypothetico-Probabilistic (HP-) method', which is, like the HD-method, functional for empirical progress and even truth approximation.
- the idea that there must be some relation between theory revision and so-called belief revision, even such that nomic truth approximation can be achieved by evidence guided belief revision, either in the form of belief *set* revision or belief *base* revision,
- finally, the idea that the thus elaborated kind of comparative (and constructive) nomic realism is the best explanation of empirical progress and hence the best response to antirealism.

However, in 2012 I discovered that my basic theory of nomic truth approximation, published in 1982 in *Erkenntnis*, was based on an unnecessarily strong assumption, viz. the assumption that theories are maximally specific in the sense that they do not only exclude certain conceptual possibilities, the exclusion claim, but also that all non-excluded possibilities are in fact claimed to be nomically possible, the inclusion claim. It turned out that the exclusion claim alone, or for that matter the inclusion claim alone, was already sufficient to motivate the formal definition of being closer to the nomic truth 'in the basic sense'. This discovery was strongly stimulated by related work of Gustavo Cevolani, Vincenzo Crupi and Roberto Festa on conjunctive theories, published in *Erkenntnis* (Cevolani et al. 2011), in particular their systematic way of defining the truth- and falsity-content of a theory.

The finding was published in 2014, entitled "Empirical progress and nomic truth approximation revisited" in *Studies in History and Philosophy of Science*, focusing on exclusion or Popperian theories (Kuipers 2014). It also suggested the possibility of nomic truth approximation by revising inclusion or representation theories, i.e., theories considered as sets of models claimed to represent nomic possibilities. Hence, it even suggested the possibility of nomic truth approximation by revising two-sided theories conceived of as tuples of models and postulates, claimed to be included and claiming to exclude, respectively, and leaving room for undecided conceptual possibilities.

This generalized perspective on the basic theory of nomic truth approximation generated not only the need to revise the new papers published so far, but also to reconsider the various concretizations of the basic theory as presented in my book of 2000, viz. the quantitative version, replacing the purely set-theoretic notions of the basic theory by quantitative notions, the refined version, taking into account that not all counterexamples are equally bad, and the stratified version, taking the observational-theoretical distinction into account.

Hence, the present book is not just a synthesis of new papers on nomic truth approximation, from 1998 onwards, but a thorough revision of them, in Chaps. 8, 9, 10, 11, 12, 13, 14, 15, and 16, starting with the indicated generalization of the main notions and claims of the 2000-book in Chaps. 2, 3, 4, 5, 6, and 7.

In the resulting book most chapters can be read independently from each other, except Chaps. 5, 6, and 7, dealing with the three mentioned concretizations, they presuppose Chaps. 3 and 4. However, they can be read independently from each other. Of course, some overlap between the other chapters, especially in the introductory sections, was unavoidable in order to make them separately readable.

Section 1.2 provides a summary of the main conclusions of *From Instrumentalism to Constructive Realism* of 2000. Section 1.3 gives an overview of the chapters of the present book. Chapters 1, 3, 5, 6, and 7 are especially written for this book. The references to the original versions of the other chapters are given in Sect. 1.4. Section 1.5 concludes with some indications for new research, in particular in order to compare the presented approach with other approaches.

1.2 The Main Conclusions of *From Instrumentalism to Constructive Realism*

The methodological themes of confirmation and empirical progress form the stepping stones to the epistemological theme of nomic truth approximation. All three themes were elaborated, in that order, in my *From Instrumentalism to Constructive Realism* (2000), henceforth sometimes abbreviated as ICR. It will be useful to start the present book with a survey of the main conclusions of ICR. It will provide its background. The most relevant conclusions will be indicated by *. The survey is an adaptation of ICR-Section 13.1. It should be kept in mind that the conclusions from the fourth conclusion onwards were assuming a strong claim of theories: what they did not claim to be nomically impossible (exclusion claim) was claimed to be nomically possible (inclusion claim). A main message of the present book is that these conclusions remain by and large valid when a theory's claim is restricted to an exclusion claim, or to an inclusion claim.

*The *first* conclusion, from Part I and II of ICR, is that a useful distinction can be made between two kinds of applying the Hypothetico-Deductive (HD-)method, viz., testing of hypotheses, in order to try to establish their truth-value, leading to falsification or confirmation, and evaluation of theories, in the first place, in order to establish their observational merits and failures.

Second, according to Part I, the notion of confirmation is primarily related to hypothesis testing, and can be given a direct qualitative explication, presented in Chap. 2, at least as far as deductive confirmation is concerned. Moreover, a quantitative, Bayesian, explication, given in Chap. 3, leads to a clear division of deductive and non-deductive confirmation. This quantitative explication generates indirectly a qualitative explication of confirmation in general and of non-deductive confirmation in particular. For the acceptance of observational hypotheses as true, fallible, inductive jumps have to be made in the sense of inductive generalizations. In Chap. 4 it is shown that inductive confirmation and inductive logic pertain to a particular kind of probability functions governing quantitative, deductive and non-deductive, confirmation.

**Third*, the evaluation of theories by the HD-method, studied in Part II, concerns, in the first place, the separate evaluation in terms of general successes and counterexamples, Chap. 5, but immediately suggests a comparative sequel, leading to the possible establishment of empirical progress, Chap. 6. The resulting evaluation methodology can be seen as an explication of the instrumentalist methodology, according to which counterexamples play a crucial role, but not the dramatic one that is associated with the falsificationist methodology. This explains and even justifies non-falsificationist, i.e., dogmatic, behavior of scientists. However, in contrast to pseudoscientists, they nevertheless aim at the improvement of their theories, which is frequently possible within a set of dogmas.

**Fourth*, empirical progress cannot be seen only as an aim in itself; it can also be seen as a means to truth approximation, as argued in Part III (Basic Truth Approximation) and Part IV (Refined Truth Approximation). From Chap. 7 onwards, 'the truth' is conceived as the nomic truth, provided by the so-called Nomic Postulate (Nomic Principle, in the present book), i.e., the strongest true theory about what is nomically possible for a given domain of reality assuming a certain vocabulary. Using the structuralist (or set-theoretic) approach to theories, the Nomic Postulate assures the existence of a subset of nomic possibilities within the set of conceptual possibilities generated by the vocabulary. Within this framework, a *prima facie* plausible, so-called basic, definition of the idea that one theory is more truthlike than another can be given. The latter can be seen as a hypothesis, the truth approximation hypothesis, which can be tested due to its prediction of empirical progress, which makes empirical progress functional for truth approximation.

Fifth, the idea of more truthlikeness and the aim of truth approximation can be shown, in Chap. 8, to be in accordance with scientific common sense about theory improvement, viz., more true consequences and more correct models. Moreover, they suggest intralevel explications of some persisting intuitions among philosophers, viz., the correspondence theory of truth and intuitions governing dialectical reasoning.

**Sixth*, the (shifting) distinction between observational and theoretical terms not only naturally leads, together with the Nomic Postulate, in Chap. 9, to the distinction between theoretical and observational truthlikeness, but also to definitions of reference, referential claims of theories, and referential truthlikeness. Whereas observational truthlikeness implies empirical progress, theoretical truthlikeness only implies observational truthlikeness as far as true consequences are concerned. This leads to a further relativization of the role of counterexamples. Whereas the loss of a true observational consequence is a genuine drawback of a new theory, an extra counterexample of a new theory may well be an accidental (observational) merit of the old one. Moreover, since there are no deductive relations between referential truthlikeness and the other comparative notions, other kinds of (theoretical and experimental) arguments may also have to be taken into account for the assessment of referential claims.

**Seventh*, although the analysis sketched above may come close to scientific common sense, it does not yet mean that the crucial notions are directly applicable to real-life scientific examples. However, Part IV, more specifically, Chap. 10, shows a refined account, according to which it is perfectly possible to take into account that the improvement of theories is usually not a matter of replacing mistaken models by correct ones, but of replacing mistaken models by less mistaken ones. This typically occurs when using the method of idealization and concretization, as for instance Van der Waals did in two concretization steps.

Eighth, Chap. 11 provides two more real-life examples. It demonstrates that the succession of the (old) quantum theories of the atom of Rutherford, Bohr and Sommerfeld provides an example of potential truth approximation, more specifically, by specialization followed by concretization. A different kind of truth approximation is illustrated by the succession of theories about the capital structure of firms, the theory of Kraus and Litzenberger, followed by that of Modigliani and Miller. Though different, it is, formally speaking, again by double concretization, as in the case of Van der Waals.

**Ninth*, and for now finally, speaking of distance from the truth and truth approximation easily suggests quantitative explications, as presented in Chap. 12. Although this is formally possible, assuming that there is some kind of distance function between the conceptual possibilities, it only seems to make sense in contexts where such a distance function is plausible.

Many other conclusions followed from the above ones. We only mention a few. Concerning confirmation, the famous paradoxes of Hempel and Goodman disappear, whereas Popper's notion of corroboration turns out to largely agree with confirmation, but suggests a straightforward definition of the severity of tests.

Concerning empirical progress, the need for systematic social explanations of non-falsificationist behavior disappears in the light of the analysis of comparative evaluation of theories.

Concerning truth approximation, our analysis throws new light on novel facts, crucial experiments, and inference to the best explanation. Moreover, it enables more precise definitions of descriptive, including inductive, research programs, on the one hand, and of theoretical or explanatory research programs on the other.

The above conclusions pertain to Parts I to IV. However, ICR opens and closes with general epistemological chapters, and I will draw a *tenth* conclusion of epistemological nature based on the parts. Chapter 1 introduces the main epistemological positions, instrumentalism,[1] constructive empiricism, referential realism, constructive (theory) realism and essentialist (theory) realism.[2] Chapter 13 provides a further articulation of constructive realism. The reason is that at the end of Part III the following conclusion was already drawn provisionally and was further strengthened in Part IV:

**Tenth*, the instrumentalist methodology provides good reasons for the transition of epistemological positions from instrumentalism, via constructive empiricism and referential realism, to constructive realism. Here, the intermediate step from constructive empiricism to referential realism turned out to be the hardest one. To be sure, the step from constructive to essentialist realism had to be rejected.

In the rest of Chap. 13 I have drawn the main lines of the resulting favorite epistemological position of *constructive realism*. It is a conceptually relative, hence non-essentialist, nomic truth approximation version of theory realism, accounting for objective truths. They can be approached by an intersubjective method, viz., the evaluation methodology for theories, in which the role of (truth-)testing of hypotheses primarily concerns testing test implications of theories as well as testing comparative success and truth approximation hypotheses of theories.

The term 'constructive realism' has earlier been used by Giere (1985), and my conception of it is rather similar, except that I include in it, of course, truth approximation, whereas Giere still focuses on the true/false dichotomy, but he fully recognizes the nomic aim of theorizing. With respect to truth approximation, my position is rather similar to that of Niiniluoto (1987, 2017). The main difference between his and my position is, besides my more cautious primarily (qualitative) comparative approach versus his straightforward quantitative approach, my emphasis on the nomic aim of theorizing. In sum, my constructive realism reflects the combination of their deviating strengths by emphasizing *comparative nomic truth approximation* as opposed to the *actual truth-value or degree of actual truthlikeness* of theories. For this reason I like to qualify my constructive realism nowadays more specifically as *comparative constructive nomic realism*.

---

[1] Instrumentalism as an epistemological position should be distinguished from the instrumentalist or evaluation methodology. The latter is claimed to serve also other epistemological positions.

[2] Note that so-called 'structural realism' is not taken into account. Apart from a few remarks (Chap. 16, Sects. 16.4.2 and 16.5) the same holds for the present book.

1.3 Overview of the Chapters

This section brings together the abstracts of all coming chapters. They will facilitate the choice of chapters to read in a first round.

Part I is devoted to the generalization of the basic approach to 'closer to the nomic truth'.

Chapter 2: Nomic Truth Approximation by Empirical Progress Revisited
In my *From Instrumentalism to Constructive Realism* (2000) I have shown how an instrumentalist account of empirical progress can be related to nomic truth approximation. However, it was assumed that a strong notion of nomic theories was needed for that analysis. In this chapter it is shown, in terms of truth- and falsity-content, that the analysis already applies when, in line with scientific common sense, nomic theories are merely assumed to exclude certain conceptual possibilities as nomic possibilities. The resulting generalized and thereby simplified approach to nomic truth approximation was strongly stimulated by the related work on conjunctive theories of Gustavo Cevolani, Vincenzo Crupi and Roberto Festa (*Erkenntnis* 2011), in particular their systematic way of defining the truth- and falsity-content of a theory. It will be shown that explicating '(nomic) truth approximation' and 'empirical progress' in these terms does justice to some basic instrumentalist/empiricist and realist conditions of adequacy.

Chapter 3: Perspectives and Guidelines for Theories of (Nomic) Truth Approximation by Empirical Progress
The account of nomic truth approximation by empirical progress presented in Chap. 2 is based on the exclusion claim of theories. It is a simplified and generalized version of the one presented in my book *From Instrumentalism to Constructive Realism* of 2000. However, its core itself, called the *basic* account, is based on several idealizations. In this chapter I will sketch a number of possible concretizations, to begin with a two-sided, otherwise also basic, version, based on an exclusion and inclusion claim, followed by quantification, refinement and stratification of this two-sided version. They will be elaborated in Chaps. 4, 5, 6, and 7.

Chapter 2 also illustrated straightforwardly how a theory of truth approximation in relation to empirical progress can be set up. The general tenet and the common elements will be summarized in this chapter. It will lead to a general outline of theories of nomic truth approximation by empirical progress. This will guide, explicitly or implicitly, the elaborating Chaps. 4, 5, 6, and 7.

Moreover, the common core of such theories of nomic truth approximation will also be summarized: the Nomic Principle, postulating a unique divide between nomic possibilities and impossibilities, the formal structure of nomic evidence, in terms of realized (nomic) possibilities and induced generalizations, and the hypothetico-deductive background of nomic evidence.

Chapter 4: Models, Postulates, and Generalized Nomic Truth Approximation The qualitative theory of nomic truth approximation, presented in Kuipers (2000), in which 'the truth' concerns the distinction between nomic, e.g. physical, possibilities and impossibilities, rests on a very restrictive assumption viz. that theories always claim to characterize the boundary between nomic possibilities and impossibilities. In Chap. 2 we have already shown that truth approximation is possible by ascribing, in Popperian spirit, theories only an excluding function. By fully recognizing two different functions of theories, viz. excluding and representing (including), this chapter drops the restrictive assumption by conceiving theories in development as tuples of postulates and models, where the postulates claim to exclude nomic impossibilities and the (not-excluded) models claim to represent nomic possibilities.

Revising such 'two-sided' theories becomes then a matter of adding or revising models and/or postulates in the light of increasing evidence, now captured by a special kind of theories, viz. 'data-theories'. Under the assumption that the data-theory is true, achieving empirical progress on its basis provides good reasons for the abductive conclusion that truth approximation has been achieved as well. As in Chap. 2, the notions of truth approximation and empirical progress are formally direct generalizations of the earlier ones. However, truth approximation is now explicitly defined in two-sided terms of increasing truth-content and decreasing falsity-content of theories, whereas empirical progress is defined in terms of lasting increased accepted-content and decreased rejected-content in the light of increasing evidence. As already noted in Chap. 2, these definitions are strongly inspired by a paper of Gustavo Cevolani, Vincenzo Crupi and Roberto Festa, viz., "Verisimilitude and belief change for conjunctive theories" (*Erkenntnis* 2011).

Part II is devoted to three concretizations of the basic two-sided approach: quantification, refinement and stratification. I start with the common part of the abstracts, followed by their specific complements.

Assuming that the target of theory oriented empirical science in general and of nomic truth approximation in particular is to characterize the boundary or demarcation between nomic possibilities and nomic impossibilities, I have presented in Chap. 4 ("Models, postulates, and generalized nomic truth approximation") the 'basic' version of generalized nomic truth approximation, starting from 'two-sided' theories. Its main claim is that nomic truth approximation can perfectly be achieved by combining two prima facie opposing views on theories: (1) the traditional (Popperian) view: theories are (sets of models of) postulates that exclude certain possibilities from being realizable, enabling explanation and prediction and (2) the model view: theories are sets of models that claim to (approximately) represent certain realizable possibilities. Nomic truth approximation, i.e., increasing truth-content and decreasing falsity-content, becomes in this way revising theories by revising their models and/or their postulates in the face of increasing evidence. The basic version of generalized nomic truth approximation is in many respects as simple as possible. At least three concretizations are in order: quantification, refinement, and stratification.

Chapter 5: Quantitative Nomic Truth Approximation
The generalized basic is qualitative in the sense that it is purely based on set-theoretic relations. The present chapter presents the straightforward quantitative concretization of it. According to the 'expected success theorem', based on some probabilistic experimental conditions, greater truthlikeness, or verisimilitude, leads to greater expected empirical success. This enables tentative nomic truth approximation conclusions by abductive reasoning.

Chapter 6: Refined Nomic Truth Approximation
The generalized basic version does not take into account that one conceptual possibility may be more similar (or closer) to another than a third one (is to that other). However, for example, one theory may include a possibility that is more similar to a wrongly not included possibility than another theory can offer. Similarly, for wrongly not excluded possibilities. In this chapter it will be shown that such 'refined' considerations can be taken into account by adapted clauses based on a ternary similarity relation between possibilities (structures). This allows again abductive conclusions about refined truth approximation if a theory is persistently more successful in the refined sense than another.

It will also be indicated and illustrated that this refined approach enables a specification to the effect that refined truth approximation can be obtained by the method of idealization and subsequent concretization. Finally, the basic and the refined approach will be evaluated with regard to some general principles and objections that have been discussed in the literature.

Chapter 7: Stratified Nomic Truth Approximation
The generalized basic version does not make a (theory-relative) distinction between an observational and a theoretical level. This raises the question to what extent in a stratified set-up (nomic) truth approximation on a theoretical level entails truth approximation on the observational level. In other words, the question whether 'closer to the truth' can be projected from the theoretical to the observational level. Of course, 'empirical success' is now explicated in observational terms, and 'closer to the observational truth' and 'more successfulness' are now related in the same way as in the unstratified case.

The chapter closes with *quantitative* and *refined* versions of stratified nomic truth approximation in line with Chaps. 5 and 6, respectively.[3]

Part III deals with two quite different extensions: variable domains and aesthetic factors.

Chapter 8: Theories Looking for Domains. Nomic Truth Approximation by Domain Revision
The theory of nomic truth approximation essentially deals primarily with truth approximation by theory revision for a fixed domain. However, variable domains and a fixed theory can also be taken into account, where the main changes concern domain extensions and restrictions. In this chapter I will present a coherent set of

[3] Referential truth approximation is dealt with in Chap. 11.

definitions of 'more truthlikeness', 'empirical progress' and 'truth approximation' due to a revision of the 'domain of intended applications'. This set of definitions seems to be the natural counterpart of the basic definitions of similar notions as far as theory revision is concerned.

The resulting formal aspects of theory and domain revision strongly suggest an analogy between truth approximation and design research, for example, drug research: a new drug (theory) may be better for a certain disease (domain) than an old one. However, it also occurs that a certain drug may be better for another disease than for the original target disease, a phenomenon which was nicely captured by the title of a study by Rein Vos (1991): *Drugs Looking for Diseases*. Similarly, truth approximation may not only take the shape of theory revision but also of domain revision, naturally suggesting the phenomenon of 'Theories looking for domains'. However, whereas Vos documented his title with a number of examples, so far, apart from plausible cases of 'truth accumulation by domain extension' and the conjecture that cases of so-called Kuhn-loss can be reconstructed as cases of 'truth approximation by domain restriction', I did not find clear-cut but more complex empirical instantiations of the analogy, only some very interesting, non-empirical examples.

Chapter 9: Beauty, a Road to the Truth?
In this chapter I give a naturalistic-cum-formal analysis of the relation between beauty, empirical success, and truth. The analysis is based on the one hand on a hypothetical variant of the so-called 'mere-exposure effect' which has been more or less established in experimental psychology regarding exposure-affect relationships in general and aesthetic appreciation in particular. It was initiated by Robert Zajonc (1968). On the other hand it is based on the formal theory of truthlikeness and truth approximation as presented in my *From instrumentalism to constructive realism* (2000).

The analysis supports the findings of James McAllister in his beautiful *Beauty and revolution in science* (1996), by explaining and justifying them. First, scientists are essentially right in regarding aesthetic criteria useful for empirical progress and even for truth approximation, provided they conceive of them as less hard than empirical criteria. Second, the aesthetic criteria of the time, the 'aesthetic canon', may well be based on 'aesthetic induction' regarding non-empirical features of paradigms of successful theories which scientists have come to appreciate as beautiful. Third, aesthetic criteria can play a crucial, schismatic role in scientific revolutions. Since they may well be wrong, they may, in the hands of aesthetic conservatives, retard empirical progress and hence truth approximation, but this does not happen in the hands of aesthetically flexible, 'revolutionary' scientists.

Part IV deals with various kinds of reasoning around nomic truth approximation.

Chapter 10: Abduction Aiming at Empirical Progress or Even Truth Approximation: A Challenge for Computational Modelling
This chapter primarily deals with the conceptual prospects for generalizing the aim of abduction from the standard one of explaining surprising or anomalous observations to that of empirical progress or even truth approximation. It turns out that the

main abduction task then becomes the instrumentalist task of theory revision aiming at an empirically more successful theory, relative to the available data, but not necessarily compatible with them. The rest, that is, genuine empirical progress as well as observational, referential and theoretical truth approximation, is a matter of evaluation and selection, and possibly new generation tasks for further improvement. The chapter concludes with a survey of possible points of departure, in AI and logic, for computational treatment of the instrumentalist task guided by the 'comparative evaluation matrix'.

Chapter 11: Inference to the Best Theory, Rather Than Inference to the Best Explanation. Kinds of Abduction and Induction

An interesting consequence of the theory of nomic truth approximation, as developed in my *From Instrumentalism to Constructive Realism* (2000), and enriched in this book, concerns so-called 'inference to the best explanation' (IBE). It can be argued that this popular rule among scientific realists can better be replaced by, various kinds of, 'inference to the best theory' (IBT). This chapter provides a survey of observational, theoretical, and referential kinds of IBT and discusses when and in what sense they may be seen as cases of 'abduction', that is, arguments for which an 'abductive' formulation and, hence, justification can be given in the minimal sense of Peirce.

Whereas it is difficult to imagine an abductive justification for IBE, especially due to its asymmetric character, it appears that abductive justifications can be given for IBT *as the closest to the relevant truth*, where 'the (relevant) truth' is defined as the strongest true claim that can be made with the relevant vocabulary about the relevant domain.

The relation of IBT with the instrumentalist or evaluative use of the hypothetico-deductive (HD-)method is also indicated. In particular, it is argued that the theoretical kind of IBT may well be seen as the proper realist extension of the instrumentalist *rule of success*, prescribing to choose the most successful theory, whether already falsified or not.

The chapter also introduces weak and strong forms of observational, theoretical, and referential induction. The weak forms merely conclude that a theory is true in the relevant sense and the strong forms can be seen as extreme special cases of the corresponding kinds of IBT, viz. 'inference to the best theory *as the relevant truth*'. Finally, it is argued that the most benevolent and defensible reading of IBE combines IBT with the corresponding weak kind of induction.

Chapter 12: Pragmatic Aspects of Truth Approximation

In this chapter I will draw a very general picture of an overall pragmatic theory of justification from the viewpoint of *truth approximation*. Crucial is the comparative definition of *success* according to which a theory $\underline{P^*}$ is more successful than a theory $\underline{P}$ if and only if (1) $\underline{P^*}$ shares all general successes of $\underline{P}$, that is all empirically confirmed general implications, and has some extra ones, and (2) $\underline{P}$ shares all empirical counter-examples of $\underline{P^*}$. The *rule of success* says that if so far $\underline{P^*}$ has been more successful than $\underline{P}$, then – for the time being – accept $\underline{P^*}$ as the better one and eliminate $\underline{P}$. It is further argued that according to the pragmatic-instrumentalist view of scientific theories, theory elimination should exclusively be guided by the rule of success rather than by Popper's *rule of falsification* – for it is pragmatically reasonable to accept the most successful theory available, even if this theory is known to be false. In the last part of the chapter, it is argued that the connection between pragmatic theory-success and truth approximation is otherwise rather close to Popper: at least 'almost' always, a theory which is closer to the truth will also turn out to be the pragmatically more successful theory. However, in the upshot, 'being false' and hence 'being falsified' are not very important for truth approximation.

Chapter 13: Empirical Progress and Nomic Truth Approximation by the 'Hypothetico-Probabilistic Method'
Three related intuitions are explicated in this chapter. The first is the idea that there must be some kind of probabilistic version of the HD-method, a 'Hypothetico-Probabilistic (HP-) method', in terms of something like probabilistic consequences, instead of deductive consequences. According to the second intuition, the comparative application of this method should also be functional for some probabilistic kind of empirical progress, and according to the third intuition this should be functional for something like probabilistic truth approximation. In all three cases, the guiding idea is to explicate these intuitions by explicating the crucial notions as appropriate 'concretizations' of their basic or deductive analogues, being 'idealizations'.

It turns out that the comparative version of the proposed HP-method amounts to a sophisticated likelihood ratio success (SLRS-) method applied to the cumulated evidence. This method turns out to be not only functional for probabilistic empirical progress but also for probabilistic truth approximation. The latter is based on a probabilistic threshold theorem constituting for this reason the analogue of the deductive success theorem.

This chapter is in fact a supplement to Chap. 5. Despite some terminological deviation, there is substantial partial overlap to enable independent reading.

A special kind of reasoning concerns 'belief revision aiming at truth approximation'. In Part V two forms are considered: belief set revision, aiming at nomic truth approximation, and belief base revision, enabling various kinds of truth approximation.

Chapter 14: Basic and Refined Nomic Truth Approximation by Evidence–Guided Belief Set Revision
Straightforward theory revision, taking into account as effectively as possible the established nomic possibilities and, on their basis induced empirical laws, is conducive for nomic truth approximation. The question this paper asks is: is it possible to reconstruct the relevant theory revision steps, on the basis of incoming evidence, in the terms of so-called belief revision and, more specifically, in the terms of belief *set* revision, also called the AGM-approach? Assuming exclusion theories, a positive answer will be given in two rounds, first for the case in which the initial theory is compatible with the established empirical laws, then for the case in which it is incompatible with at least one such a law.

Chapter 15: Dovetailing Belief Base Revision with Truth Approximation
Gustavo Cevolani, Vincenzo Crupi and Roberto Festa (2011) have shown that their account of verisimilitude of 'conjunctive theories' of a finite propositional language can be nicely linked to a variant of AGM belief *set* revision, viz. belief *base* revision, in the sense that the latter kind of revision is functional for truth approximation according to the conjunctive account. In the present chapter I offer a generalization of these ideas to the case of approaching any divide of a (finite or infinite) universe, allowing several interpretations, besides true (false) atomic propositions, notably nomic states (not) in equilibrium, nomic (im-)possibilities, (non-)instantiated 'Q-predicates' of a monadic language. It shows how and why approximation of 'the true boundary' takes place by belief base revision guided by evidence.

In the nomic (im-)possibilities interpretation this chapter essentially deals with a belief *base* revision perspective on *basic* and *quantitative* nomic truth approximation of two-sided theories in the sense of Chaps. 4 and 5. The previous chapter dealt with a belief *set* revision (i.e., AGM-) perspective on *basic and refined* nomic truth approximation of (one-sided) exclusion theories. This chapter will leave the challenge open of a belief base revision perspective on refined nomic truth approximation of two-sided theories, and, more generally, a belief base perspective on refined approximation of 'the true boundary' belonging to whatever interpretation.

In the final chapter, constituting Part VI, we draw the main epistemological conclusion that corresponds best to the theory of nomic truth approximation, guided by empirical and aesthetic progress, elaborated in this book, that is, comparative constructive nomic realism.

Chapter 16: Comparative Realism as the Best Response to Antirealism
Arguments for and against scientific realism usually presuppose as the main epistemic claim about theories that we may have good reasons to conclude that they are true, or at least approximately true. The antirealist charges against this claim are not easy to counter. In this chapter it is argued that the defense of realism is much easier if we relativize its epistemic claim in the light of (theories of) truth approximation. From this *comparative realist* perspective the main epistemic claim becomes that we may have good reasons to conclude that successor theories are closer to the truth than their predecessors. For example, although Einstein's theory of general relativity may still be false, and not even approximately true, we have good reasons to assume that it is closer to the truth than Newton's theory of gravitation. A similar relativization of claims that theoretical terms refer to things in the world is argued for in terms of 'being closer to the referential truth'. For both purposes it is also plausible to relativize 'being empirically successful' to the comparative notion of 'being (persistently) empirically more successful'.

Comparative realism hence is realism guided by the comparative perspective on success and on truth approximation, that is, the notions of 'more successful' and 'closer to the (observational, referential, and theoretical) truth', and their mutual relations. This approach is defended against the antirealist charges and compared with the main other realist responses. The main positive claim of comparative realism is that (theoretical) truth approximation provides the (stratified) default explanation and prediction of empirical progress between non-empirically equivalent theories and of 'aesthetic progress' between empirically equivalent theories. Here 'aesthetic progress' is based on the comparative notion of 'aesthetically more successful' and specified in terms of the prevailing 'aesthetic canon' in the relevant field and period, that is, the prevailing non-empirical virtues of theories.

1.4 Origin of the Chapters[4]

Chapter 1	Introduction *New*
Chapter 2	Nomic Truth Approximation by Empirical Progress Revisited *Revised version of*: "Empirical Progress and Truth Approximation Revisited", Proceedings Tilburg conference on Progress in Science (April, 2012), *Studies in History and Philosophy of Science*, special section *Progress in Science*, 46, 2014, 64–72.
Chapter 3	Perspectives and guidelines for theories of (nomic) truth approximation by empirical progress *New*
Chapter 4	Models, postulates, and generalized nomic truth approximation *Revised version of*: "Models, postulates, and generalized nomic truth approximation", *Synthese*, 193 (10), 2016, 3057–3077.
Chapter 5	Quantitative nomic truth approximation *New. An adapted version* appeared as: "Quantitative nomic truth approximation by revising models and postulates", *European Studies in Philosophy of Science*, EPSA15 Selected papers, Eds. M. Massimi, J.W. Romeijn & G. Schurz, Dordrecht, Springer 2017, pp. 309–320.
Chapter 6	Refined nomic truth approximation *New. An adapted version*, entitled *Refined nomic truth approximation by revising models and postulates*, will appear in a special issue of *Synthese* in honor of Gerhard Schurz.
Chapter 7	Stratified nomic truth approximation *New*
Chapter 8	Nomic truth approximation by domain revision *Revised version of*: "Theories looking for domains. Fact or fiction? Reversing Structuralist Truth Approximation" In: L. Magnani (ed.), *Model-Based Reasoning in Science and Engineering*, Studies in Logic, Vol. 2, College Publications, King's College, London, 2006, pp. 33–50.

[4]The (strongly) revised versions of all previously published papers, and one forthcoming paper (Chap. 6), appear here with the consent of the publishers concerned.

Chapter 9	Beauty, a road to the truth? *Revised version of*: "Beauty, a road to the truth", *Synthese*, **131**.3, 2002, 291–328.
Chapter 10	Abduction aiming at empirical progress or even truth approximation *Revised version of*: "Abduction aiming at empirical progress or even at truth approximation, leading to challenge for computational modelling", *Scientific Discovery and Creativity*, eds. J. Meheus, T. Nickles, special issue of *Foundations of Science*. Vol. **4**.3, 1999, 307–23.
Chapter 11	Inference to the best theory, rather than inference to the best explanation *Revised version of:* "Inference to the best theory, rather than inference to the best explanation. Kinds of abduction and induction". In: F. Stadler (ed.), *Induction and Deduction in the Sciences*, Proceedings of the ESF-workshop *Induction and Deduction in the Sciences*, Vienna, July, 2002, Dordrecht, Kluwer Academic Publishers, pp. 25–51, followed by a commentary of Adam Grobler, pp. 53–56.
Chapter 12	Pragmatic aspects of truth approximation *Revised version of*: "Pragmatic aspects of truth approximation", in: P. Weingartner, G. Schurz and G. Dorn, *The role of pragmatics in contemporary philosophy*, Hölder-Pichler-Temsky, Vienna, (Proceedings of the 20th International Wittgenstein -Symposium, August 1997), 1998, 288–300.
Chapter 13	Empirical progress and nomic truth approximation by the 'Hypothetico-Probabilistic Method' *Strongly revised version of*: "Empirical progress and Truth Approximation by the 'Hypothetico-Probabilistic Method'", *Erkenntnis, 70*.3, 2009, 313–330.
Chapter 14	Basic and refined nomic truth approximation by evidence-guided belief set revision *Strongly revised version of:* "Basic and refined nomic truth approximation by evidence-guided belief revision in AGM-terms", *Erkenntnis*, 75, 2, 2011, pp. 223–236.
Chapter 15	Dovetailing belief base revision with truth approximation *Revised version of:* "Dovetailing belief base revision with (basic) truth approximation", in E. Weber, D. Wouters, and J. Meheus, *Logic, Reasoning, and Rationality*, proceedings-conference in Ghent, 2010, Springer, 2014, pp. 77–93.
Chapter 16	Comparative realism as the best response to antirealism

Revised version of: "Comparative realism as the best response to antirealism". In Logic, Methodology and Philosophy of Science. Proceedings of the 13th International Congress (August 9–15, 2007, Beijing), eds. Clark Glymour, Wang Wei and Dag Westerstahl. College Publications, King's College, London, 2009, pp. 221–250.

1.5 Comparison with Other Approaches Needed

My theory of nomic truth approximation was and remains a specific comparative approach to truth approximation, with several points of contact to other approaches, such as that of Ilkka Niiniluoto (1987) (quantitative likeness approach), Graham Oddie (1986) (a variant of the likeness approach), and Gustavo Cevolani, Vincenzo Crupi and Roberto Festa (2011) (the conjunctive approach).[5] Since the main focus of the other approaches is the actual (or historical) truth, they are, though mutually competing, *prima facie* compatible with my nomic approach, which hence remains an autonomous contribution to the relevant literature.

However, for this reason it is also a challenging touchstone for all 'actual truth' approaches to find out how a redirection of their other leading ideas to nomic truth approximation would look like, and whether these adapted approaches would be competing or compatible with my nomic approach. For the conjunctive approach this has already been done and it appears that (the basic version of) the nomic approach can be seen as an application of the conjunctive one, as indicated at relevant places in this book and already published in a (with hindsight) premature version in *Synthese* (with Gustavo Cevolani and Roberto Festa), entitled "Verisimilitude and belief change for nomic conjunctive theories" (Cevolani et al. 2013). It was premature because it could have been formulated much more transparent in terms of the two-sided approach of Chap. 4.

Triggered by a plea of Jonathan Cohen (1980, 1987) for the study of 'legisimilitude' besides verisimilitude or, in my terms, approximation of the nomic truth besides approximation of the actual truth, Ilkka Niiniluoto (1983, 1987, Chap. 11) and Graham Oddie (1982) have already elaborated how their likeness approaches could be applied to 'nomic constituents' and nomic theories in general. Niiniluoto (1987, p. 382) comes to the conclusion that "there is a remarkable convergence between two quite independently developed approaches", meaning his and mine. There is certainly convergence in spirit, but my approach at the time (Kuipers 1982) was still restricted to the basic (set-theoretic) version and assuming the strong claim of theories. It would be interesting to compare the present refined, two-sided version (Chap. 6) with the best nomic versions of the likeness approaches of Niiniluoto and Oddie.

[5] See also Cevolani and Festa (to appear) as the most recent presentation, including a significant extension to the case of arbitrary theories, notably disjunctive theories.

There is also a couple of approaches to actual truthlikeness that are restricted to propositional languages: the relevant-consequence approach of Gerhard Schurz and Paul Weingartner (1987), see also Schurz (2005), the refined approach of Sjoerd Zwart (2001, Chap. 6), and the algebraic approach of Chris Brink, Isabella Burger and Johannes Heidema, see Brink and Heidema (1987) and Burger and Heidema (2005). Their mutual comparison and the comparison with the likeness approaches of Niiniluoto and Oddie and with the basic and refined version of my nomic approach deserves an overarching research project. There are two survey papers and a book taking a number of them into account. They provide a good point of departure: Niiniluoto (1998), Oddie (2016) and Zwart (2001).

It is important to note that, besides my focus on nomic truth, there is one technical difference between my approach and all others, except that of David Miller (1978). I prefer to present the analysis in terms of conceptual possibilities or, more technically, in terms of set-theoretic structures. But, as a rule, it is possible to nevertheless compare the results with the more usual syntactic approaches, assuming some specific language. The conceptual possibilities then correspond to the models of the language. In a propositional language the so-called constituents can even be identified with its models, but in a (first order) predicate language the relation between constituents and models is more complex.

In several chapters it is suggested that the conceptual possibilities have to be set-theoretic structures. This is particularly plausible in case the scientific theories concerned have some formal, usually mathematical, character. However, as becomes particularly clear in Chap. 15, the formal machinery developed in this book for nomic truth approximation, notably in Chap. 4, can be used for approximation of a binary divide within all kinds of universes of conceptual possibilities. Besides the divide between nomic possibilities and nomic impossibilities, the target may be, for example, the divide between true and false atomic propositions and that between instantiated and non-instantiated 'Q-predicates' of a monadic predicate language. In short, the target may be the approximation of 'the true boundary' generated by a binary division of any set of conceptual possibilities. Sometimes examples will be mentioned that allude to such other interpretations.

Finally, regarding epistemological positions, the last chapter will conclude with characterizing my position, as stated before, as comparative constructive nomic realism, after comparing it, on the basis of detailed analyses in my 2000-book, summarized in Chap. 11, with the main positions in the literature, notably constructive empiricism, referential realism and essentialistic or metaphysical realism. However, so-called structural realism (Russell 1927; Worrall 1989) is a position that still lacks an analysis from my comparative, constructive and nomic point of view. The main question is whether the transition from constructive empiricism via structural realism to constructive realism is as plausible as that from constructive empiricism via referential realism to constructive realism?

References

Brink, C., & Heidema, J. (1987). A verisimilar ordering of theories phrased in a propositional language. *The British Journal for the Philosophy of Science, 38*, 533–549.

Burger, I., & Heidema, J. (2005). For better, for worse: Comparative orderings on states and theories. In A. Aliseda, R. Festa, & J. Peijnenburg (Eds.), *Confirmation, empirical progress and truth approximation, Essays in debate with Theo Kuipers* (Vol. 1, pp. 459–488). Amsterdam: Rodopi Reply by T. Kuipers, 489–492.

Cevolani, G., & Festa, R. (To appear). *To appear. A partial consequence account of truthlikeness.* To appear in a special issue of *Synthese* in honor of Gerhard Schurz.

Cevolani, G., Crupi, V., & Festa, R. (2011). Verisimilitude and belief change for conjunctive theories. *Erkenntnis, 75*(2), 183–202.

Cevolani, G., Festa, R., & Kuipers, T. (2013). Verisimilitude and belief change for *nomic* conjunctive theories. *Synthese, 190*, 3307–3324.

Cohen, L. J. (1980). What has science to do with truth? *Synthese, 45*, 489–510.

Cohen, L. J. (1987). Verisimilitude and Legisimilitude. In T. Kuipers (Ed.), *What is closer-to-the-truth? (Poznan studies in the philosophy of science and the humanities Vol. 10)* (pp. 129–144). Amsterdam: Rodopi.

Giere, R. (1985). Constructive realism. In P. Churchland & C. Clifford (Eds.), *Images of science* (pp. 75–98). Chicago: The University of Chicago Press.

Kuipers, T. (1982). Approaching descriptive and theoretical truth. *Erkenntnis, 18*(3), 343–378.

Kuipers, T. (2000). *[ICR] From instrumentalism to constructive realism. On some relations between confirmation, empirical progress, and truth approximation. Synthese library 287.* Dordrecht: Kluwer Academic Publishers.

Kuipers, T. (2014). Empirical progress and truth approximation revisited. Proceedings Tilburg conference on progress in science (April, 2012). *Studies in History and Philosophy of Science, special section Progress in Science, 46*, 64–72.

Miller, D. (1978). On distance from the truth as a true distance. In J. Hintikka et al. (Eds.), *Essays on mathematical and philosophical logic* (pp. 415–435). Dordrecht: Reidel.

Niiniluoto, I. (1983). Verisimilitude vs Legisimilitude. *Studia Logica, 17*, 315–329.

Niiniluoto, I. (1987). *Truthlikeness. Synthese library 185.* Dordrecht: Reidel.

Niiniluoto, I. (1998). Verisimilitude: The third period. *The British Journal for the Philosophy of Science., 49*(1), 1–29.

Niiniluoto, I. (2017). Optimistic realism about scientific progress. *Synthese, 194*, 3291–3309.

Oddie, G. (1982). Cohen on verisimilitude and natural necessity. *Synthese, 51*, 355–379.

Oddie, G. (1986). *Likeness to truth*. Dordrecht: Reidel.

Oddie, G. (2016). Truthlikeness. In E. N. Zalta (Ed.), *The stanford encyclopedia of philosophy* (Winter 2016 Edition). https://plato.stanford.edu/archives/win2016/entries/truthlikeness

Russell, B. (1927). *The analysis of matter*. London: Kegan Paul.

Schurz, G. (2005). Bayesian HD-confirmation and structuralistic truthlikeness: Discussion and comparison with the relevant-element and content-part approach. In A. Aliseda, R. Festa, & J. Peijnenburg (Eds.), *Confirmation, empirical progress and truth approximation, Essays in debate with Theo Kuipers* (Vol. 1, pp. 141–195). Amsterdam: Rodopi Reply by T. Kuipers, 160–166.

Schurz, G., & Weingartner, P. (1987). Verisimilitude defined by relevant consequence-elements. A new reconstruction of Popper's original idea. In T. Kuipers (Ed.), *What is closer-to-the-truth? (Poznan studies in the philosophy of science and the humanities Vol. 10)* (pp. 47–78). Amsterdam: Rodopi.

Vos, R. (1991). *Drugs looking for diseases*. Dordrecht: Kluwer Academic Publishers.

Worrall, J. (1989). Structural realism: The best of both worlds? *Dialectica, 43*, 99–124.

Zajonc, R. (1968). Attitudinal effects of mere exposure. *Journal of Personality and Social Psychology Monograph Supplement, 9*, 1–27.

Zwart, S. (2001). *Refined verisimilitude. Synthese library 307.* Dordrecht: Kluwer Academic Publishers.

Part I
Closer to the Nomic Truth, Basic Version

Chapter 2
Nomic Truth Approximation by Empirical Progress Revisited

Abstract In my *From Instrumentalism to Constructive Realism* (2000) I have shown how an instrumentalist account of empirical progress can be related to nomic truth approximation. However, it was assumed that a strong notion of nomic theories was needed for that analysis. In this chapter it is shown, in terms of truth- and falsity-content, that the analysis already applies when, in line with scientific common sense, nomic theories are merely assumed to exclude certain conceptual possibilities as nomic possibilities. The resulting generalized and thereby simplified approach to nomic truth approximation was strongly stimulated by the related work on conjunctive theories of Gustavo Cevolani, Vincenzo Crupi and Roberto Festa (*Erkenntnis*, 2011), in particular their systematic way of defining the truth- and falsity-content of a theory. It will be shown that explicating '(nomic) truth approximation' and 'empirical progress' in these terms does justice to some basic instrumentalist/empiricist and realist conditions of adequacy.

Keywords Empirical progress · Nomic truth approximation · Conceptual possibilities · Nomic possibilities · Truth-content · Falsity-content · Success theorem · Instrumentalism · Realism · Exclusion claim

---

Revised version of: "Empirical Progress and Truth Approximation Revisited", Proceedings Tilburg conference on Progress in Science (April, 2012), *Studies in History and Philosophy of Science*, special section *Progress in Science*, 46, 2014, 64–72. *Acknowledgements*: I like to thank Gustavo Cevolani, Roberto Festa, Jan Sprenger and two anonymous referees for their constructive remarks on the draft.

T. A. F. Kuipers, *Nomic Truth Approximation Revisited*, Synthese Library 399,
https://doi.org/10.1007/978-3-319-98388-2_2

2.1 Introduction[1]

The intuitive idea underlying the notion of truth approximation can be expressed as follows: one theory is closer to the truth than another when the first says more things about the domain under investigation and more of all things said are true. The first formal definition of 'closer to the truth', or 'more truthlike' or 'more verisimilar', was proposed by Popper (1963). According to Popper, a theory is more truthlike than another if the former implies more true sentences and fewer false sentences than the latter. Notwithstanding its intuitive appeal, Popper's definition was shown to be untenable by Tichý (1974) and Miller (1974), who independently proved that, according to this definition, a false theory can never be closer to the truth than another (true or false) theory. The Tichý-Miller theorem opened the way to the post-Popperian approaches to truthlikeness, which emerged since 1975. Such approaches escape the strictures pointed out by Tichý and Miller, allowing for a comparison of at least some false theories with regard to their closeness to the truth. Excellent surveys of most post-Popperian accounts of verisimilitude can be found in Niiniluoto (1998), Zwart (2001) and Oddie (2016).

In Kuipers (1982, 1984, 2000) I have developed the so-called *nomic* account. The intuitive idea underlying the notion of *nomic (verisimilitude or) truthlikeness* can be expressed as follows. Given a domain of inquiry and a vocabulary, let U be the set of mutually exclusive and together exhaustive *conceptual possibilities* which might occur within this domain. Hence, U may be construed as the conceptual frame of a given scientific inquiry, specifying the relevant kinds of objects, events or states of natural systems or artifacts under investigation. As an example, U may contain four kinds of object: 'black raven', 'black non-raven', 'non-black raven' and 'non-black non-raven'. One may assume that there is a unique subset **T** of U, including precisely all *nomic possibilities*, i.e., all conceptual possibilities which are 'really' possible in the domain of inquiry. Here, 'nomic possibilities' may assume different meanings, depending on the particular context: e.g., **T** may concern the physical, chemical, biological, psychological or socio-economical possibilities of the domain. Of course, the set U−**T** is then the set of the *nomic impossibilities*, i.e., the set of those conceptual possibilities which are impossible as a matter of fact. For example, **T** might contain the conceptual possibilities 'black raven', 'black non-raven', and 'non-black non-raven', whereas U−**T** might contain the conceptual possibility 'non-black raven'. Since theory-oriented scientists usually aim at understanding and discovering the nomic features of the world, **T** can be construed as the target, or 'the great unknown', of the scientific inquiry of the domain. Accordingly, **T** may be construed as *the* (whole) *nomic truth* about the domain and the *nomic truthlikeness* of a (nomic) theory may be defined as its closeness or simi-

[1] With their consent, I start the introduction of the present chapter with a few paragraphs derived from the joint paper with Cevolani and Festa (Cevolani et al. 2013), viz. of the beginning of the paper and of Sect. 2.2.

larity to the nomic truth **T**. The *logical* problem of nomic truthlikeness thus amounts to the explication of the idea that a given theory is closer to **T** than another.

As long as **T** is unknown, solving the logical problem of nomic truthlikeness by an explication does not enable direct applications. The *empirical* problem of nomic truthlikeness is to relate the explication to empirical evidence in a convincing way, i.e., in such a way that we may have good empirical reasons to conclude, for the time being, that a new theory is closer to the nomic truth than an old one. That is, that we approximate the nomic truth better or more by the new one than by the old. It is plausible to try to link nomic truth approximation in this tentative sense to (an explication of the idea of) empirical progress, a notion that is used by realists and instrumentalists.

More specifically, explicating 'empirical progress' and '(nomic) truth approximation' should do justice to some basic instrumentalist/empiricist and realist Conditions of Adequacy.

- *CA-instrumentalist*: the explication of 'empirical progress' (1) should at most make use of inductive (generalization) steps and (2) should not be laden by realist notions, notably, 'the truth' and 'closer to the truth'.
- *CA-realist*: the explication of 'truth approximation' and 'empirical progress' should be such that (1) 'truth approximation' explains 'empirical progress' and (2) 'empirical progress' supports the 'truth approximation'-hypothesis.

The first condition is important in order to convince instrumentalists that the realist intentions in the second condition pertain to their crucial notion of empirical progress. The notion of 'estimated progress' of Niiniluoto (1987, 2015) cannot work in this respect. It evidently does not satisfy the first condition because it is defined as (the degree of) increase of estimated verisimilitude, a highly plausible, but typical realist notion for it is based on a measure of verisimilitude, viz. Niiniluoto's well-known quantitative measure of verisimilitude.

After 30 years, I discovered in 2012 that my qualitative approach to 'nomic truth approximation by empirical progress' (Kuipers 1982, 1984, 2000) can be presented in a much more general way than I always thought. The definition of 'closer to the truth' can already be conceptually motivated by assuming that the claim of a theory only excludes certain conceptual possibilities as nomic possibilities, i.e., the 'exclusion claim'. I always thought that the 'inclusion claim' had to be added that the not excluded possibilities were nomically possible. The new, generalized and thereby simplified approach to nomic truth approximation was strongly stimulated by the related work on conjunctive theories of Gustavo Cevolani, Vincenzo Crupi and Roberto Festa (2011), in particular their systematic way of defining the truth- and falsity-content of a theory.

In Sect. 2.2 the generalized story based only on the exclusion claim will be presented in its 'basic' form, that isthe explication of nomic truth approximation and its relation to empirical progress assuming that there is just one language that generates the conceptual possibilities. In Sect. 2.3 the story will be concretized by taking the crucial (theory-relative) distinction between an observational language and a theoretical language into account. In this case, nomic truth approximation is of course explicated in terms of the theoretical language and empirical progress in terms of the observational language.

Section 2.3 deals with one possibility of concretization of the basic account presented in Sect. 2.2. In Chap. 3 we will start with indicating several other possibilities of concretization.

2.2 The Basic Story

Let me start with my favorite toy example of theory oriented science (Kuipers 2000, p. 143). To represent an electric circuit with several switches and bulbs one may use a language with elementary propositions that enable to indicate which switches are on and which are off and also to indicate which bulbs give light and which do not. Several of the conceptually possible states will be physically possible, and one of them will be the actual state.

Referring to Fig. 2.1, let p_i for $1 \leq i \leq 4$ indicate that switch *i* is on (⇔) and $\neg p_i$ that it is off (⇕). Let q(¬q) indicate that the bulb lights (does not light). It is assumed that the bulb is not defective and that there is enough voltage. A possible state of the circuit can be represented by a conjunction of negated and un-negated p_i's. It is clear that there is just one true description of the actual state of the circuit as it is depicted, $q\&p_1\&\neg p_2\&p_3\&p_4$, according to the standard propositional representation. Hence, the example nicely illustrates, among others, that we consider 'the actual world' primarily as something partial and local, i.e., one or more aspects of a small part of the actual universe. However, it need not be restricted to a momentary state; it may also concern an actual trajectory of states in a certain time interval. In sum, the actual world is the actual world in a certain context.

To represent the physically possible states by one proposition or theory one will have to design a complex proposition, the nomic truth, in the above case:

$$\mathrm{q} \leftrightarrow \left[\left(\mathrm{p1} \,\&\, p2\right) \mathrm{or}\, \mathrm{p3}\right] \&\, p4$$

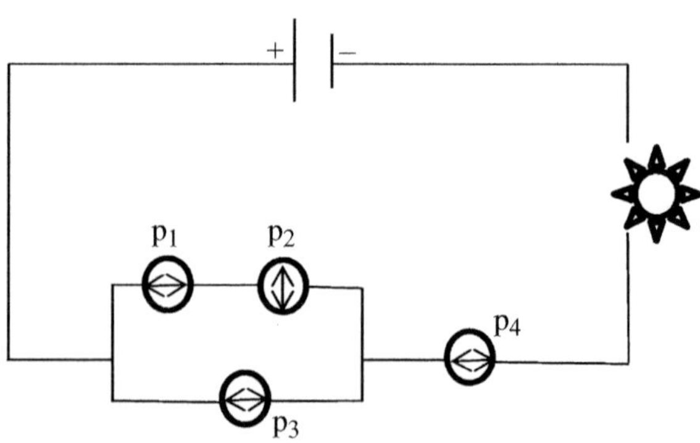

Fig. 2.1 An electric circuit

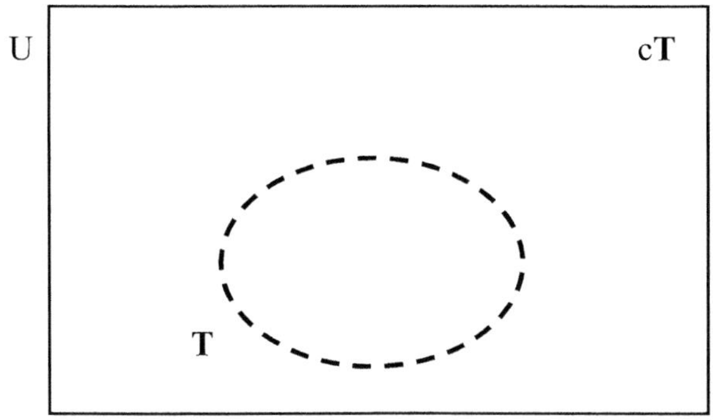

Fig. 2.2 The set of conceptual possibilities U and the (unknown) subset of nomic possibilities **T**

All states in which this proposition is true are physically possible, all others are not. All propositions that can be formulated in the indicated language may be considered as candidates for being this nomic truth and, at least intuitively, one proposition may be closer to the truth than another, viz. when it captures more physically possible states and less impossible ones. This may be a toy example for representing theory oriented science, but in present day epigenetics there is a close analogy: in fact, genes are considered as switches that may be on or off. However this may be, only the general tenet of the example is relevant: theory oriented science is ultimately aiming at characterizing 'the nomic target', that is, what is e.g. physically, biologically or economically possible, and theories are tested by experiments which are realizations of possibilities.

Let U, as above, indicate the set of *conceptual possibilities* in a given domain D (e.g. the possible states or trajectories of a system, or possible kinds of systems), generated by a descriptive vocabulary V in which U is characterized, and subsets of U, e.g. X, Y, P, R, S, can be characterized, technically as sets of set-theoretic structures of a certain similarity type. Let (bold) **T** indicate the subset of *nomic*, e.g. physical, *possibilities*, and hence c**T** (the complement of **T**, U−**T**) the subset of nomic *im*possibilities. The assumption that such a set exists, we call the *Nomic Principle*.[2] By the bold '**T**' we indicate that we do not (yet) dispose of a characterization of it in terms of V. Figure 2.2 indicates this further by depicting **T** as an interrupted ellipse. Note that, whereas U only depends on V, **T** depends on both the domain D and the vocabulary V, hence formally we might write **T**(D,V) instead of **T**, but we shall not do so. The *target of research* is identifying, if possible, **T**'s boundary in V-terms, indicated by (non-bold) T, hence **T** = T, assuming such a characterization exists, which I will do throughout. **T** is called 'the nomic truth' and T may be called 'the *explicit* nomic truth', both for reasons that will become clear

[2] In (Kuipers 2000) this principle and some other general principles were called postulates, but this term will get in this book a more specific use.

soon.[3] From now on, 'the truth' and 'truth approximation' will always refer to the nomic truth and nomic truth approximation, respectively.

In a nomic context attempts to characterize **T** in V-terms are primarily done by theories that exclude certain conceptual possibilities as nomic possibilities. This exclusion function of theories is typically associated with speaking of the axioms, principles or postulates of theories. They have to be satisfied and hence they exclude together everything which does not satisfy them all. Let subset P of U, defined in V-terms, indicate the set of conceptual possibilities (technically, structures) satisfying all postulates of a theory; they are usually called the models of the postulates. Let theory $\underline{P}$, or simply $\underline{P}$, indicate this subset P with the (exclusion) claim "$\mathbf{T} \subseteq P$", or equivalently "$cP \subseteq c\mathbf{T}$", i.e., all non-members of P are *excluded* (by the postulates) as nomic possibilities.[4] Note that this claim is a conjunctive claim in the sense that e.g. "$cP \subseteq c\mathbf{T}$" states that all items in cP are in c**T**, which means that the claim can be seen as a 'conjunctive theory' in the sense of Cevolani et al. (2011).

It is now plausible to define that $\underline{P}$ is *true* iff (if and only if) its claim is true, i.e., iff $cP \subseteq c\mathbf{T}$, that is, iff $cP - c\mathbf{T} = \varnothing$, and hence also iff $\mathbf{T} - P = \varnothing$, and $\underline{P}$ is *false* otherwise. If $P^{\#}$ is a subset of P, the claim of $\underline{P^{\#}}$ is of course stronger than that of $\underline{P}$. Hence, it is easy to see that there is at most one strongest true theory, viz. the characterization of **T** in V-terms, *if it exists*, as before indicated by T, with non-bold 'T', i.e., the target of research! This logical fact justifies calling **T** 'the (nomic) truth' and T 'the explicit (nomic) truth'.

Now it is also plausible to define:

- the *truth-content* of $\underline{P}$, TC($\underline{P}$): the largest subset of cP which is also a subset of c**T**: $cP \cap c\mathbf{T}$, that is, the subset of all members x of cP for which the claim "$x \in c\mathbf{T}$" is true,
- the *falsity-content* of $\underline{P}$, FC($\underline{P}$): the largest subset of cP which is also a subset of **T**: $cP \cap \mathbf{T} = cP - c\mathbf{T}$, that is, the subset of all members x of cP for which the claim "$x \in c\mathbf{T}$" is false.

Figure 2.3 depicts the defined notions.

Note that the truth-content of $\underline{P}$ as well as its falsity-content is 'outside' P due to the fact that they derive from an exclusion claim. Their union equals the comple-

[3] Note already here that, in view of the yes or no matter of being a member of **T**, the nomic truth is of a deterministic nature. This will be assumed in the whole book.

[4] In the structuralist spirit, we might define $\underline{P}$ more formally as the set-theoretic predicate <U, P, **T**> with the claims that P and **T** are subsets of U and that **T** is a subset of P. The following standard set-theoretic notation will be used. For sets A and B, the intersection will be designated by '$A \cap B$', which means the set of all common elements; '$A - B$' indicates the difference, which is the set of elements that are elements of A but not elements of B. '$A \subseteq B$' indicates that A is a *subset* of B and '$A \subset B$' that A is a *proper subset* of B. The first case does not exclude '$A = B$', the second does. '$\varnothing$' indicates the *empty* set.

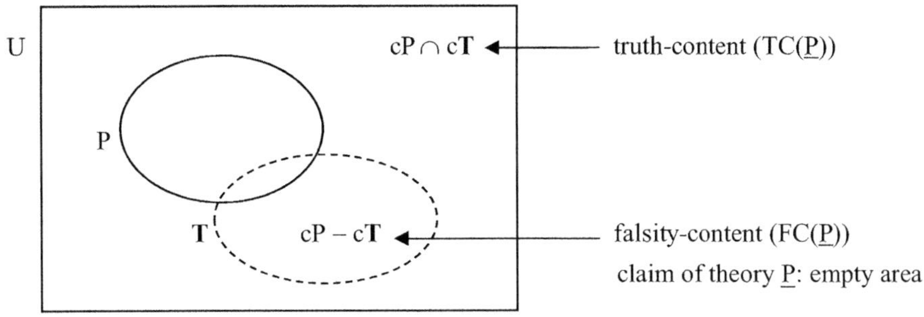

Fig. 2.3 The truth-content and the falsity-content of theory P

ment of P, cP, which I like to call the Popperian content of theory P, for it contains all conceptual possibilities that are excluded for the status of nomic possibility.

Finally it is plausible to define, for theory P and another theory P*:

P* is *at least as close to* ***T***, or to the truth, as P, iff:

TC-clause: the truth-content of P is a subset of that of P*:
TC(P) ⊆ TC(P*), that is, $cP \cap c\mathbf{T} \subseteq cP^* \cap c\mathbf{T}$

&

FC-clause: the falsity-content of P* is a subset of that of P:
FC(P*) ⊆ FC(P), that is, $cP^* - c\mathbf{T} \subseteq cP - c\mathbf{T}$

P* is *closer to* ***T***, or to the truth, than P, iff:
P* is at least as close to **T** as P
& one of the clauses may be strengthened to a proper subset claim.

Of course, we speak of *truth approximation* when a new theory is closer to the truth than an old one.

It is easy to check that the TC- and the FC-clause of 'at least as close' are equivalent to:

$c(P\cup\mathbf{T}) \subseteq c(P^*\cup\mathbf{T})$ resp. $cP^* \cap \mathbf{T} \subseteq cP \cap \mathbf{T}$

$P^*\cup\mathbf{T} \subseteq P\cup\mathbf{T}$ resp. $\mathbf{T} - P^* \subseteq \mathbf{T} - P$ ⟵

⟶ $P^* - \mathbf{T} \subseteq P - \mathbf{T}$ resp. $\mathbf{T} \cap P \subseteq \mathbf{T} \cap P^*$

Hence, the single claim "$cP \subseteq c\mathbf{T}$" already generates by appropriate comparison with its P*-version "$cP^* \subseteq c\mathbf{T}$" the two difference clauses (indicated by arrows), and hence the symmetric difference definition, viz. $P^*\Delta\mathbf{T} \subseteq P\Delta\mathbf{T}$ (Kuipers 1982,

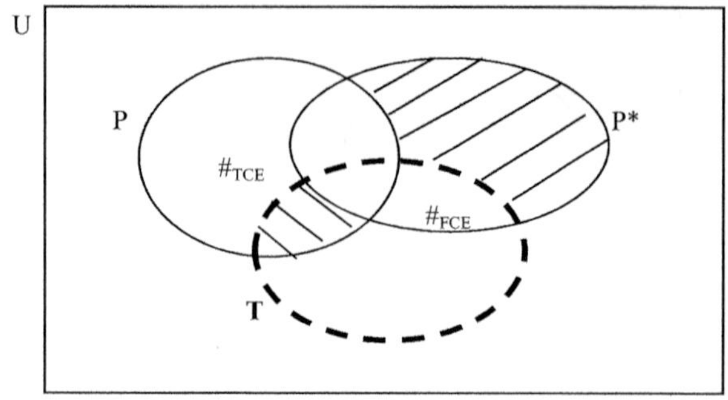

Fig. 2.4 Theory $\underline{P}^*$ is closer to the truth than theory $\underline{P}$

2000).[5,6] To motivate the symmetric difference definition of 'closer to the truth' I always thought to also have to assume the 'inclusion' claim, "$P \subseteq \mathbf{T}$", hence together the 'strong claim': "$P = \mathbf{T}$", but as we have shown this turns out to be formally unnecessary. That is, the symmetric difference definition can be conceptually motivated by defining 'closer to the truth' in terms of truth-content and falsity-content with respect to the exclusion claim alone. Consequently, the so far added 'inclusion' claim "$P \subseteq \mathbf{T}$", leading to the claim "$P = \mathbf{T}$", is totally redundant.

Figure 2.4 represents the situation that theory $\underline{P}^*$ is closer to the truth than theory $\underline{P}$ as follows: shaded areas indicate empty sets and at least one of the two #- areas has to be non-empty. As is easy to check, the ///–shaded areas are empty due to the TC-clause, which is equivalent to $TC(\underline{P}) - TC(\underline{P}^*) = \varnothing$, while the \\\-shaded areas are empty due to the FC-clause, which is equivalent to $FC(\underline{P}^*) - FC(\underline{P}) = \varnothing$. Moreover, $\#_{TCE}$ represents the extra truth-content of $\underline{P}^*$ relative to $\underline{P}$ and $\#_{FCE}$ the extra falsity-content of $\underline{P}$ relative to $\underline{P}^*$.

A propositional example, easily interpretable in terms of a circuit, we can provide for the case that the conceptual possibilities exist of the propositional constituents generated by a finite number n of elementary propositions $p_1, p_2, \ldots p_n$: $(\pm)p_1\&(\pm)p_2\&\ldots(\pm)p_n$, where each $(\pm)$ is either a negation sign or no sign at all. In Table 2.1 with $n = 3$ it is made clear by truth-tables, coupling models or states to propositions, that theory $\underline{P}^*$ '=' $p_1 \Leftrightarrow p_3$, formally: $\underline{P}^* = \underline{Mod(p_1 \Leftrightarrow p_3)}$, is closer to $\mathbf{T}$ '=' $(p_1 \vee p_2) \Leftrightarrow p_3$ than theory $\underline{P}$ '=' $(p_1\&p_2) \Leftrightarrow p_3$. For we get: $Mod(P) = \{s_1, s_6, s_7, s_8\}$, $Mod(P^*) = \{s_1, s_3,$

[5]The symmetric difference of sets A and B, $A\Delta B$, is standardly defined as $(A-B) \cup (B-A)$. It is easy to check that the conditions $P^* - \mathbf{T} \subseteq P - \mathbf{T}$ and $\mathbf{T} - P^* \subseteq \mathbf{T} - P$ together can be abbreviated to $P^*\Delta\mathbf{T} \subseteq P^*\Delta\mathbf{T}$.

[6]Soon after 1982, I came to know of David Miller's symmetric difference definition (Miller 1978), which he however did not put in the nomic context, but in the context of 'the actual or descriptive truth'.

Table 2.1 A propositional example: $\underline{P^*} = \underline{Mod(p_1 \Leftrightarrow p_3)}$ is closer to $\mathbf{T} = \underline{Mod((p_1 v p_2) \Leftrightarrow p_3)}$ than $\underline{P} = \underline{Mod((p_1 \& p_2) \Leftrightarrow p_3)}$

Models/states	p_1	p_2	p_3	P: $(p_1\&p_2)\Leftrightarrow p_3$	P*: $p_1\Leftrightarrow p_3$	**T**: $(p_1 v p_2)\Leftrightarrow p_3$
$s_1 = p_1\&p_2\&p_3$	1	1	1	1	1	1
$s_2 = \neg p_1\&p_2\&p_3$	0	1	1	0	0	1
$s_3 = p_1\&\neg p_2\&p_3$	1	0	1	0	1	1
$s_4 = p_1\&p_2\&\neg p_3$	1	1	0	0	0	0
$s_5 = \neg p_1\&\neg p_2\&p_3$	0	0	1	0	0	0
$s_6 = \neg p_1\&p_2\&\neg p_3$	0	1	0	1	1	0
$s_7 = p_1\&\neg p_2\&\neg p_3$	1	0	0	1	0	0
$s_8 = \neg p_1\&\neg p_2\&\neg p_3$	0	0	0	1	1	1

s_6, $s_8\}$ and Mod(**T**) = $\{s_1,s_2,s_3,s_8\}$. Hence $P^*\Delta\mathbf{T} = \{s_2, s_6\} \subset P\Delta\mathbf{T} = \{s_2, s_3, s_6, s_7\}$.[7] Finally, it is easy to check that $\{s_7\} = \{p_1\&\neg p_2\&\neg p_3\}$ represents the extra truth-content (extra rightly excluded) of $\underline{P^*}$ relative to $\underline{P}$ and $\{s_3\} = \{p_1\&\neg p_2\&p_3\}$ the extra falsity-content (extra wrongly excluded) of P relative to $\underline{P^*}$.

So far, we have been engaged with the logical problem of verisimilitude, that is, explicating 'being closer to the truth', assuming that we dispose of the truth in one way or another. Now we turn to the epistemic problem of specifying empirical conditions that support the conclusion, at least for the time being, that one theory is closer to the unknown truth than another. As we will see, empirical progress is such a condition.

In the nomic context the empirical data are *asymmetric* in the following sense. We can establish by experiments nomic possibilities. In fact, every experiment realizes by definition a nomic possibility. However, it is evident that we cannot establish nomic impossibilities in such a direct way. But in the empirical sciences we use to 'induce' nomic impossibilities indirectly by inductive (empirical) generalizations that we accept, for the time being, on the basis of 'sufficient' experimentation, notably by trying to realize counterexamples. For example, the observation of a black raven is the realization of a nomic possibility, whereas concluding that non-black ravens do not exist is an inductive generalization, viz. all ravens are black.

We indicate the (asymmetric) *data* at a certain moment by R/S, where R indicates the set of realized nomic possibilities (e.g. realized physical possibilities) and S indicates the strongest law induced on the basis of R, that is, the set of conceptual possibilities that are not excluded by any of the accepted inductive generalizations.[8] In this setup cS indicates the set of induced nomic impossibilities. Of course, we

---

[7] Note that in this example exclusion claims are supposed to be associated with theories $\underline{P}$ and $\underline{P^*}$, but it might also be strong claims leading to maximal theories, i.e., theories also claiming $P \subseteq \mathbf{T}$, the inclusion claim. In Chap. 4 we will see that the theories may also be interpreted in terms of an inclusion claim only.

[8] Note that in fact the claim "$\mathbf{T} \subseteq S$", i.e., "$cS \subseteq c\mathbf{T}$", is associated with S, and hence that S is just a special kind of theory, which might hence also be indicated by $\underline{S}$, but we will not do so, for it will always be clear whether we mean the set S or the (strongest) law S, i.e., S with the claim.

may always assume that R ⊆ S, for if not, any element in R−S would represent a realized counterexample to S and hence to at least one of its constituting inductive generalizations.

The following definitions of accepted- and rejected-content are now plausible:

- the *accepted-content* of P, AC(P): the set of 'induced examples' of P: cP∩cS = c(P∪S), that is, the subset of those members of cP which are also excluded by the accepted evidence,
- the *rejected-content* of P, RC(P): the set of realized counterexamples of P: cP∩R = R−P, that is, the subset of those members of cP for which the exclusion claim is rejected by the accepted evidence.

Figure 2.5 depicts the corresponding sets, lying of course again outside P.

Analogous to the previous truth-oriented comparative definitions, we can now define success-oriented comparative notions:

P* is *(empirically) at least as successful* as P, relative to R/S, iff:

AC-clause: the accepted-content of P is a subset of that of P*:
AC(P) ⊆ AC(P*), that is, cP ∩ cS ⊆ cP* ∩ cS, or equivalently
S ∪ P* ⊆ S ∪ P
In words: all induced examples of P are (induced) examples of P*, or, equivalently, all induced laws entailed by P are entailed by P*[9]

&

RC-clause: the rejected-content of P* is a subset of that of P:
RC(P*) ⊆ RC(P), that is, R − P* ⊆ R−P

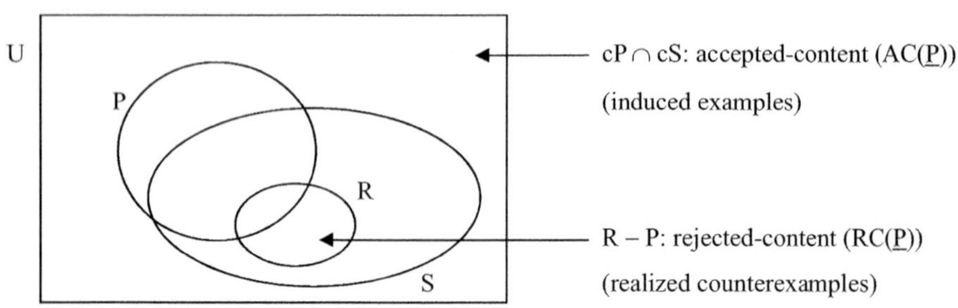

Fig. 2.5 The accepted-content and the rejected-content of theory P relative to evidence R/S

[9] Note first that cP ∩ cS = c(P ∪ S) and hence cP ∩ cS ⊆ cP* ∩ cS is equivalent to c(P ∪ S) ⊆ c(P* ∪ S), hence to S ∪ P* ⊆ S ∪ P. Now if L is a superset of S ∪ P (S ∪ P ⊆ L) it is a superset of both S (S ⊆ L) and P (P ⊆ L). The former implies that L is at least implicitly an induced law, the latter that it is entailed by P. If S ∪ P* ⊆ S ∪ P, as assumed by the AC-clause, any superset L of S ∪ P will be a superset of S ∪ P*, which explains the 'induced law' version of the clause.

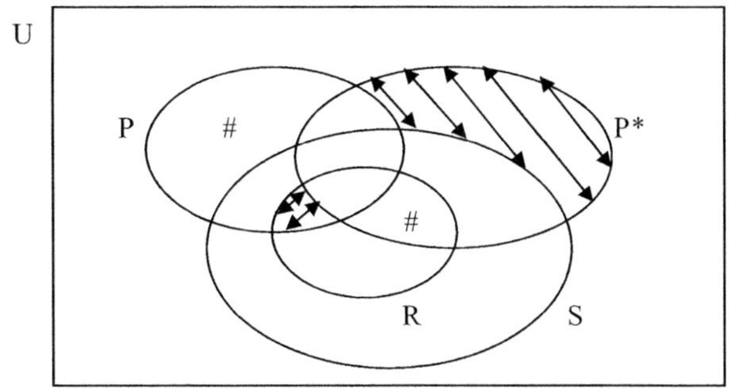

Fig. 2.6 Theory P* is more successful than theory P relative to evidence R/S

In words: all realized counterexamples of P* are (realized) counter examples of P

P* is *more successful than* P, relative to R/S:
P* is at least as successful as P
& one of the clauses may be strengthened to a proper subset claim.

Figure 2.6 represents the situation that theory P* is more successful than theory P as follows: shaded areas indicate empty sets, the ↘ – shaded area due to the AC-clause, which is equivalent to AC(P) – AC(P*) = ∅, and the ↙ – shaded area due to the RC-clause, which is equivalent to RC(P*) – RC(P) = ∅. Moreover, at least one of the two #-areas has to be non-empty.

Now we can turn to the connection between 'closer to the truth' and 'more successful'.[10] To begin with, it is conceptually trivial that, if R is correctly described and if S is correctly induced, $R \subseteq \mathbf{T} \subseteq S$. This will be called the *correct data* (CD-) hypothesis, surely a very strong assumption, in particular the inductive component. However, now it is easy to prove the first clause of the following:

Success Theorem: Assuming the Correct Data (CD-) hypothesis:

(a) if P* is at least as close to the truth as P, P* is at least as successful,
(b) if P* is closer to the truth than P, P* is or will become more successful in the long run.

[10] From here on the section presents an enriched, adapted and simplified version of Sects. 7.3.3 and 7.3.4 of (Kuipers 2000).

Proof (a): as suggested it is formally almost trivial to prove this; however, it is conceptually interesting to write out a proof in some detail. We start with a lemma, following from the CD-hypothesis.

Lemma:	$R \subseteq \mathbf{T} \subseteq S$ i.e., (CD-i) $R \subseteq \mathbf{T}$ & (CD-j) $\mathbf{T} \subseteq S$, entails		
	(i1) $RC(\underline{P}) \subseteq FC(\underline{P})$:		$cP - cR \subseteq cP - c\mathbf{T}$
	(i2) $RC(\underline{P^*}) \cap FC(\underline{P}) \subseteq RC(\underline{P})$		$(cP^* - cR) \cap (cP - c\mathbf{T}) \subseteq cP - cR$
	(j1) $AC(\underline{P}) \subseteq TC(\underline{P})$:		$cP \cap cS \subseteq cP \cap c\mathbf{T}$
	(j2) $AC(\underline{P}) \cap TC(\underline{P^*}) \subseteq AC(\underline{P^*})$:		$(cP \cap cS) \cap (cP^* \cap c\mathbf{T}) \subseteq cP^* \cap cS$

In words: the accepted-/rejected-content of a theory is a subset of its truth-/falsity-content and the accepted-/rejected-content of one theory belonging to the truth-/falsity-content of another also belongs to the accepted-/rejected-content of that other.

Assuming the CD-hypothesis and the TC- and FC-clause, (a) is proven by:

$AC(\underline{P}) \subseteq_{(j1)} TC(\underline{P}) \subseteq_{\text{TC-clause}} TC(\underline{P^*})$ and hence, by (j2), $AC(\underline{P}) \subseteq AC(\underline{P^*})$
&
$RC(\underline{P^*}) \subseteq_{(i1)} FC(\underline{P^*}) \subseteq_{\text{FC-clause}} FC(\underline{P})$ and hence, by (i2), $RC(\underline{P^*}) \subseteq RC(\underline{P})$

Proof (b): here we only give an informal indication. In view of the fact that $\underline{P^*}$ is closer to the truth than $\underline{P}$, there must exist a nomic possibility in $(P^* \cap \mathbf{T})-P$ or a nomic impossibility in $P-(P^* \cup \mathbf{T})$. Hence we may assume that by sufficient experimentation sooner or later such a nomic possibility in $(P^* \cap \mathbf{T})-P$ will be realized or such a nomic impossibility in $P - (P^* \cup \mathbf{T})$ will be induced (by accepting a new inductive generalization). To get a straightforward proof more specific probabilistic conditions about the experimentation have to be assumed (see the note after the proof of the Expected Success Theorem in Sect. 5.2).

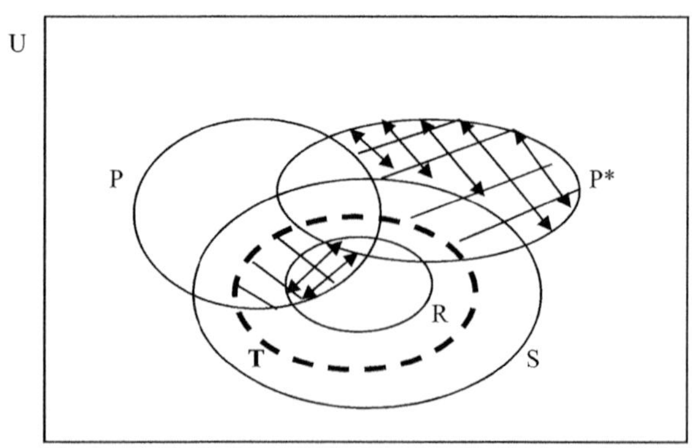

Fig. 2.7 A visualization of the proof of part (a) of the Success Theorem

Figure 2.7 makes the proof of (a) visible. Assuming $R \subseteq \mathbf{T} \subseteq S$, the emptiness of the simple shaded areas, representing "theory $\underline{P}^*$ is at least as close to the truth as theory $\underline{P}$", guarantees the emptiness of the areas shaded with double arrows, representing "theory $\underline{P}^*$ is at least as successful as theory $\underline{P}$".

Now we are in the position to explicate the notion of *empirical progress*. If a new theory, $\underline{P}^*$, is more successful than the present one, $\underline{P}$, relative to the available data R/S, this suggests the

Empirical Progress (EP-)hypothesis,
$\underline{P}^*$ (is and) remains more successful than $\underline{P}$, i.e.,

- all past and future induced examples of $\underline{P}$ are examples of $\underline{P}^*$,
- all past and future realized counterexamples of $\underline{P}^*$ are counterexamples of $\underline{P}$,
- $\underline{P}^*$ will have at least some extra induced examples, at least in the past,
- or $\underline{P}$ will have at least some extra realized counterexamples, at least in the past.

Note that this is a comparative hypothesis, which can be tested by designing tests that may confirm or falsify the EP-hypothesis. If it is falsified, it is a first illustration of 'divided success', i.e., the situation that a theory is more successful in some respects than another, *and vice versa*.

The core idea of *Empirical Progress* now is the acceptance of the EP-hypothesis when it has been 'sufficiently confirmed' to be accepted as true, at least for the time being. Of course, being 'sufficiently confirmed' will always be a matter of debate.

Recall that the explication of 'empirical data' and of 'more successful' only made use of descriptive and inductive steps. Moreover, the definition of Empirical Progress is based on two (comparative) inductive steps, viz. the two inductive generalizations of the EP-hypothesis, that is, all (past) and future induced examples of $\underline{P}$ are examples of $\underline{P}^*$, and all (past) and future realized counterexamples of $\underline{P}^*$ are counterexamples of $\underline{P}$. Finally, the definition does not depend on that of **T**, let alone on that of 'closer to **T**'. Hence, the instrumentalist condition of adequacy (CA-instrumentalist) is satisfied.

This brings us to the crucial point of the analysis: the connection between Truth Approximation (TA) and Empirical Progress (EP), i.e., the satisfaction of the realist condition of adequacy consisting of two requirements, TA should explain EP and EP should justify TA.

The first is simple, according to the Success Theorem, TA entails EP, and hence it explains EP, at least in a basic sense. The immediate consequence of this is that EP suggests the Truth Approximation (TA-)hypothesis: $\underline{P}^*$ is closer to the truth than $\underline{P}$. The way of suggesting is straightforwardly abductive in the Peircean sense: given EP and given that TA explains EP, "hence, there is reason to suspect that [the TA-hypothesis] is true", to quote Peirce's famous conclusion.

The TA-hypothesis is to be tested by testing the EP-hypothesis, for the former entails the latter. The EP-hypothesis even specifies all its empirical consequences.

Moreover, the *reverse consequences* of the Success Theorem are such that EP not only suggests the TA-hypothesis, they also justify it to a substantial extent:

- the Success Theorem not only makes it perfectly possible that P* is closer to the truth than P, it would even explain the greater success,
- it is impossible that P* is further from the truth than P (and hence that P is closer to the truth than P*), for otherwise, so shows the Success Theorem, P* could not be more successful,
- it is also possible that P* is neither closer to, nor further from the truth than P, in which case, however, another specific explanation has to be given for the fact that P* has so far proven to be more successful, e.g. by a biased choice of experiments.

Note that the TA-hypothesis provides a typical *default* explanation of EP, that is, an adequate explanation *unless* there turns out to be reason for another diagnosis. The third reverse consequence provides the room for 'unless' conditions and hence for future 'divided success'.

Our conclusion from the reverse consequences is that Empirical Progress justifies "Inference to the Best Theory (among the available theories) as the closest to the truth" (IBT), i.e., acceptance of the TA-hypothesis, at least for the time being. IBT is a sophisticated version of IBE (Inference to the Best Explanation), see (Kuipers 2004; Chap. 11 of this book). Note that IBT is not just an inductive generalization step, for it explicitly refers to '(closer to) the truth'.

In view of the foregoing analysis we may conclude that our explication of nomic truth approximation satisfies the realist condition of adequacy (CA-realist). In combination with our explication of empirical progress, satisfying the instrumentalist condition of adequacy (CA-instrumentalist), we may conclude that 'nomic truth approximation by empirical progress' is perfectly possible by only assuming an exclusion claim of theories.[11] However, we did not yet make a distinction between observational and theoretical terms, for which reason we call the above analysis the (generalized) basic story.

2.3 The Stratified Story

The stratified story is based on the assumption that there is a theory-relative distinction between observational and theoretical terms, that is, although the observational terms may well be laden with underlying theories, they are not laden with the

[11] One of the relieving consequences of the possibility to base nomic truth approximation merely on the exclusion claim, is that my paper (Kuipers 2011) on truth approximation by refined AGM-style belief *set* revision is not at all as *ad hoc* as I thought: after the AGM-revision of a theory by evidence, I wrongly assumed to have to add the inclusion (representation) claim in order to prove nomic truth approximation. Chap. 14 presents a revised version of the paper of 2011.

theories we are interested in. Like the excluding and nomic nature of the claim of theories, the distinction of an observational and a theoretical level (o- and t-level) is also typical for theory-oriented science.[12] In the electric circuit example, at the beginning of Sect. 2.2, the distinction arises by assuming that some or even all connections between the switches and the bulbs are hidden and hence invisible.

I will restrict the attention to the relevant notions of truth- and falsity-content, 'closer to the truth' and 'more successful' and the relevant relations. Apart from some qualifications, the rest of the story remains the same: the EP-hypothesis, Empirical Progress, the TA-hypothesis and 'inference to the best theory'.

So, let there be for the t-level an encompassing vocabulary V_t with a sub-vocabulary V_0 on the o-level, with the consequence that there are two universes or levels of conceptual possibilities, U_t and U_0, the first based on V_t, the second only on V_0. Of course, every member of U_t has a unique counterpart in U_0, but not vice versa. Hence, let π indicate the (many-one) projection function of members and subsets of U_t to U_0. For the projection of subsets P of U_t we will replace the standard notation $\pi(P)$ by πP.

The realist will now assume the Nomic Principle on the t-level, that is, that there is a subset $\mathbf{T_t}$ of $\mathbf{U}_t$, of nomic possibilities. From this realist point of view the primary target of research is to find a characterization of $\mathbf{T}_t$ in terms of V_t, i.e., (non-bold) T_t. Whether or not there is such a characterization, the realist will also assume that there is a subset $\mathbf{T}_0$ of U_0, if possible to be characterized in V_0, i.e., by some T_0. The instrumentalist has at most to assume that the target of research is $\mathbf{T}_0$. The realist will claim that this must be $\pi\mathbf{T}_t$, i.e., $\mathbf{T}_0 = \pi\mathbf{T}_t$, called the *Truth Projection (TP-) Principle*, and hence $T_0 = \pi T_t$.[13] If $\mathbf{T}_0$ has no characterization only using V_0, the instrumentalist will also be happy with an indirect characterization πT_t, based on (non-bold) T_t, and hence V_t, but of course merely as an instrument.

Assuming theories are aiming at $\mathbf{T}_t$, hence on the t-level, with the corresponding exclusion claim, there is again at most one strongest true theory, 'the truth', viz. T_t (non-bold). Hence, we can also take over the definitions of truth- and falsity-content, of 'closer to the truth' and of truth approximation; all with the adjective 'theoretical', when confusion might arise otherwise.

On the o-level we can do all the same, now with the adjective 'observational'. Moreover, the notions of 'more successful' and 'empirical progress' have now to be localized on this level. Hence, on the o-level the full basic story of Sect. 2.2 applies, including the connection between empirical progress and (observational) truth approximation.

The crucial question is: what is the relation between theoretical and observational truth approximation and, hence, indirectly, what is the relation between theoretical truth approximation and empirical progress? To study this we need to focus on the relation between 'closer to the truth' on the two levels. It is not difficult to

[12] This section is heavily based on (Kuipers 2000, Sect. 9.1.1, and 2004, Sect. 3.1).

[13] See (Kuipers 2000, pp. 215–7) for a detailed motivation of the TP-principle.

check that the TC-clause on the t-level is projected or preserved on the o-level. That is, if the TC of $\underline{P^*}$ includes the TC of $\underline{P}$, the TC of $\underline{\pi P^*}$ includes that of $\underline{\pi P}$. This can be transformed into: "$P^* - (P \cup \mathbf{T}_t) = \varnothing$" implies "$\pi P^* - (\pi P \cup \pi \mathbf{T}_t) = \varnothing$", which is easy to check. However, the FC-clause is not simply preserved by projection. That is, if the FC of $\underline{P^*}$ is a subset of the FC of $\underline{P}$, the FC of $\underline{\pi P^*}$ need not be a subset of the FC of $\underline{\pi P}$. This can be transformed into: "$(\mathbf{T}_t \cap P) - P^* = \varnothing$" does not imply "$(\pi \mathbf{T}_t \cap \pi P) - \pi P^* = \varnothing$".

This means that if $\underline{P^*}$ is closer to the truth than $\underline{P}$ on the theoretical level, there still exists the possibility on the theoretical level of a 'lucky miss' of $\underline{P}$ relative to a mistaken exclusion of both $\underline{P}$ and $\underline{P^*}$, leading to a counterexample to the claim that $\underline{P^*}$ is at least as close to the truth as $\underline{P}$ on the observational level. That is, there may be a theoretical (nomic) impossibility *x*, that is rightly excluded by $\underline{P^*}$, but not by $\underline{P}$ (hence, missed by $\underline{P}$), however such that there is also a theoretical possibility *z*, wrongly excluded by both, that has the same observational projection as *x*.

A picture will be illuminating. Figure 2.8 represents on the t-level "theory $\underline{P^*}$ is at least as close to the truth as theory $\underline{P}$" by two empty shaded areas. Regarding the TC-clause, it is easy to see that if the ///–shaded area on the t-level is empty, the projections are such that the similarly shaded area on the o-level must also be empty. Regarding the FC-clause, however, the emptiness of the \\\-shaded area on the t-level, leaves room for the non-emptiness of the formally corresponding area, witness the indicated possibility that the projections of *x* (the lucky miss of $\underline{P}$) and *z* (the common mistaken exclusion) are the same, viz. *a*.

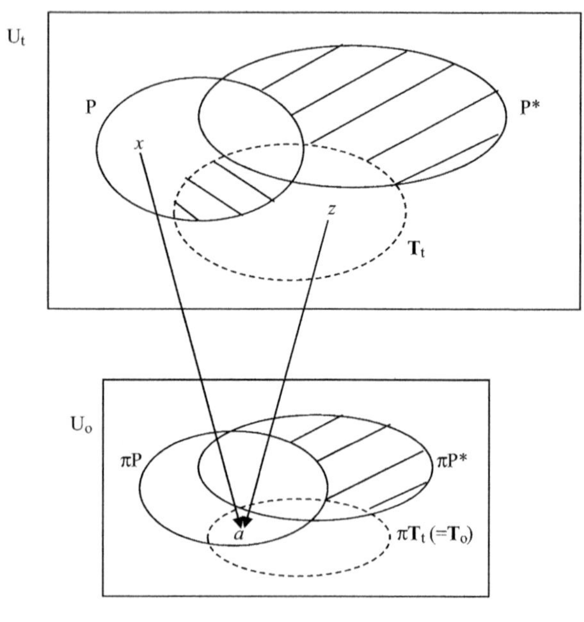

Fig. 2.8 Explanation, see text

Such potential lucky misses can be excluded in a useful general way by assuming that theory $\underline{P}$ is relatively correct, that is, $\pi(P\cap\mathbf{T}_t)$ is not only, by the nature of projection, a subset of $\pi P\cap\pi\mathbf{T}_t$, but also vice versa, and hence:

$\pi(P\cap\mathbf{T}_t) = \pi P\cap\pi\mathbf{T}_t$ theory $\underline{P}$ is *relatively correct*

The name of the assumption is based on the fact that it precisely guarantees that (relative to $\pi\mathbf{T}_t$) correct items of theory $\underline{\pi P}$ (i.e., in the sense of correctly not excluded by theory $\underline{\pi P}$) can be extended to correct items of theory $\underline{P}$ (relative to $\mathbf{T}_t$, in the corresponding sense). In other words, it guarantees that $\underline{P}$ has theoretically correct reproductions of all its observationally correct items. How strong is this assumption? It means something like: theory $\underline{P}$ is on the right track, as far as V_t relative to V_0 is concerned. Apart from this, it is not a very strong condition, as argued in some detail in Kuipers (2000, p. 213).

Now it is easy to prove the following

Projection Theorem:
If theory $\underline{P^*}$ is at least as close to the theoretical truth as theory $\underline{P}$, and if $\underline{P}$ is relatively correct, then theory $\underline{\pi P^*}$ is at least as close to the observational truth as theory $\underline{\pi P^*}$.

Together with the Success Theorem, applied on the o-level, we get the conditioned consequence that theoretical truth approximation entails more successfulness.

Success Corollary:
If theory $\underline{P^*}$ is at least as close to the theoretical truth as theory $\underline{P}$, *and if $\underline{P}$ is relatively correct*, then theory $\underline{\pi P^*}$ is always at least as successful as theory $\underline{\pi P^*}$.

For the opposite direction, we assume of course that empirical progress has been established in the sense described in Sect. 2.2, now on the o-level, hence it remains to satisfy the instrumentalist condition of adequacy (CA-instrumentalist). In the stratified situation empirical progress not only suggests the observational truth approximation (oTA-)hypothesis, but also the theoretical truth approximation (tTA-) hypothesis, having the same overall test implication, viz. the empirical progress (EP-)hypothesis, however now with the proviso that the discarded theory ($\underline{P}$) is relatively correct. Hence, the empirical progress conclusion now not only justifies 'inference to the best theory (among the available theories) as the closest to the *observational* truth', i.e., acceptance of the oTA-hypothesis, at least for the time being. It also justifies, though evidently in a weaker sense, 'inference to the best theory as the closest to the *theoretical* truth', i.e., acceptance of the tTA-hypothesis, at least for the time being. In which sense 'weaker' is difficult to spell out. However, it is at least possible to construe a case of theoretical truth approximation by some revision of the two theories.[14]

Note that potential lucky misses may wrongly block the theoretical truth approximation conclusion. It may occur that the AC (accepted-content) of theory $\underline{P^*}$ exceeds the AC of theory $\underline{P}$, but that the former's RC (rejected-content) is not a subset of the latter's RC, that is, that $\underline{P^*}$ has *extra* counterexamples due to the fact that these items are lucky misses of $\underline{P}$. This possibility gives reason for the realist to introduce the notion of one-sided or *explanatory* empirical progress: the AC of theory $\underline{P}$ is a proper subset of the AC of theory $\underline{P^*}$, that is, all induced laws entailed by $\underline{P}$ form a *proper* subset of the induced laws entailed by $\underline{P^*}$, even if the latter has extra counterexamples. The explanatory progress might be due to theoretical truth approximation, and the extra counterexamples of $\underline{P^*}$ might be cases of lucky misses of $\underline{P}$.

However, this may be, since the tTA-hypothesis explains EP, at least when assuming relative correctness, and since EP supports the tTA-hypothesis, the realist condition (CA-realist) is also satisfied.

2.4 Conclusion

To be sure, the analysis in Sect. 2.2 is based on the simplest assumptions about the further nature of theories and their claims. In Sect. 2.3 we concretized one such an assumption. In the next chapter I will indicate a number of perspectives for concretization, including stratification, as far as relevant, in line with those of Kuipers (2000). They will be elaborated in later chapters.

References

Cevolani, G., Crupi, V., & Festa, R. (2011). Verisimilitude and belief change for conjunctive theories. *Erkenntnis, 75*(2), 183–202.

Cevolani, G., Festa, R., & Kuipers, T. (2013). Verisimilitude and belief change for *nomic* conjunctive theories. *Synthese, 190*, 3307–3324.

Kuipers, T. (1982). Approaching descriptive and theoretical truth. *Erkenntnis, 18*(3), 343–378.

Kuipers, T. (1984). Approaching the truth with the rule of success. In P. Weingartner & C. Pühringer (Eds.), *Philosophy of science – History of science, selection 7th LMPS Salzburg 1983* (pp. 244–253, Philosophia Naturalis 21.2–4).

Kuipers, T. (2000). *From instrumentalism to constructive realism. On some relations between confirmation, empirical progress, and truth approximation. Synthese Library 287*. Dordrecht: Kluwer Academic Publishers.

[14] Assuming that $\underline{\pi P^*}$ is closer to the truth than $\underline{\pi P}$ and defining $P^\# =_{def} P - \{x \in P - P^* / \pi(x) \in \pi P^* \cap To\}$ and $P^{*\#} =_{def} P \cap P^*$ (see Kuipers 2000, p. 214), then $\underline{P^{*\#}}$ is closer to the truth than $\underline{P^\#}$.

Kuipers, T. (2004). Inference to the best theory, rather than inference to the best explanation. Kinds of abduction and induction. In F. Stadler (Ed.), *Induction and deduction in the sciences. Proceedings of the ESF-workshop induction and deduction in the sciences, Vienna, July, 2002* (pp. 25–51). Dordrecht: Kluwer Academic Publishers.

Kuipers, T. (2011). Basic and refined nomic truth approximation by evidence-guided belief revision in AGM-terms. *Erkenntnis, 75*(2), 223–236.

Miller, D. (1974). Popper's qualitative theory of verisimilitude. *The British Journal for the Philosophy of Science, 25*(2), 166–177.

Miller, D. (1978). On distance from the truth as a true distance. In J. Hintikka et al. (Eds.), *Essays on mathematical and philosophical logic* (pp. 415–435). Dordrecht: Reidel.

Niiniluoto, I. (1987). *Truthlikeness. Synthese Library 185*. Dordrecht: Reidel.

Niiniluoto, I. (1998). Verisimilitude: The third period. *The British Journal for the Philosophy of Science, 49*(1), 1–29.

Niiniluoto, I. (2015). Scientific progress. *The Stanford encyclopedia of philosophy* (Summer 2015 ed., E. N. Zalta, Ed.). https://plato.stanford.edu/archives/sum2015/entries/scientific-progress/

Oddie, G. (2016). Truthlikeness. *The Stanford encyclopedia of philosophy* (Winter 2016 ed., E. N. Zalta, Ed.). https://plato.stanford.edu/archives/win2016/entries/truthlikeness

Popper, K. R. (1963). *Conjectures and refutations: The growth of scientific knowledge* (3rd ed.). London: Routledge and Kegan Paul.

Tichý, P. (1974). On Popper's definitions of verisimilitude. *The British Journal for the Philosophy of Science, 25*(2), 155–160.

Zwart, S. (2001). *Refined verisimilitude. Synthese Library 307*. Dordrecht: Kluwer Academic Publishers.

Chapter 3
Perspectives and Guidelines for Theories of (Nomic) Truth Approximation by Empirical Progress

Abstract The account of nomic truth approximation by empirical progress presented in Chap. 2 is based on the exclusion claim of theories. It is a simplified and generalized version of the one presented in my book *From Instrumentalism to Constructive Realism* of 2000. However, its core itself, called the *basic* account, is based on several idealizations. In this chapter I will sketch a number of possible concretizations, to begin with a two-sided, otherwise also basic, version, based on an exclusion and inclusion claim, followed by quantification, refinement and stratification of this two-sided version. They will be elaborated in Chaps. 4, 5, 6, and 7.

Chapter 2 also illustrated straightforwardly how a theory of truth approximation in relation to empirical progress can be set up. The general tenet and the common elements will be summarized in this chapter. It will lead to a general outline of theories of nomic truth approximation by empirical progress. This will guide, explicitly or implicitly, the elaborating Chaps. 4, 5, 6, and 7.

Moreover, the common core of such theories of nomic truth approximation will also be summarized: the Nomic Principle, postulating a unique divide between nomic possibilities and impossibilities, the formal structure of nomic evidence, in terms of realized (nomic) possibilities and induced generalizations, and the hypothetico-deductive background of nomic evidence.

Keywords Guidelines for theory of truth approximation · Nomic principle · Nomic evidence · Truth approximation · Empirical progress · Exclusion claim · Inclusion claim · Two-sided theories · Quantification · Refinement · Stratification · Hypothetico-deductive method background of nomic evidence

3.1 Introduction

The account of truth approximation by empirical progress presented in Chap. 2 is based on the exclusion claim of theories. It is a simplified and generalized version of the one presented in my book *From Instrumentalism to Constructive Realism* of 2000. However, its core itself, called the *basic* account, is based on several idealizations. In Sect. 3.2 I will sketch a number of possible concretizations, to begin with

T. A. F. Kuipers, *Nomic Truth Approximation Revisited*, Synthese Library 399,
https://doi.org/10.1007/978-3-319-98388-2_3

a two-sided, otherwise also basic, version, based on an exclusion and inclusion claim, followed by quantification, refinement and stratification of this two-sided version. They will be elaborated in Chaps. 4, 5, 6, and 7.

Chapter 2 also illustrated straightforwardly how a theory of truth approximation in relation to empirical progress can be set up. The general tenet and the common elements will be summarized in Sect. 3.3. It will lead to a general outline of theories of nomic truth approximation by empirical progress. This will guide, explicitly or implicitly, the elaborating Chaps. 4, 5, 6, and 7.

Moreover, the common core of such theories of nomic truth approximation will be summarized in Sect. 3.4: the Nomic Principle, postulating a unique divide between nomic possibilities and impossibilities, the formal structure of nomic evidence, in terms of realized (nomic) possibilities and induced generalizations, and the hypothetico-deductive background of nomic evidence.

3.2 Perspectives for Concretization

In this section I will briefly indicate the following perspectives for concretization of basic (one-sided) truth approximation by empirical progress:

- a two-sided approach, based on an exclusion and an inclusion claim
- quantification, based on some measure function
- refinement, based on an underlying, ternary similarity relation
- stratification, based on a distinction between observational and theoretical terms

As far as relevant, the concretizations will largely be in line with those dealt with in Kuipers (2000). They concern quantification, refinement and stratification. However, the first concretization, the two-sided approach, is new.

3.2.1 The Two-Sided Approach

The possibility of the generalized (basic) approach presented in Chap. 2, opens the way for an alternative. In fact, it is not difficult to check that if we replace in Sect. 3.2.2 the exclusion claim ("$\mathbf{T} \subseteq$ P") by the inclusion claim ("P $\subseteq \mathbf{T}$") and proceed conceptually otherwise in the same way, we get formally the same symmetric difference definition of 'closer to the truth'. However, the formal versions of the truth- and falsity-content clauses have been interchanged.

The consequence of these two roads to the symmetric difference definition is that we can distinguish three types of theories and methods of (nomic) truth approximation, viz. truth approximation by theories based on an exclusion claim, from now on called 'exclusion theories', by theories based on an inclusion claim, called 'inclusion theories', and by theories based on a combination of such claims, called 'two-sided' theories. Assuming a set U of conceptual possibilities, two-sided theories are theories consisting of two subsets of U, <M, P>, a set of models M being a subset of P (the set of models of some postulates), with the two-sided claim "$M \subseteq \mathbf{T} \subseteq P$". An extreme case of a two-sided theory, called *maximal* theory, results when M = P, with the resulting (strong) claim "M (= P) = **T**". As explained in the previous chapter, I wrongly assumed for 30 years (1982–2012) that this strong claim was necessary for conceptually motivating the symmetric difference definition.

All three methods of truth approximation fit in the conjunctive approach to truth approximation (Cevolani et al. 2011; Kuipers 2014), which is based on an idea of Roberto Festa. Crucial is the way of defining the truth- and the falsity-content. In (Cevolani et al. 2013) it is already shown that my nomic approach (Kuipers 2000) can be presented as a special case of the conjunctive approach. Without stating this clearly, that paper in fact generalizes the nomic approach to two-sided theories, with on the one hand maximal theories as extreme special case and on the other hand two otherwise extreme kinds, viz. two one-sided theories, that is, theories with merely an inclusion claim (inclusion theories) or merely an exclusion claim (exclusion theories).

In Chap. 4 I will elaborate the basic version of the two-sided approach in some detail. It will already be clear that this theory of nomic truth approximation becomes conceptually in complete harmony with two *prima facie* opposing functions of theories. On the one hand, the Popperian or *exclusion view* that theories exclude by their postulates certain (conceptual) possibilities from occurring or being realizable. On the other hand, there is the inclusion or *representation view* that theories represent, by specification of models, certain possibilities as realized or realizable, at least approximately, due to idealizations. To keep these two functions of theories in mind, the two-sided approach will also provide the point of departure for the concretizations in Chaps. 5, 6, and 7: quantification, refinement, and stratification, to be indicated now.

3.2.2 Quantification

It is important to note that 'closer to the truth' and 'more successful' in the basic sense are partial order relations. Hence, theories will frequently not be comparable in either direction. There are at least two plausible way-outs. From a methodological point of view it seems important to have a strategy to deal with cases of 'divided success', that is, when the one theory is more successful in some respects and the other in other respects. The qualitative ideal suggests to try to apply in this situation a kind of 'principle of dialectics', that is: *try to improve both theories in one stroke*. In other words, try to design a new theory that is and remains more successful than both, that is, try to achieve genuine empirical progress, and hence, presumably, truth approximation with respect to both theories.

Another way to deal with the non-comparability problem is to design a quantitative concretization. In a finite context it is even plausible to just count number of elements, leading to the quantitative symmetric difference definition of 'closer to the truth', viz. $|P^*\Delta T| \leq |P\Delta T|$. However, as soon as one wants to differentiate between the weight of 'successes' and 'failures' or if U is infinite, ad hoc elements, e.g. weighing factors and other parameters, are unavoidable, witness Niiniluoto's otherwise impressive approach (Niiniluoto 1987). In Chap. 5 I will nevertheless present a general quantitative (two-sided basic) approach, viz. a measure-theoretical one. It is largely in the spirit of Kuipers (2000, Chap. 12). It leads almost always to an ordering of two theories, however, with a non-deductive 'success theorem', in terms of expectation values, about the relation between truth approximation and the corresponding quantitative notion of empirical progress. But after sufficient confirmation of the quantitative empirical progress hypothesis, the theorem will support the abductive quantitative 'closer to the truth' conclusion substantially.

The quantitative account will be given a probabilistic turn in Chap. 13, linking it to a probabilistic version of the hypothetico-deductive method, that is, the hypothetico-*probabilistic* method.

3.2.3 Refinement

There are two plausible qualitative concretizations of the basic version of the two-sided approach presented in Chap. 4, refinement and stratification (cf. Kuipers 2000, Chaps. 10 and 9, respectively). The first qualitative concretization, refinement, takes into account that one counterexample is less severe than another, e.g. in the sense that it is less idealized than the other. In (Kuipers 2000, Chap. 10) I have presented such a refined approach to empirical progress and (nomic) truth approximation based on an underlying ternary similarity relation between possibilities, that is, structures, and hence called a structurelikeness relation. Of course, the point of departure was then the strong claim, combining the inclusion and the exclusion method in an extreme complementary way.

From the present two-sided perspective, the refined definitions of 'closer to the truth' and 'more successful' seem to be primarily concretizations of the clauses corresponding to the representation function of inclusion theories. In the refined inclusion method new theories revise old theories by changing the models of the theories to some extent, e.g. by taking new factors into account that have been neglected before. The new models are claimed to be more similar to 'the true ones' than those before. Hence, although the exclusion method may appeal to Popperian intuitions, revising theories in the suggested refined way seems to reflect the representation part of scientific common sense. Therefore, it is plausible to think that an adequate modeling of scientific common sense has to take into account both intuitions. This is perfectly possible from the present perspective: two-sided theories for which the basic clauses are used for the exclusion subtheories, whereas the refined clauses are used for the inclusion subtheories. In terms of (Zwart 2001), this is a way of combining the (Popperian) 'content-approach', with the 'similarity approach' to truth approximation, à la Niiniluoto (1987) and Oddie (1986).

The suggested asymmetric way of dealing with refinement may seem plausible; in the practice of science refinement occurs on both sides. E.g. Einstein's postulates refine, relative to Newton's postulates, which possibilities are excluded from being realizable.

Since it is perfectly possible to translate the refinement of the model-side of theories to their postulates, by taking suitable complements, leading to the refinement of the postulate side, we prefer to conceive the refinement of truthlikeness of two-sided theories primarily in a symmetric way (see Chap. 6). In this symmetric version, 'closer to the truth' is, roughly, defined by requiring on both sides a larger truth-content and a less problematic falsity-content, the latter in the sense that the falsity-content of the one theory is, in terms of the structurelikeness relation, more similar to (part of) the truth than the falsity-content of the other. From this symmetric version it is easy to derive the above suggested asymmetric version by 'idealizing' the exclusion side in order to get its basic version back. This can be done by assuming a 'trivial' similarity relation.

Finally, it turns out that 'more successful' can best be refined in a somewhat weaker, but more plausible way than before, with the attractive consequence that the Success Theorem remains unconditionally valid: truth approximation in the refined sense entails at least as successfulness in the refined sense.

3.2.4 *Stratification*

The second important qualitative concretization of the basic version of the two-sided approach deals with stratification in terms of an observational and a theoretical level. It is more or less crucial for the realism/instrumentalism debate. In Sect. 3.2.3 I have already presented stratification for exclusion theories, in line with Kuipers (2000, Chap. 9). As we saw, it leads to some substantial weakening of the connection between empirical progress and truth approximation, but the connection remains remarkable. In Chap. 7 I will elaborate this stratification for two-sided theories. The crucial question for both sides is to what extent 'closer to the truth' on the theoretical level is projected on the observational level. Though formally similar, the possible exceptions have different methodological impact for the two sides, due to the asymmetric nature of evidence in the nomic context. It will also be indicated how stratification is possible for the refined and quantified versions of 'closer to the truth', with specific limits to its projection on the observational level.

The observational/theoretical distinction not only suggests the problem of truth approximation on the theoretical level, but also the problem of the reference of theoretical terms. However, although it is not difficult to adapt the analysis of *referential* truth approximation, given in Kuipers (2000, Chap. 9) for maximal theories to exclusion theories, it is not easy to see how to take inclusion theories, and hence genuinely two-sided theories into account. For this and some other reasons the adapted version for exclusion theories is postponed to Chap. 11.

3.2.5 *Revising the Vocabulary*

There is one type of concretization that is too difficult to take into account in some systematic way: a change of vocabulary. So far a given and fixed vocabulary was assumed. However, in practice, guided by empirical findings formulated within a given vocabulary, vocabularies may be enriched or changed in some other more or less rigorous way. However this may be, comparing theories formulated within different vocabularies is difficult, even if the relevant observational vocabulary is not changed. Although some kind of fundamental incommensurability may even be claimed in theory, some kind of pragmatic comparability usually remains possible, not only regarding empirical success, but also regarding truth approximation. We refer to (Kuipers 2000, Sects. 6.3 and 9.3) for a number of relevant considerations, that largely remain to apply to (concretizations of) two-sided theories.

3.3 Guidelines for Theories of (Nomic) Truth Approximation by Empirical Progress

The previous chapter illustrated straightforwardly how a theory of truth approximation in relation to empirical progress can be set up. The general tenet and the common elements will be summarized in this section. It will lead to a general outline of theories of nomic truth approximation by empirical progress. It will guide, explicitly or implicitly, the Chaps. 4, 5, 6, and 7, elaborating the indicated concretizations.

On the basis of previous work (Kuipers 2000; Cevolani et al. 2011) and the previous chapter the following crucial steps in setting up any specific theory of nomic truth approximation indicated by empirical progress can be discerned.

I General logico-semantic step: Intentional description of the target

- Assuming a domain of research and a vocabulary, clarify the target you are aiming at, that is, about which you aim to approach the truth. In the nomic case, the target is '*the nomic truth*', i.e., which conceptual possibilities are nomically possible in the domain of research and which are not, a dichotomy that is based on the *Nomic Principle*.

II Specific logico-semantic steps: Define '*closer to the truth*'[1]

- Define the notion of theory and define next the notion of true (false) theories and hence, if possible, of the strongest true theory, i.e., '*the truth*'.
- Define the notions of the '*truth-content*' and the '*falsity-content*' of a theory.

[1] Also called 'greater truthlikeness' or 'greater verisimilitude'.

- Define *'closer to the truth'* in terms of suitably specified notions of a larger truth-content and a smaller, or otherwise less problematic, falsity-content.

III General epistemological step: Identify the kind of evidence

- Identify the kind of evidence that experiments provide. In the nomic case, the evidence will consist of realized nomic possibilities and inductive generalizations 'based' on them.

IV Specific epistemological steps: Define *'more successful'*

- Assuming that the evidence is accepted, define the therewith *'accepted-content'* and *'rejected-content'* of a theory.
- Define *'more successful'* in terms of a larger accepted-content and a smaller, or otherwise less problematic, rejected-content.

V Theoretical step: From verisimilitude to success

- Assuming the truth of the (accepted) evidence, prove the strongest *'success theorem'*. Ideally this theorem amounts to: 'closer to the truth' unconditionally entails 'at least as successful' and in the long run even 'more successful'. However, it may well turn out that only a weaker theorem can be proven, but still of great interest.

VI Methodological steps: From success to verisimilitude

- Assuming that a new theory is at a certain moment more successful than the old one, propose and test the *'empirical progress (EP-)hypothesis'*: the new theory (is and) remains more successful than the old one.
- Assuming that after 'sufficient confirmation' the empirical progress hypothesis is accepted (for the time being), argue on the basis of the relevant success theorem to what extent the *'truth approximation (TA-)hypothesis'*, that is, the new theory is closer to the truth than the old one, the best explanation is for this case of empirical progress, i.e., that this is a case of truth approximation.
- Abductively conclude (for the time being) that the new theory is closer to the truth than the old one, i.e., that truth approximation has been achieved.

3.4 Common Core of Theories of Nomic Truth Approximation by Empirical Progress

In this section we will deal with the nomic version of the two common general steps (I), the nomic target, and (III), nomic evidence. This will largely be based on the relevant parts of Chap. 2. Next, the hypothetico-deductive (HD-) background of nomic evidence will be indicated.

3.4.1 Ad I: General Logico-Semantic Step: Intentional Description of the Target

Recall my favorite toy example of theory oriented science in Sect. 3.2.2. To represent an electric circuit with several switches and bulbs, but possibly invisible connections, the language will have elementary propositions that enable to indicate which switches are on and which are off and also to indicate which bulbs give light and which do not. Among the conceptually possible states there will at any moment be one actual state, but several other states are also physically possible. To represent these states by one proposition or theory one will have to design a complex proposition, the nomic truth. All states in which this proposition is true are physically possible, all others are not. All propositions that can be formulated in the indicated language may be considered as candidates for being this nomic truth and, at least intuitively, one proposition may be closer to the truth than another. This is a toy example for representing theory oriented science, but only the general tenet of the example is relevant: 'the nomic target' of theory oriented science is characterizing what is physically, biologically or economically possible and theories are tested by experiments, which are realizations of possibilities.

Let U indicate the set of mutually exclusive and together exhaustive *conceptual possibilities* in a given context (e.g. the possible states, trajectories or transformations[2] of a system, or possible kinds of systems), generated by a descriptive vocabulary V. Formally, U is characterized as V's set of set-theoretic structures of the corresponding similarity type. Subsets of U, e.g. X, Y, R, S, can also be characterized in terms of V. Complements of sets will be indicated by a preceding 'c': e.g. cX. By the way, U should not be taken as a set of possible worlds in the standard 'there is only one, all-inclusive, world' sense. Our possibilities are relative to a certain context or (type of) system(s), sometimes called 'small worlds'. They are only mutually exclusive and together exhaustive in the same case in the given context, e.g. in one state of a system.

Let (bold) **T** indicate the subset of *nomic*, e.g. physical, *possibilities*, and hence c**T** the subset of nomic *im*possibilities. The assumption that such a set exists, we call the *Nomic Principle*. By the bold '**T**' we indicate that it is an intentional description. We do not (yet) dispose of a characterization of it in terms of V, indicated by the *interrupted* ellipse in Fig. 3.1. Note that, whereas U only depends on V, **T** depends on both the domain D and the vocabulary V, hence formally we might write **T**(D,V) instead of **T**, but we shall not do so. The *nomic target* of research is identifying, if possible, **T**'s boundary in V-terms, indicated by (non-bold) T, hence T = **T**, assuming that such a characterization exists, which I will do in this book, except when otherwise stated. T will be called 'the (explicit) (nomic) truth', for reasons that will become clear in any specific case.

[2] Deutsch (2013) is drawing attention, in terms of 'constructor theory', to the importance in physics and elsewhere of the distinction between possible and impossible 'transformations'. In these terms, the present chapter is dealing with two-sided truth approximation regarding the true boundary between nomically possible and nomically impossible (but conceptually possible) transformations.

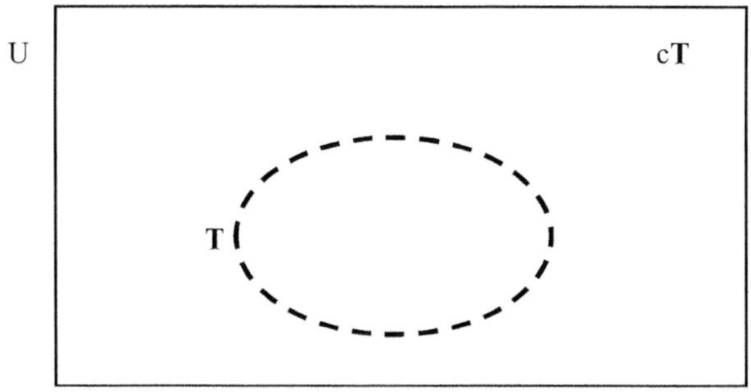

Fig. 3.1 The set of conceptual possibilities U and the (unknown) subset of nomic possibilities **T**

3.4.2 Ad. III: General Epistemological Step: Identify the Kind of Evidence

In the toy example, experiments consist of course in putting every switch in the on or off state and see which bulbs are lighting and which are not. In this way, every experiment realizes a nomic possibility. On the basis of several experiments we may also come to the (inductive) conclusion that e.g. a certain switch must be on for a certain bulb to give light.

In general, in the nomic context the empirical data are *asymmetric* in the following sense (cf. Kuipers 2000, p. 157). We can establish by experiments nomic possibilities. In fact, every experiment realizes by definition a nomic possibility. However, it is evident that we cannot establish nomic impossibilities in such a direct way. But in the empirical sciences we use to 'induce' nomic impossibilities indirectly by inductive (empirical) generalizations that we accept, for the time being, on the basis of 'sufficient' experimentation, notably by trying to realize counterexamples of theories and potential inductive generalizations. For example, the observation of a black raven is the realization of a nomic possibility, whereas concluding that non-black ravens do not exist is an inductive generalization, viz. all ravens are black. That is, non-black ravens are induced inductive *im*possibilities.

We indicate the (asymmetric) *data* at a certain moment by R/S, where R indicates the set of *realized nomic possibilities* (e.g. realized physical possibilities) and S indicates the strongest law induced on the basis of R, more precisely, the set of conceptual possibilities that are not excluded by (the combination of) the accepted inductive generalizations. In this setup cS indicates the set of *induced nomic impossibilities*. Of course, we may always assume that R $\subseteq$ S, for, if not, any element in R−S would represent a realized counterexample to S and hence to at least one of its constituting inductive generalizations. Moreover, if the experiments are correctly described by R (relative to the vocabulary) and if, in addition, S is correctly induced, we may conclude that R $\subseteq$ **T** $\subseteq$ S, whatever **T** is. Hence, accepting the evidence R/S amounts to concluding that the claims R $\subseteq$ **T** and **T** $\subseteq$ S are true. Of course, the correctness assumptions are very substantial, in particular that of correct inductive generalizations. Fig. 3.2 displays correct empirical data R/S.

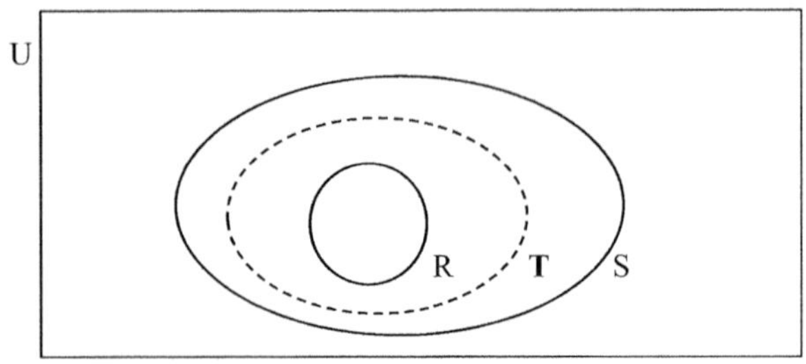

Fig. 3.2 Correct empirical data R/S

In addition to the above elaborations of the two general steps (I and III), in the chapters to come, I will put the formal emphasis on the specific steps II (Define '*closer to the* truth') and IV (Define '*more successful*'), because they are conceptually new. In most cases essentially similar versions of the remaining steps V and VI have been elaborated in Kuipers (2000) and the previous chapter. I will focus in these steps on specific aspects and conditions.

3.4.3 Hypothetico-Deductive Background of Nomic Evidence

As suggested, the background of R/S typically is that of theory testing and evaluation. Here we will elaborate this somewhat further. As far as theories exclude, as is the case when one looks for 'counterexamples', the standard story of Hypothetico-Deductive (HD-)testing and (HD-)evaluation (Kuipers 2000, Chaps. 5 and 6; Kuipers 2001, Chaps. 7 and 8) can be applied. One derives general test implications (GTI's) from a theory, and from them one derives individual test implications (ITI's). Ideally, performing an individual test, by realizing the relevant initial conditions, leads either to a success (of the GTI and hence) of the theory or to a counterexample. In both cases, the realized possibility is added to the existing set of relevant realized nomic possibilities R. In the case of an individual success the typical continuation is to test other ITI's of the same GTI by varying the initial conditions within the relevant range. If after many trials of various relevant kinds no counterexample shows up, the inductive jump to the conclusion that the GTI is true may be made, of course, always for the time being. This GTI may now not only be booked as a general success of the theory, but also as another inductive generalizations that may be 'added' to the existing set of relevant induced generalizations, and hence to their conjunction S. Hence, R grows and S shrinks (while cS grows) in the course of time, that is, as long as no previous additions are withdrawn for some reason or another. Recall that, whereas the members of R are called realized nomic possibilities, the members of cS are called induced nomic impossibilities. And, if no descriptive and inductive mistakes have been made, we may conclude: $R \subseteq \mathbf{T} \subseteq S$.

In our previous work we have stressed that the simple occurrence of a counterexample of a theory is no good reason to throw the theory away. It is crucial to look first for a better theory in two stages. In the first stage it is crucial to find a new theory that is (empirically) more successful relative to the existing data R/S, in the plausible sense that it entails all inductive generalizations the old one entails and that all its counterexamples are counterexamples of the old one, and is superior in at least one of these respects. In the second stage it is important to the test the 'comparative success hypothesis' or 'empirical progress (EP-)hypothesis' claiming that this relation remains to hold whatever new tests we perform. Hence, instead of HD-*testing* of a theory, aiming at establishing its truth value, we are more interested in separate and comparative HD-*evaluation*, in the suggested way. HD-testing and -evaluation are of course based on deductive consequences of theories.

Comparative evaluation is typical for the instrumentalist methodology as opposed to the (naïve) falsification methodology. The irony is that this instrumentalist methodology turns out to be functional for truth approximation. This is the topic of the last step (VI: Methodological steps: From success to verisimilitude) in the above described set-up of theories of specific truth approximation (indicated) by empirical progress.

References

Cevolani, G., Crupi, V., & Festa, R. (2011). Verisimilitude and belief change for conjunctive theories. *Erkenntnis, 75*(2), 183–202.

Cevolani, G., Festa, R., & Kuipers, T. (2013). Verisimilitude and belief change for *nomic* conjunctive theories. *Synthese, 190*, 3307–3324.

Deutsch, D. (2013). Constructor theory. *Synthese, 190*, 4331–4359.

Kuipers, T. (2000). *From instrumentalism to constructive realism. On some relations between confirmation, empirical progress, and truth approximation. Synthese library 287*. Dordrecht: Kluwer Academic Publishers.

Kuipers, T. (2001). *Structures in science. Heuristic patterns based on cognitive structures. An advanced textbook in neo-classical philosophy of science. Synthese library 301*. Dordrecht: Kluwer Academic Publishers.

Kuipers, T. (2014). Dovetailing belief base revision with (basic) truth approximation. In E. Weber, D. Wouters, & J. Meheus (Eds.), *Logic, reasoning, and rationality, Proceedings-conference in Ghent, 2010* (pp. 77–93). Dordrecht: Springer.

Niiniluoto, I. (1987). *Truthlikeness. Synthese library 185*. Dordrecht: Reidel.

Oddie, G. (1986). *Likeness to truth*. Dordrecht: Reidel.

Zwart, S. (2001). *Refined verisimilitude. Synthese library 307*. Dordrecht: Kluwer Academic Publishers.

Chapter 4
Models, Postulates, and Generalized Nomic Truth Approximation

Abstract The qualitative theory of nomic truth approximation, presented in Kuipers (From instrumentalism to constructive realism. On some relations between confirmation, empirical progress, and truth approximation. Synthese library, vol. 287. Kluwer Academic Publishers, Dordrecht, 2000), in which 'the truth' concerns the distinction between nomic, e.g. physical, possibilities and impossibilities, rests on a very restrictive assumption viz. that theories always claim to characterize the boundary between nomic possibilities and impossibilities. In Chap. 2 we have already shown that truth approximation is possible by ascribing, in Popperian spirit, theories only an excluding function. By fully recognizing two different functions of theories, viz. excluding and representing (including), this chapter drops the restrictive assumption by conceiving theories in development as tuples of postulates and models, where the postulates claim to exclude nomic impossibilities and the (not-excluded) models claim to represent nomic possibilities.

Revising such 'two-sided' theories becomes then a matter of adding or revising models and/or postulates in the light of increasing evidence, now captured by a special kind of theories, viz. 'data-theories'. Under the assumption that the data-theory is true, achieving empirical progress on its basis provides good reasons for the abductive conclusion that truth approximation has been achieved as well. As in Chap. 2, the notions of truth approximation and empirical progress are formally direct generalizations of the earlier ones. However, truth approximation is now explicitly defined in two-sided terms of increasing truth-content and decreasing falsity-content of theories, whereas empirical progress is defined in terms of lasting increased accepted-content and decreased rejected-content in the light of increasing evidence. As already noted in Chap. 2, these definitions are strongly inspired by a paper of Gustavo Cevolani, Vincenzo Crupi and Roberto Festa, viz., "Verisimilitude and belief change for conjunctive theories" (*Erkenntnis*, 2011).

Revised version of: "Models, postulates, and generalized nomic truth approximation", *Synthese*, 193 (10), 2016, 3057–3077. *Acknowledgement*: I like to thank David Atkinson, Erik Krabbe, and Jan-Willem Romeijn for their helpful remarks at a first presentation in Groningen (PCCP), Fred Muller as commentator and the rest of the audience in Rotterdam (EIPE), the audience in Cambridge (CamPoS) and Roberto Festa and, above all, Gustavo Cevolani for their constructive comments.

T. A. F. Kuipers, *Nomic Truth Approximation Revisited*, Synthese Library 399,
https://doi.org/10.1007/978-3-319-98388-2_4

Keywords Models · Postulates · Nomic truth approximation · Empirical progress · Success theorem · Theory revision · Two-sided theories · Truth-content · Falsity-content

4.1 Introduction

The qualitative theory of nomic truth approximation, as presented in Kuipers (2000), aims at defining when a theory is closer than another to 'the nomic truth', i.e., the true boundary between nomic, e.g. physical, biological, etc., possibilities and impossibilities within a target domain. This theory rests on a very restrictive assumption viz. that theories claim to characterize the boundary between nomic possibilities and impossibilities, i.e., a theory amounts to a classification of all conceptual possibilities as either nomically possible or nomically impossible. In Chap. 2 we have already shown that truth approximation is possible by ascribing, in Popperian spirit, theories only an excluding function. The present chapter drops the very restrictive assumption in such a way that the theory of nomic truth approximation becomes conceptually in complete harmony with two *prima facie* opposing functions of theories. On the one hand, the Popperian or *exclusion view* that theories exclude certain (conceptual) possibilities from occurring or being realizable. On the other hand, there is the inclusion or *representation view* that theories represent, in relevant respects, certain possibilities as realized or realizable, at least approximately, due to idealizations. The exclusion function is typically associated with speaking of the axioms, principles or postulates of theories. They have to be satisfied and hence they exclude together everything which does not satisfy them all. The representation function on the other hand is typically associated with speaking of (specified) models of theories.[1]

Now in Kuipers (2000) I just assumed that the set of (representing) models coincides with the set of models, in the formal sense, of the (excluding) postulates of the theory. Although such a 'maximal' theory may be the ultimate aim of nomic theorizing, viz. the strongest true theory, it is not realistic to assume this of 'theories in development'. However, it is perfectly possible to separate models and postulates, in order to fully recognize the twofold function of theories in general and in aiming at truth approximation in particular. That is, although (sets of) specified models of a (consistent) theory will have to satisfy the postulates of the theory, there is no need to assume in general that they exhaust the set of all models of the theory's postulates in the formal sense. This kind of completeness or maximality would only be appropriate if the postulates are supposed to exclude together precisely everything which is not realizable. Hence, in this chapter a theory will in general be taken as possibly 'non-maximal' in the following two-sided sense, viz. as a combination of a set of specified (intended and interpreted) models M and a set of postulates Π, where the former have to satisfy the latter, but need not exhaust the set of models of the joint

[1] Note that as far as postulates represent, they represent unrealizable possibilities.

postulates. As done in Chap. 2, it will be convenient to represent the postulates by this set of all its models, i.e., models(Π), and we will indicate this set simply by P. Hence, a theory becomes a tuple of the following form:

$$<M, P> \qquad \text{with } P =_{def} \text{models}(\Pi) \text{ and } M \text{ a subset of } P$$

As suggested, the set of models of a non-maximal theory is a *proper* subset of all models of the joint postulates (M⊂P). The specification of the associated inclusion and exclusion claim will follow in Sect. 4.3.

The two-sided approach to theories can do justice to three different views in philosophy of science, viz. theorizing is mainly a matter of (1) formulating and revising postulates, or (2) designing and redesigning models, or (3) the two-sided combination of them. The generalized theory of truth approximation to follow will provide additional support for the two-sided view, but will leave room for both one-sided views.

Note, as an aside, that the two-sided view on theories is also in perfect agreement with the hypothetical-deductive (HD) and deductive-nomological (DN) views on prediction and explanation, respectively. For the prediction and explanation of an event we start with representing (modelling) the situation in the relevant terms, as far as possible, but without the crucial event. The occurrence of the event is derived by applying the relevant postulates, which amounts to completing or closing the model as far as required. Hence, prediction and explanation naturally appear as co-production of (partial) models and postulates, that is, of representation and exclusion.

The generalized theory of truth approximation presented below is based on the two-sided view on theories and technically reduces to that of Kuipers (2000) by assuming M = P throughout. The representation claim will be associated with M and the exclusion claim with P. The two-sided construction of truth approximation is conceptually strongly stimulated by a paper of Gustavo Cevolani et al. (2011).[2] Truth approximation can thus be construed in terms of increasing truth-content and decreasing falsity-content by adding or revising models and/or postulates in the light of increasing evidence.

The generalized theory further follows Kuipers (2000) in reconstructing evidence in terms of realized, hence nomic, possibilities R and the strongest inductive generalization S (with cS as the set of induced nomic *im*possibilities), which can now be seen as a special kind of theories, 'data-theories': <R, S>. The evidence will guide the comparative assessment of the success of theories and the subsequent planning of new experiments to be performed leading to increasing evidence. Ultimately, the comparative success assessment may give good reasons for the inductive conclusion that empirical progress has been achieved by a revised version of the theory relative to the original. Here, empirical progress is defined in terms of lasting increased accepted- and decreased rejected-content in the light of increasing

[2] See also Cevolani and Festa (to appear) as the most recent presentation, including a significant extension to the case of arbitrary theories, notably disjunctive theories.

evidence. In addition, and of crucial importance, empirical progress may provide good reasons for the abductive conclusion that the revised theory is closer to the truth than the original and hence that truth approximation has been achieved.

We start with presenting an example of a two-sided theory (Sect. 4.2) and then, along the guidelines of the previous chapter, a relatively detailed presentation (Sects. 4.3, 4.4, and 4.5) with emphasis on what is conceptually new relative to (Kuipers 2000) and we close by indicating a number of perspectives not yet captured in the previous chapter.

4.2 An Example of a Two–Sided Theory

The following, simplified, example illustrates the idea of a two-sided theory. Newton's theory can be represented on the one hand by the three laws of motion, its (general) postulates, and on the other by various sets of models, e.g. (cpm-)models of *c*lassical *p*articles *m*echanics (cpm) and models of classical rigid body mechanics, with specific sub-classes.

Focusing on the set or universe of conceptually possible systems of classical particle mechanics, U_{cpm}, the general postulates determine together the set of all models satisfying them, P_{cpm} ($\subseteq U_{cpm}$). Various sub-classes of P_{cpm} are based on specific assumptions about the nature of such systems, notably so-called special force laws, e.g. the set of models P_{gcpm} satisfying the law of gravitation, but also on many system specific assumptions. Hence, a specific model for a specific system is built up by starting *bottom-up* with system specific details and is *top-down* completed by applying the *g*eneral and the *s*pecial *p*ostulates, here called the *GSP-closure*. Such a specific model may or may not be generalized to a subset of models for a specific subtype of cpm-systems.[3] For example, let M_{fo} indicate the GSP-closure of objects of medium size and weight, falling from a medium distance on the earth.[4]

As soon as a particular model is claimed to represent a particular system (e.g. a falling brick) it predicts behavior of the system in accordance with that model.[5] If that model is disconfirmed, some specific (bottom-up) assumption may have to be

[3] Note that such a set of specified (intended and interpreted) models is comparable to a set of *exemplars* in Kuhn's sense or a *set* (of GSP-closures of tentative representations) *of the intended applications* in Sneed's sense. Examples in the present context are falling objects, pendulae, ballistic and planetary curves.

[4] Subsets of P_{cpm} corresponding to systems with different kinds of forces may overlap, notably in the case of two kinds of forces, but they need not exhaust P_{cpm}, e.g. there may be room for as yet unknown (formally different) kinds of forces. Moreover, although it is (claimed to be) excluded that non-members of the subset of, say gravitational cpm-models, are realizable, it may, but need not be claimed that all members are realizable. It seems plausible that physicists before Einstein assumed that e.g. all gravitational cpm-models are (in principle) physically realizable by natural systems, but this had to be revised by revising (concepts and) postulates.

[5] This also provides a potential explanation of that behavior, where, in agreement with the DN-model of explanation, the bottom-up data function as initial conditions and the postulates as laws.

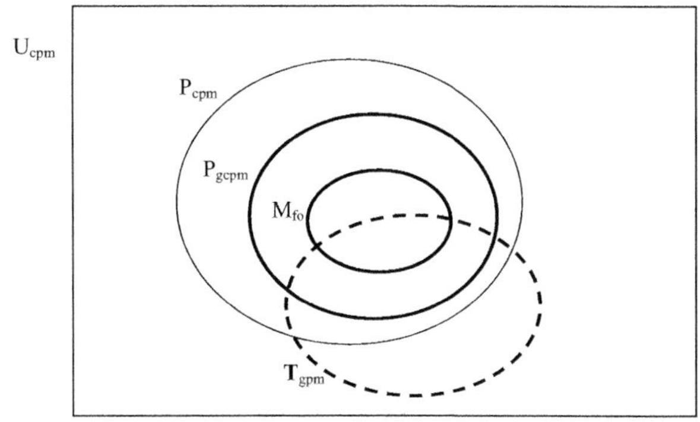

Fig. 4.1 The classical theory of gravitational particle mechanics (still) restricted to falling objects

revised, but if similar models for similar systems are also disconfirmed, the whole relevant class of particular models may have to be revised, e.g. by revising the particular force law (top-down). In this way we see that revising a two-sided theory may be a matter of adding or revising postulates that exclude things (top-down) or of adding or revising models that are supposed to represent and hence to be included (bottom-up), or both.

Let us now assume that there is a sharp, but not yet specified, distinction or boundary between the set $\mathbf{T}_{gpm}$ of *physically possible systems* of gravitational *p*article *m*echanics, as represented in U_{cpm}, and the set of *physically impossible systems* of gravitational particle mechanics, that is, the complement of $\mathbf{T}_{gpm}$. Figure 4.1 illustrates the four subsets of U_{cpm} introduced: P_{cpm}, the subset P_{gcpm}, and the latter's subset M_{fo}, and finally the, unknown, target set $\mathbf{T}_{gpm}$.

Now we may reconstruct truth approximation as a matter of aiming at establishing the true boundary between nomic possibilities and impossibilities, or the truth about such a boundary, in the present case, starting from the two-sided theory $<M_{fo}, P_{gcpm}>$ with the claim $M_{fo} \subseteq \mathbf{T}_{gpm} \subseteq P_{gcpm}$, and revising it by adding or revising the models and/or adding or revising the postulates in the light of increasing evidence, with the ideal end the 'maximal and true' or the strongest true theory $<M^{\#}, P^{\#}>$, i.e., the theory for which $M^{\#} = \mathbf{T}_{gpm} = P^{\#}$.

4.3 II Specific Logico-Semantic Steps: Define *'Closer to the Truth'*

In Chap. 3 we have given an outline of specific theories of truth approximation. Here we will follow the steps, with emphasis on the specific aspects of two-sided theories. The first step (I) was a general step, the intentional description of the target, and was already presented in Sect. 3.4. The basic idea was that the target of theory oriented research is, in a given context, the characterization of the subset of nomic possibilities within the set U of mutually exclusive and together exhaustive conceptual possibilities generated by the chosen vocabulary (formally, a set of set-theoretic structures of a certain similarity type). The existence of a unique divide between nomic possibilities and nomic impossibilities was postulated by the Nomic Principle. Hence, we turn to the first of the specific logico-semantic steps, preparing the definition of 'closer to the truth'.

4.3.1 *Theories, Their Claims, and 'The Truth'*

For the logico-semantic steps we start with defining the notion of theories. In the present nomic context, theories are intended to (at least partially) characterize **T**. Recall that for that purpose a theory is conceived as a tuple <M, P> of subsets of U, defined in V-terms, with P = models (postulates Π). Now we can specify its two claims:

- "$M \subseteq \mathbf{T}$", the inclusion or representation claim: all members of M are nomic possibilities,
- "$\mathbf{T} \subseteq P$", i.e., "$cP \subseteq c\mathbf{T}$", the exclusion claim: all non-members of P are excluded from being nomic possibilities (are nomic *im*possibilities), or, equivalently, no nomic possibility is excluded by the postulates of the theory.

The members of M are called the models *simpliciter* and those of P the models of the postulates of the theory, or even, formally incorrect, simply postulates.[6]

A theory is *consistent* if $M \subseteq P$, i.e., when its two claims are compatible; it is *inconsistent* otherwise. Note that this notion of inconsistency is not the standard logical one, but here plausible and of course related.[7] This chapter is restricted to consistent theories, leaving inconsistent ones for future research, for they seem interesting as well. A theory is *maximal* if M = P; it is *non-maximal* otherwise. Kuipers (2000) is restricted to maximal theories. Maximal theories in this sense seem also characteristic for the model-theoretic and structuralist[8] or semantic view on theories.

---

[6]Although we suggest that P is determined as the set of models of a given set of postulates Π, and M is just given as a set of members of U, it may well be that in some contexts both P and M are directly specified as subsets of U. Alternatively, whereas P may be based on a set of exclusion postulates, M may be based on a set of inclusion postulates. However, I prefer to speak of models because the term 'model' suggests to represent something and hence to be included in some target set, whereas the term 'postulate' suggests to exclude things by imposing a necessary condition to be satisfied in order to be included.

[7]A theory is inconsistent in the syntactic (logical) sense when it is possible to derive a contradiction or in the semantic (logical), but equivalent, sense when there is no model satisfying it.

[8]In the first view, the postulates (or axioms) are explicitly formulated in a specified language and the models are those structures of the language that satisfy all of them. In the second view, the language is not specified but the models are directly specified in such a way that the postulates are built into them. Note that in the basic format of a theory in the structuralist sense <U, M, I>, U corresponds with our U, M with our M and P, with M = P, and I with **T**. As suggested, both views on theories seem to conceive them as maximal. However, one may also argue that they are purely dealing with an exclusion claim, viz. ($\mathbf{T}$=) $I \subseteq M$ (=P), in the structuralist case.

Note that, assuming $M \subseteq P$, the two claims "$M \subseteq \mathbf{T}$" and "$cP \subseteq c\mathbf{T}$" are compatible conjunctive claims in the sense that e.g. "$M \subseteq \mathbf{T}$" states that all items in M are in **T**, which means that the conjunction of these claims can be seen as a 'conjunctive theory' in the sense of Cevolani et al. (2011).

It will be useful to also define two other extreme kinds of theories, besides maximal ones, so-called pure (or one-sided) theories. A theory <M, P> is a *pure* theory of postulates, or a pure exclusion theory, if M=∅, i.e., <∅, P>, and it is a *pure* theory of models, or a pure inclusion theory, if P=U, i.e., <M, U>. In these terms, a (non-pure or two-sided) theory <M, P> is a combination of a pure theory of models <M, U> and a pure theory of postulates <∅, P>, also called the M-side (or inclusion) theory and the P-side (or exclusion) theory, respectively.

Finally, a theory <M, P> is *true* if both claims are true, i.e., $M \subseteq \mathbf{T} \subseteq P$, *false* otherwise. Now, if $P^{\#}$ is a *subset* of P, the claim associated with $P^{\#}$ is of course stronger than that associated with P. Conversely, if $M^{\#}$ is a *superset* of M, the claim associated with $M^{\#}$ is of course stronger than that associated with M. Hence, it is now easy to see that there is at most one true maximal or strongest true theory, called *the true(nomic) theory* or simply *the (nomic) truth*, viz. the one for which $M = \mathbf{T} = P$. From now on, 'the truth' will refer to 'the nomic truth', except when otherwise stated. It results from the characterization of **T** in V-terms, *if it exists*. It will be indicated by <T, T>, or simply T, with non-bold 'T'. This T is the target of (theory-oriented) research.

It is also easy to check that there is at most one strongest true pure theory of models, viz. <T, U> and that there is at most one strongest true pure theory of postulates, viz. <∅, T>. Hence, <T, T> is the strongest true (two-sided) theory.

4.3.2 *Truth- and Falsity-Content*

Turning to the second specific logico-semantic step, it is, in view of the two claims of a theory, plausible how to define the truth-content and the falsity-content of a theory. Consider the M-side claim $M \subseteq \mathbf{T}$. As far as $M \cap \mathbf{T}$ is concerned the sub-claim $M \cap \mathbf{T} \subseteq \mathbf{T}$ is true and as far as $M - \mathbf{T}$ is concerned the remaining sub-claim $M - \mathbf{T} \subseteq \mathbf{T}$ is false. Hence we define $M \cap \mathbf{T}$ as its *truth-content* and $M-\mathbf{T}$ as its *falsity-content*. Note that their union equals M, which can be seen as the (total) content of the M-side claim.

Similarly, regarding the P-side claim $\mathbf{T} \subseteq P$, that is, $cP \subseteq c\mathbf{T}$, the sub-claim $cP \cap c\mathbf{T} \subseteq c\mathbf{T}$ is true and the additional sub-claim $cP - c\mathbf{T} \subseteq c\mathbf{T}$ is false. Hence we define $cP \cap c\mathbf{T}$ as its truth-content and $cP - c\mathbf{T}$ as its falsity-content. Their union equals cP, in which Popper's idea of the empirical content can be recognized easily: the set of possibilities which are excluded by the postulates.

Table 4.1 summarizes these definitions.

In combination with Table 4.1, Fig. 4.2 makes graphically clear which subsets are involved.

Table 4.1 Truth- and falsity-content

Theory <M, P>	M-side: inclusion/ representation	P-side: exclusion/ explanation&prediction
M- resp. P-claim	$M \subseteq \mathbf{T}$	$\mathbf{T} \subseteq P$ ($cP \subseteq c\mathbf{T}$)
Truth-content (TC)	$M \cap \mathbf{T}$	$cP \cap c\mathbf{T}$
Falsity-content (FC)	$M - \mathbf{T}$	$cP - c\mathbf{T}$
M- resp. P-content	M	cP

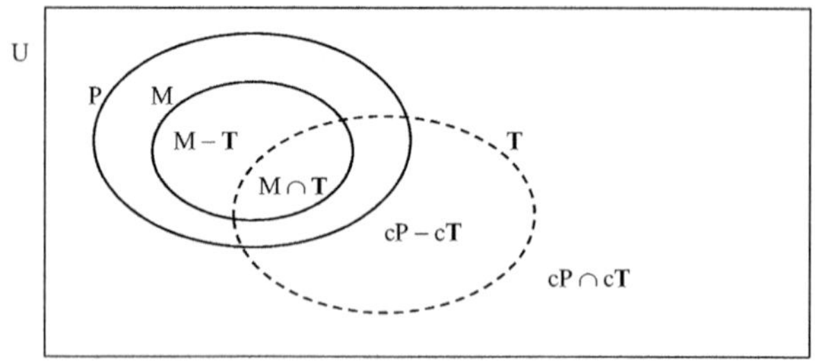

Fig. 4.2 Truth- and falsity-content. The indicated intersections constitute the relevant truth-content and the indicated difference sets the relevant falsity-content

4.3.3 Closer to the Truth (Greater Verisimilitude or Greater Truthlikeness)

Turning to the third and final specific logico-semantic step, the notions of truth- and falsity-content just defined allow for a straightforward definition of 'closer to the truth' in terms of greater truth-content and smaller falsity-content. Focusing first on pure M-side theories, for example, one can define:

<M*, U> is *at least as close to the truth* as the M-side theory <M, U > iff
TC-clause: the truth-content of <M, U> is a subset of the truth-content of <M*, U> and
FC-clause: the falsity-content of <M*, U> is a subset of the falsity-content of <M, U>

A similar definition can of course be given for pure P-side theories.

Table 4.2 presents the combined definition of "<M*, P*> is at least as close to the truth as <M, P>".

It is not difficult to check that the four single clauses are independent as long as the theories are non-maximal. Moreover, as indicated, on both sides the two single difference clauses can be combined into the corresponding combined symmetric-difference clauses. It is important to note that for maximal theories (M = P) the two combined clauses are formally equivalent.[9]

[9] This leads to the original 'symmetric difference'- definition (Miller 1978; Kuipers 1982), assuming the strong claim of maximal theories: M = P = **T**. By the way, Miller's definition was focused

Table 4.2 <M*, P*> is at least as close to the truth as <M, P>

<M*, P* > is at least as close to the truth as <M, P>	M-side	P-side
TC-clauses: TC is a subset of TC*	$M \cap \mathbf{T} \subseteq M^* \cap \mathbf{T}$ $\equiv \mathbf{T} - M^* \subseteq \mathbf{T} - M$	$cP \cap c\mathbf{T} \subseteq cP^* \cap c\mathbf{T}$ $\equiv P^* - \mathbf{T} \subseteq P - \mathbf{T}$ $\equiv P^* \cup \mathbf{T} \subseteq P \cup \mathbf{T}$
FC-clauses: FC* is a subset of FC	$M^* - \mathbf{T} \subseteq M - \mathbf{T}$	$cP^* - c\mathbf{T} \subseteq cP - c\mathbf{T}$ $\equiv \mathbf{T} - P^* \subseteq \mathbf{T} - P$ $\equiv P \cap \mathbf{T} \subseteq P^* \cap \mathbf{T}$
Combined clauses[a]	$M^* \Delta \mathbf{T} \subseteq M \Delta \mathbf{T}$	$cP^* \Delta c\mathbf{T} \subseteq cP \Delta c\mathbf{T}$ $\equiv P^* \Delta \mathbf{T} \subseteq P \Delta \mathbf{T}$

[a]The symmetric difference $A\Delta B$ is standardly defined as $(A-B)\cup(B-A)$

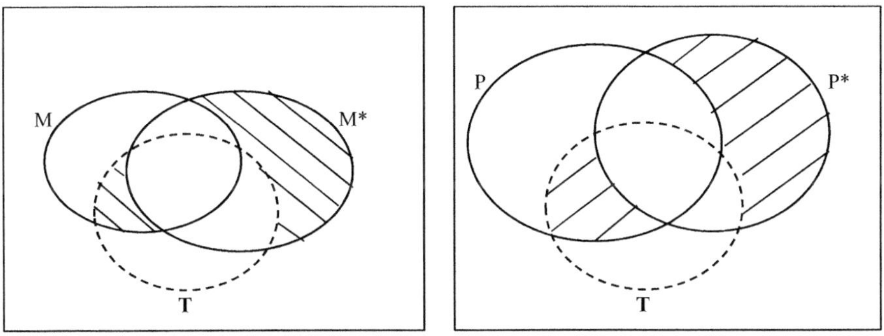

Fig. 4.3 <M*, P*> is at least as close to the truth as <M, P>

In Chap. 2 we have given a propositional example, in present terms, of pure exclusion theories 'closer to the truth' (see Table 2.1). In view of the combined clauses these theories may also be seen as pure inclusion theories, such that <M*, ∅> with M*'=' $p_1 \Leftrightarrow p_3$ (formally: M* = Mod($p_1 \Leftrightarrow p_3$)) is closer to **T**'=' $(p_1 \vee p_2) \Leftrightarrow p_3$ than <M, ∅> with M'=' $(p_1 \& p_2) \Leftrightarrow p_3$. Now it is easy to check that $\{p_1 \& \neg p_2 \& p_3\}$ represents the extra truth-content (extra rightly included) of <M*, ∅> relative to <M, ∅> and $\{s_3\} = \{p_1 \& \neg p_2 \& \neg p_3\}$ the extra falsity-content (extra wrongly included) of <M, ∅> relative to <M*, ∅>. Of course, the example theories may also be interpreted as maximal theories.

Figure 4.3 illustrates "<M*, P*> is at least as close to the truth as <M, P>". In combination with Table 4.2 the separated figures make graphically clear by shading which subsets have to be empty at the M-side (left) and at the P-side (right). Of

---

on 'the actual truth', i.e., one particular possibility. However, both share the characteristic that they are intuitively similar to Popper's original but faulty definition in terms of more true and less false consequences. The symmetric difference definition for maximal theories may be summarized as: (1) more correctly and less incorrectly included possibilities (as nomic) or, equivalently, (2) more correctly and less incorrectly excluded possibilities (from being nomic). In the generalized theory, applied to non-maximal theories, (1) characterizes 'closer to' on the M-side, and (2) on the P-side.

course, the figures have to be conceived as combined, taking into account that $M \subseteq P$ and $M^* \subseteq P^*$. Hence, in a combined figure there would appear single and double shaded subareas.

Although the two figures look formally similar, it is important to note that corresponding shaded areas are empty due to 'opposite' clauses. E.g. in the left figure the shaded area on the left is empty due to the TC-clause for the M-side, viz. $M \cap \mathbf{T} \subseteq M^* \cap \mathbf{T}$, whereas in the right figure it is empty due to the FC-clause for the P-side, viz. $cP^* - c\mathbf{T} \subseteq cP - c\mathbf{T}$, or, equivalently $\mathbf{T} - P^* \subseteq \mathbf{T} - P$.

It is also not difficult to check that the P-side can be summarized by 'having at least as many true consequences (for $P^* \cup \mathbf{T}$ is included in $P \cup \mathbf{T}$[10]) and allowing at least as many correct models ($P^* \cap \mathbf{T}$ includes $P \cap \mathbf{T}$) and, conversely, the M-side by 'having at least as many correct models ($M^* \cap \mathbf{T}$ includes $M \cap \mathbf{T}$) and allowing at least as many true consequences (for $M^* \cup \mathbf{T}$ is included in $M \cup \mathbf{T}$)'.

Finally, it is plausible to define 'closer to the truth' if, in addition to 'at least as close to the truth', at least one of the four required single set-theoretic inclusions can be replaced by a proper inclusion and hence if at least one of the two required combined inclusions in terms of symmetric differences is proper.

4.4 IV Specific Epistemological Steps: Define 'More Successful'

The first epistemological step (III) was a general step, viz. identifying the kind of evidence, and was, like step I, already presented in Sect. 3.4, followed by an indication of the hypothetico-deductive (HD-) background of this kind of evidence. For the nomic case, the evidence R/S could be characterized by subsets R and S of U, indicating, respectively, the set of realized nomic possibilities and the strongest inductive generalization 'based' on them.[11] In the present context of two-sided theories, it is interesting to note that R/S is in fact such a two-sided theory, viz. <R, S>, now also called a data-theory. Recall, that if the experiments are correctly described by R (relative to the vocabulary) and if, in addition, S is correctly induced, we may conclude that $R \subseteq \mathbf{T} \subseteq S$, whatever $\mathbf{T}$ is. Hence, we not only have that <R, S> is a

[10] A true consequence of <M, P> at the P-side can be represented as a superset of both P and $\mathbf{T}$, i.e., a set having both P and $\mathbf{T}$ as subsets, for the following reason. Let P, $\mathbf{T}$, and C (subset of U) represent the models of the (complex) sentences p, **t**, and c, respectively, and let $P \cup \mathbf{T} \subseteq C$. Then c is a consequence of p and of **t**, and hence a true consequence of p, for c is true in all models of P and $\mathbf{T}$. The condition that $P^* \cup \mathbf{T}$ is included in $P \cup \mathbf{T}$ now guarantees that all supersets of $P \cup \mathbf{T}$ are supersets of $P^* \cup \mathbf{T}$, and hence that all true consequences of <M, P> (at the P-side) are (true) consequences of <M*, P*>. For a detailed comparison between Popper's failing consequence-based approach and the 'model-based' approach, both for maximal theories, the reader is referred to Kuipers (2000, Chap. 8.1), where the latter is also translated in terms of consequences, leading to the identification of Popper's bad luck.

[11] Formally more neat, we should represent the set of inductive generalizations e.g. by Σ, such that S is the set of models of Σ.

theory, but by accepting the data we even assume that it is a true theory, that is, we have accepted the claims $R \subseteq \mathbf{T}$ and $\mathbf{T} \subseteq S$ ($\equiv cS \subseteq c\mathbf{T}$). Again we have to stress that the correctness assumptions are very substantial, in particular that of correct inductive generalizations.

4.4.1 Accepted- and Rejected-Content

Turning to the first of the specific epistemological steps (IV), to get plausible definitions of accepted-content and rejected-content of a theory <M, P> in the light of an accepted data-theory <R, S> we have to confront the claims of the former, that is, $M \subseteq \mathbf{T}$ and $cP \subseteq c\mathbf{T}$, with the accepted claims of the latter, that is, with the claims $R \subseteq \mathbf{T}$ and $cS \subseteq c\mathbf{T}$. Let us first look at the M-side. Of the theory claim $M \subseteq \mathbf{T}$, recall, with content M, we have, by accepting $R \subseteq \mathbf{T}$, accepted the sub-claim $M \cap R \subseteq \mathbf{T}$, and hence the *accepted M-content* is $M \cap R$, which may be called the set of *realized examples.*[12] However, by accepting $cS \subseteq c\mathbf{T}$, we have also rejected the sub-claim $M \cap cS \subseteq \mathbf{T}$, and hence the *rejected M-content* is $M \cap cS$ ($=M-S$), which may be called the set of *induced counterexamples.*

Similarly for the P-side. Of the theory claim $cP \subseteq c\mathbf{T}$, recall, with content cP, we have, by accepting $cS \subseteq c\mathbf{T}$, accepted the sub-claim $cP \cap cS \subseteq c\mathbf{T}$, and hence the *accepted P-content* is $cP \cap cS$, which may be called the set of *induced examples.* However, by accepting $R \subseteq \mathbf{T}$, we have also rejected the sub-claim $cP \cap R \subseteq c\mathbf{T}$, and hence the *rejected P-content* is $cP \cap R$ ($=R-P$), which may be called the set of *realized counterexamples.*

Table 4.3 gives a full survey, adding the also plausible terminology of 'true/false positives/negatives',[13] the undecided content on both sides,[14] and a numbering of the most relevant subsets.

Figure 4.4 illustrates all these concepts graphically, using the numbering of subsets in Table 4.3.

---

[12] This and the following '(counter-) example terminology' is a variation of similar terminology in Kuipers (2000, p. 158).

[13] True (false) positives are truly (falsely) claimed nomic possibilities, true (false) negatives are truly (falsely) claimed nomic impossibilities. This terminology assumes that we take the claim that an item in U is a nomic possibility as the claim to be 'a positive (result)' and that it is a nomic impossibility as the claim to be 'a negative (result)'. Now, for example, regarding the members of $M \cap R$, theory <M, P> truly claims that they are positives; hence they may be called 'true positives'.

[14] The undecided M-content is that part of the content of M about which <R, S> is silent, that is, the M-content apart from its accepted and rejected parts: 'M-content – (accepted M-content $\cup$ rejected M-content) $= M - [(M \cap R) \cup (M \cap cS)] = M-(R \cup cS)$. Similarly for the undecided P-content.

Table 4.3 Accepted and rejected M- and P-content of theory <M, P> in the light of accepted data-theory <R, S>

<M, P > in the light of <R, S>	M-side	P-side
M- resp. P-claim	M ⊆ **T**	**T** ⊆ P (cP ⊆ c**T**)
Accepted claims	R ⊆ **T** **T** ⊆ S (cS ⊆ c**T**)	
M-resp. P-content (of the theory)	M	cP
Accepted (M- resp. P-) content	1: M∩R realized examples (true positives)	4: cP∩cS induced examples (true negatives)
Rejected (M- resp. P-) content	2: M−S (M∩cS) induced counterexamples (false positives)	5: cP∩R (R−P) realized counterexamples (false negatives)
Undecided (M-resp. P-) content	3: M−(R∪cS)	6: cP−(R∪cS)

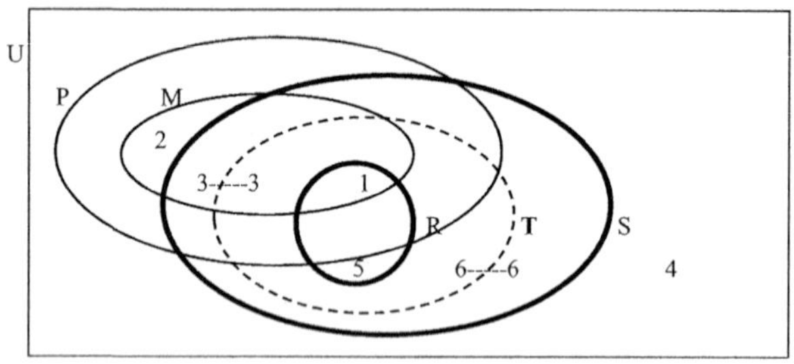

Fig. 4.4 Accepted and rejected M- and P-content of theory <M, P> in the light of accepted data-theory <R, S>. See Table 4.3 for the legend. (Note that P−S represents the wrongly not excluded nomic impossibilities, of which the subset M−S (= 2) represents the wrongly included nomic impossibilities)

4.4.2 At Least as Successful Relative to <R, S>

Turning to the second and final specific epistemological step, it is rather plausible to define now the idea that (the revised or new) theory <M*, P*> is *at least as successful* as (the initial) theory <M, P>, relative to accepted data-theory <R, S>, by requiring, that all successes (realized and induced examples) of <M, P> are successes of <M*, P*> and all failures (induced and realized counterexamples) of <M*, P*> are failures of <M, P>. Or, equivalently, at least as much accepted-content (AC-clauses) and at most as much rejected-content (RC-clauses) of the *-theory on both sides, of course, not in terms of numbers but of subset conditions. The result is given in Table 4.4.

Of course, 'more successful' is defined by requiring, in addition, that at least one of the four single clauses is a proper inclusion.

Table 4.4 <M*, P*> is at least as successful as <M, P>, relative to <R, S>

<M*, P*> is at least as successful as <M, P>, relative to <R, S>	M-side	P-side
AC-clauses: AC is a subset of AC*	$M \cap R \subseteq M^* \cap R$ $\equiv R - M^* \subseteq R - M$ all realized examples of <M, P> are examples of <M*, P*>	$cP \cap cS \subseteq cP^* \cap cS$ $\equiv P^* - S \subseteq P - S$ all induced examples of <M, P> are examples of <M*, P*>[a]
RC-clauses: RC* is a subset of RC	$M^* - S \subseteq M - S$ all induced counterexamples of <M*, P*> are counterexamples of <M, P>	$cP^* - cR \subseteq cP - cR$ $\equiv R - P^* \subseteq R - P$ all realized counterexamples of <M*, P*> are counterexamples of <M, P>

[a]Since the clause is also equivalent to $S \cup P^* \subseteq S \cup P$ it can also be paraphrased by "all induced laws entailed by <M, P> are entailed by <M*, P*>", for any superset of S∪P is a superset of S∪P*

4.5 Remaining Steps: From Verisimilitude to Success, and Vice Versa

At this point we just repeat the remaining steps in setting up a theory of nomic truth approximation and briefly give some particular features of the specific case.

V. Theoretical step: From verisimilitude to success

- Assuming the truth of the (accepted) evidence, prove the strongest 'success theorem'. Ideally this theorem amounts to: 'closer to the truth' unconditionally entails 'at least as successful' and in the long run even 'more successful'. However, it may well turn out that only a weaker theorem can be proven, but still of great interest.

  In the present case it is not difficult to check that 'ideally' applies, i.e., if <M*, P*> is closer to the truth than <M, P> and if the data-theory <R, S> is true, this entails that <M*, P*> is at least as successful as <M, P>relative to <R, S>, and, under some probabilistic test assumptions,[15] that it will become more successful in the long run.

VI. Methodological steps: From success to verisimilitude

- Assuming that a new theory is at a certain moment more successful than an old one, propose and test the *empirical progress (EP-)hypothesis*: the new theory (is and) remains more successful than the old one.

[15] E.g. suppose U is finite and let <M*, P*> be closer to the truth than <M, P> (partly) due to the 'strong' TC-clause (see Table 4.2) that $M \cap \mathbf{T}$ is a *proper* subset of $M^* \cap \mathbf{T}$. If R results from random sampling in **T**, sooner or later one of the extra items in $M^* \cap \mathbf{T}$ (i.e., items in $(M^* \cap \mathbf{T}) - M$) will show up, proving 'more successful'. Similarly, for the possible strong TC-clause on the P-side and the two possible strong FC-clauses.

Note that the two claims of a theory lead to different kinds of predictions: whereas the "$T \subseteq P$" claim leads to 'this is impossible' (hence, 'that must happen') predictions, the "$M \subseteq T$"-claim leads to 'this is possible/may happen'-predictions. With plausible consequences for differential predictions between <M, P> and <M*, P*>.

- Assuming that after 'sufficient confirmation' the EP-hypothesis is accepted (for the time being), which is an inductive conclusion, argue on the basis of the success theorem that the best explanation for this case of empirical progress is the *truth approximation (TA-)hypothesis* that the new theory is closer to the truth than the old one, i.e., that this is a case of truth approximation.

The reverse ('from success to verisimilitude') consequences of the success theorem, i.e., the consequences of that theorem in view of an accepted EP-hypothesis (and the data-theory on which it is based), are such that this situation not only suggests the TA-hypothesis, they also justify it to a substantial extent:

If <M*, P*> is, in view of <R, S>, accepted as empirically progressive relative to <M, P>, then

1. the success theorem not only makes it perfectly possible that <M*, P*> is closer to the truth than <M, P>, due to it the TA-hypothesis would even explain the greater success,
2. it is impossible that <M*, P*> is further from the truth than <M, P> (and hence <M, P> closer to the truth than <M*, P*>), for otherwise, so shows the success theorem, <M*, P*> could not be more successful,
3. it is also possible that <M*, P*> is neither closer to nor further from the truth than <M, P>, in which case, however, another specific explanation has to be given for the fact that <M*, P*> has so far proven to be more successful, e.g. due to a biased choice of experiments.

Note that the TA-hypothesis provides a typical *default* explanation of EP, that is, an adequate explanation unless there turns out to be reason for another diagnosis. The third reverse consequence provides the room for 'unless' conditions and hence for future 'divided success', i.e., the old theory may get extra successes relative to the new one. For example, the experiments so far may turn out to have been biased in favor of the new theory, and hence new experiments breaking this bias may turn out to be in favor of the old one.

- Abductively conclude (for the time being) that the new theory is closer to the truth than the old one, i.e., that truth approximation has been achieved.

This final step will now be no surprise. In fact, it is a special case of a sophisticated form of 'inference to the best explanation' or 'inference to the best theory', viz. not as a, or even the, true theory, but as the theory which is the closest to the truth of the available theories. Of course, 'inference to the best theory' should be read as: inference to the best theory of all available theories beyond the data-theory <R, S>.[16,17]

4.6 Some Perspectives for Further Research

In the next three chapters, the three announced concretizations of the two-sided approach will be presented: quantification, refinement, and stratification. Here we confine ourselves to some suggestions for further research that will not be further elaborated in this book.

4.6.1 Inconsistent Theories

Inconsistent two-sided theories, that is, theories where M is not a subset of P and hence some models are excluded by the postulates, may be based on good reasons for the models as well as for the postulates. Such theories may well be very useful for truth approximation. To begin with, starting with inconsistent theory <M, P> one may be heading for a consistent theory <M*, P*> such that the latter is 'side-wise' closer to the truth than the former. And even an inconsistent two-sided theory may be side-wise closer to the truth than another one, and hence be a step in the direction of the truth. This is formally perfectly possible. However, the question is to what extent <M, P> can still be meaningfully considered to be one theory, though inconsistent. How can M and P still be substantially related, i.e., share more than the vocabulary, when the models do not satisfy (all of) the postulates? Of course, one option is that the models may satisfy some approximate version of the postulates or only the most fundamental postulates. This is clearly something which needs to be investigated further.

4.6.2 Research Programs

Lakatos's distinction between hardcore and auxiliary postulates can easily be made on the P-side. On the M-side we can make the distinction between prototypical and tentative models. Whereas this leads on the P-side to a playground for revising the auxiliary postulates it leads on the M-side to a similar playground of revising (the set of) tentative models. Hence, Lakatosian empirical progress and truth

---

[16] The latter has to be excluded, for it is not only assumed to be true, it would provide a circular explanation of being the best theory. Moreover, opting for it would also amount to the conclusion that after all there was nothing of interest in the original theory.

[17] See Chap. 11 for an analysis of this kind of inference to the best explanation.

approximation can then be achieved when such revisions respect the prototypical models and the hardcore postulates.

4.6.3 Disjunctive Nomic Theories

A further challenge is to generalize the present two-sided approach restricted to nomic theories based on conjunctive claims to nomic theories in general, based on disjunctions of conjunctive nomic claims. It may well be that such a generalization is only possible in quantitative terms. More specifically, it seems plausible to try to use the recent generalization of Cevolani and Festa (to appear) of their quantitative approach of conjunctive theories to arbitrary theories, notably disjunctive theories. If this succeeds, it will also become possible to compare this extended nomic approach with the outline of Ilkka Niiniluoto (Niiniluoto 1987, Chap. 11) of a theory of nomic truth approximation based on distances between 'nomic constituents', where nomic theories in general are construed as disjunctions of nomic constituents.

References

Cevolani, G., & Festa, R. (To appear). *A partial consequence account of truthlikeness*. To appear in a special issue of *Synthese* in honor of Gerhard Schurz.

Cevolani, G., Crupi, V., & Festa, R. (2011). Verisimilitude and belief change for conjunctive theories. *Erkenntnis, 75*(2), 183–202.

Kuipers, T. (1982). Approaching descriptive and theoretical truth. *Erkenntnis, 18*(3), 343–378.

Kuipers, T. (2000). *From instrumentalism to constructive realism. On some relations between confirmation, empirical progress, and truth approximation. Synthese library* (Vol. 287). Dordrecht: Kluwer Academic Publishers.

Miller, D. (1978). On distance from the truth as a true distance. In J. Hintikka et al. (Eds.), *Essays on mathematical and philosophical logic* (pp. 415–435). Dordrecht: Reidel.

Niiniluoto, I. (1987). *Truthlikeness. Synthese library* (Vol. 185). Dordrecht: Reidel.

Part II
Concretizations

Chapter 5
Quantitative Nomic Truth Approximation

Abstract Assuming that the target of theory oriented empirical science in general and of nomic truth approximation in particular is to characterize the boundary or demarcation between nomic possibilities and nomic impossibilities, I have presented in Chap. 4 ("Models, postulates, and generalized nomic truth approximation") the 'basic' version of generalized nomic truth approximation, starting from 'two-sided' theories. Its main claim is that nomic truth approximation can perfectly be achieved by combining two prima facie opposing views on theories: (1) the traditional (Popperian) view: theories are (sets of models of) postulates that exclude certain possibilities from being realizable, enabling explanation and prediction and (2) the model view: theories are sets of models that claim to (approximately) represent certain realizable possibilities. Nomic truth approximation, i.e., increasing truth-content and decreasing falsity-content, becomes in this way revising theories by revising their models and/or their postulates in the face of increasing evidence.

The basic version of generalized nomic truth approximation is in many respects as simple as possible. Among other things, it is qualitative in the sense that it is purely based on set-theoretic relations. The present chapter presents the straightforward quantitative concretization of it. According to the 'expected success theorem', based on some probabilistic experimental conditions, greater truthlikeness, or verisimilitude, leads to greater expected empirical success. This enables tentative nomic truth approximation conclusions by abductive reasoning.

Keywords Degree of verisimilitude · Degree of truthlikeness · Degree of empirical success · Two-sided theories · Amount of truth-content · Amount of falsity-content · Degree of truth · Degree of falsity · Amount of accepted-content · Amount of rejected-content · Expected success theorem

---

Adapted version appeared as: "Quantitative nomic truth approximation by revising models and postulates, *European Studies in Philosophy of Science*, EPSA15 Selected papers, Eds. M. Massimi, J.W. Romeijn & G. Schurz, Dordrecht, Springer 2017, pp. 309–320. *Acknowledgement*: I like to thank an anonymous referee for some requests for clarification.

T. A. F. Kuipers, *Nomic Truth Approximation Revisited*, Synthese Library 399,
https://doi.org/10.1007/978-3-319-98388-2_5

5.1 Introduction

It is not difficult to present a simple (normalized) measure-theoretical account of nomic truth approximation, along with the (qualitative) basic version. It is important to note that whereas the basic version is only a partial ordering, a purely quantitative version will lead to complete orderings. However, we will also formulate hybrid versions, using quantitative ingredients for a couple of comparative judgments, leading to 'partially complete' orderings.

In Sect. 5.2, we will first define (M-/P-)sidewise and combined amounts of truth- and falsity-content of theories, of united content and sidewise and combined degrees of truthlikeness or verisimilitude. In passing, we will also define sidewise and combined degrees of truth and falsity of theories. Then we will define sidewise and combined amounts of accepted- and rejected-content and sidewise and combined degrees of empirical success.

Next we will present partially and purely quantitative definitions of greater verisimilitude and greater empirical success between theories. Then we will investigate how they are related. It will turn out that we can prove, assuming random experimentation, an 'Expected success theorem', according to which 'quantitatively closer to the truth' may be expected to lead to more empirical success. The theorem is crucial for making abductive truth approximation claims.

In Sect. 5.3 we will generalize the purely quantitative definitions of verisimilitude and empirical success, with three parameters, leaving room for various considerations and calculations. Finally, Sect. 5.4 concludes, among other things, with the perspective of a straightforward probabilistic reading of the present (neutral) quantitative approach in relation to a 'hypothetico-probabilistic method', all of which will be elaborated in Chap. 13.

5.2 Basic Quantitative Versions

All our quantitative definitions will presuppose a so-called measure function, more specifically, a probability measure on the set of subsets of U. Let m be a so-called normalized finitely additive real-valued measure function on the set of measurable subsets of U, that is, for all subsets X, Y of U: $m(\emptyset) = 0 \leq m(X) \leq 1 = m(U)$ and $m(X \cup Y) = m(X) + m(Y) - m(X \cap Y)$. It is important to note that e.g. m(M) is no prior in the sense of the probability that the M-claim of theory <M, P> is true, i.e., m("M $\subseteq$ **T**"). For this reason we prefer here 'm' for the probability function, where m(M) is the absolute probability (measure) of M as a subset of the total outcome space U.

5.2.1 *Amounts of Truth-Content, Falsity-Content and Verisimilitude*

Starting from theory <M, P>, plausible definitions of sidewise and combined amounts of truth- and falsity-content and of united content of theories are given in Table 5.1.[1] It also includes sidewise and combined degrees of verisimilitude. Here, and below, the corresponding basic definitions are first inserted.

As is easy to check, the amounts of truth- and falsity-content vary from 0 to the amounts of the united content. The sidewise degrees of verisimilitude vary from $-m(\mathrm{M})$ to $+m(\mathrm{M})$ and from $-m(\mathrm{cP})$ to $+m(\mathrm{cP})$, respectively, and hence the combined degree of verisimilitude varies from $-[m(\mathrm{M}) + m(\mathrm{cP})]$ to $m(\mathrm{M}) + m(\mathrm{cP})$, and hence ultimately, for maximal theories (M = P), from −1 to +1.

As announced, it is also easy to define, Table 5.2, sidewise and combined degrees of truth and falsity. Of course, the corresponding degrees of truth and falsity add up

Table 5.1 Sidewise and combined amounts of truth- and falsity-content of theories and of degrees of verisimilitude

Theory <M, P>	M-side	P-side	Combined
M- resp. P-claim	$\mathrm{M} \subseteq \mathbf{T}$	$\mathbf{T} \subseteq \mathrm{P}\ (\mathrm{cP} \subseteq \mathrm{c}\mathbf{T})$	$\mathrm{M} \subseteq \mathbf{T} \subseteq \mathrm{P}$
Basic: Truth-content (TC)	$\mathrm{M} \cap \mathbf{T}$[a]	$\mathrm{cP} \cap \mathrm{c}\mathbf{T}$	$(\mathrm{M} \cap \mathbf{T}) \cup (\mathrm{cP} \cap \mathrm{c}\mathbf{T})$
Amount of truth-content mTC $\in [0, 1]$	$m(\mathrm{M} \cap \mathbf{T})$	$m(\mathrm{cP} \cap \mathrm{c}\mathbf{T})$	$m(\mathrm{M} \cap \mathbf{T}) + m(\mathrm{cP} \cap \mathrm{c}\mathbf{T})$
Basic: Falsity-content (FC)	$\mathrm{M} - \mathbf{T}$	$\mathrm{cP} - \mathrm{c}\mathbf{T}$	$(\mathrm{M} - \mathbf{T}) \cup (\mathrm{cP} - \mathrm{c}\mathbf{T})$ $\equiv (\mathrm{M} - \mathbf{T}) \cup (\mathbf{T} - \mathrm{P})$
Amount of falsity-content mFC $\in [0, 1]$	$m(\mathrm{M} - \mathbf{T})$	$m(\mathrm{cP} - \mathrm{c}\mathbf{T})$	$m(\mathrm{M} - \mathbf{T}) + m(\mathrm{cP} - \mathrm{c}\mathbf{T})$ $= m(\mathrm{M} - \mathbf{T}) + m(\mathbf{T} - \mathrm{P})$
Basic: United content (UC) $= \mathrm{TC} \cup \mathrm{FC} = \mathrm{C}$	M	cP	$\mathrm{M} \cup \mathrm{cP}$
Amount of united content mUC $=_{\mathrm{def}} m\mathrm{TC} + m\mathrm{FC}$	$m(\mathrm{M})$	$m(\mathrm{cP})$	$m(\mathrm{M}) + m(\mathrm{cP})$
Degrees of verisimilitude (Ver) $=_{\mathrm{def}} m\mathrm{TC} - m\mathrm{FC}$	M-side		P-side
Sidewise	$m(\mathrm{M} \cap \mathbf{T}) - m(\mathrm{M} - \mathbf{T})$		$m(\mathrm{cP} \cap \mathrm{c}\mathbf{T}) - m(\mathrm{cP} - \mathrm{c}\mathbf{T})$
Combined	$[m(\mathrm{M} \cap \mathbf{T}) - m(\mathrm{M} - \mathbf{T})] + [m(\mathrm{cP} \cap \mathrm{c}\mathbf{T}) - m(\mathrm{cP} - \mathrm{c}\mathbf{T})]$ $\equiv [m(\mathrm{M} \cap \mathbf{T}) + m(\mathrm{cP} \cap \mathrm{c}\mathbf{T})] - [m(\mathrm{M} - \mathbf{T}) + m(\mathrm{cP} - \mathrm{c}\mathbf{T})]$		

[a]The set $\mathrm{M} \cap \mathbf{T}$ is the set of items about which the relevant claim, $\mathrm{M} \subseteq \mathbf{T}$, is true and the set $\mathrm{M} - \mathbf{T}$, below, is the set of items about which this claim is false

[1]The term 'combined' will always refer to some combination of sidewise notions and 'separated clauses' of the same kind, not necessarily as a strict conjunction. Separated clauses deal with truth- and falsity-content, respectively, and with accepted- and rejected-content, respectively. The term 'united' will refer to the union of such separated clauses and clause-wise notions of both relevant kinds.

Table 5.2 Sidewise and combined degrees of truth and falsity

Theory <M, P>	M-side	P-side	Combined
Degree of truth $\in [0, 1] =_{def} m\text{TC}/m\text{C}$	$\frac{m(\text{M} \cap \mathbf{T})}{m(\text{M})}$	$\frac{m(\text{cP} \cap \text{c}\mathbf{T})}{m(\text{cP})}$	$\frac{m(\text{M} \cap \mathbf{T}) + m(\text{cP} \cap \text{c}\mathbf{T})}{m(\text{M}) + m(\text{cP})}$
Degree of falsity $\in [0, 1] =_{def} m\text{FC}/m\text{C}$	$\frac{m(\text{M} - \mathbf{T})}{m(\text{M})}$	$\frac{m(\text{cP} - \text{c}\mathbf{T})}{m(\text{cP})}$	$\frac{m(\text{M} - \mathbf{T}) + m(\text{cP} - \text{c}\mathbf{T})}{m(\text{M}) + m(\text{cP})}$

Table 5.3 Sidewise and combined amounts of evidence, degrees of truth/falsity of evidence, amounts of accepted, rejected and undecided content, degrees of empirical success

Theory <M, P>	M-side	P-side	Combined
<M, P> in the light of evidence: data-theory <R, S>			
Accepted claims	$\text{R} \subseteq \mathbf{T}$	$\mathbf{T} \subseteq \text{S}$	$\text{R} \subseteq \mathbf{T} \subseteq \text{S}$
Basic: Content of evidence	R	cS	R ∪ cS
Amount of evidence	$m(\text{R})$	m(cS)	$m(\text{R})$ + m(cS)
Degrees of truth/falsity of <R,S>, if accepted claims are true	1/0	1/0	1/0
Basic: Accepted-content (AC)	M∩R	cP∩cS	(M∩R) ∪ (cP∩cS)
Amount of accepted-content (mAC)	m(M∩R)	m(cP∩cS)	m(M∩R) + m(cP∩cS)
Basic: Rejected-content (RC)	M−S	cP−cR	(M−S) ∪ (cP−cR)
Amount of rejected-content (mRC)	m(M−S)	m(cP−cR)	m(M−S) + m(cP−cR)
Amount of undecided content	m(M−(R∪cS))	m(cP−(R∪cS))	m((M∪cP)−(R∪cS))
Degrees of empirical success (Suc) $=_{def}$ mAC − mRC	M-side		P-side
Sidewise	m(M∩R) − m(M−S)		m(cP∩cS) − m(cP−cR)
Combined	[m(M∩R) − m(M−S)] − [m(cP∩cS) − m(cP−cR)] ≡ [m(M∩R) + m(cP∩cS)] − [m(M−S) + m(cP−cR)]		

to 1. However, besides an application to evidence, these degrees will not play a role in what follows.

Assuming a theory <M, P> and evidence <R, S>, plausible definitions of sidewise and combined amounts of evidence, degrees of truth/falsity of evidence, accepted, rejected and undecided content, and sidewise and combined degrees of empirical success are given in Table 5.3.

As is easy to check, the sidewise and combined degrees of empirical success vary between the same boundaries as the corresponding degrees of verisimilitude.

5.2.2 *Quantitative Comparative Definitions of Verisimilitude and Empirical Success*

Turning to the comparison of theories <M, P> and <M*, P*>, the first intended comparative judgment is "<M*, P*> is quantitatively at least as close to the truth as <M, P>", abbreviated as "<M*, P*> is at least as q-close to the truth as <M, P>". Table 5.4 starts with definitions in terms of the relevant TC- and FC-clause, first basic (0 = 0a&0b), then quantitatively, both sidewise (1 = 1a&1b) and combined (2 = 2a&2b), and in the sidewise cases an *un*conditional and an at least as transparent conditional, but equivalent variant. The quantitative definitions are typically partial or hybrid in the sense that they use quantitative content measures for sidewise and combined comparative definitions.

Table 5.4 concludes with 'united' sidewise quantitative (3) and combined (or purely) quantitative (4) definitions of 'at least as q-close to the truth', where the TC-

Table 5.4 <M*, P*> is at least as q-close to the truth as <M, P>, 4 versions, see text. Sidewise definitions (0, 1, 3) should be read as conjunctions of the sidewise specified clauses

	M-side	P-side
<M*, P* > is at least as q-close to the truth as <M, P > iff		
0a) Basic: TC-clauses: $TC \subseteq TC^*$	$M \cap \mathbf{T} \subseteq M^* \cap \mathbf{T} \equiv$ $\mathbf{T} - M^* \subseteq \mathbf{T} - M$	$cP \cap c\mathbf{T} \subseteq cP^* \cap c\mathbf{T} \equiv$ $c\mathbf{T} - cP^* \subseteq c\mathbf{T} - cP \equiv$ $P^* \cap c\mathbf{T} \subseteq P \cap c\mathbf{T}$
1a) TC-clauses: $mTC \leq mTC^*$	$m(M \cap \mathbf{T}) \leq m(M^* \cap \mathbf{T}) \equiv$ $m(\mathbf{T} - M^*) \leq m(\mathbf{T} - M)$	$m(cP \cap c\mathbf{T}) \leq m(cP^* \cap c\mathbf{T}) \equiv$ $m(c\mathbf{T} - cP^*) \leq m(c\mathbf{T} - cP)$
Conditional	$m(M \mid \mathbf{T})^a \leq m(M^* \mid \mathbf{T})$	$m(cP \mid c\mathbf{T}) \leq m(cP^* \mid c\mathbf{T}) \equiv$ $m(P^* \mid c\mathbf{T}) \leq m(P \mid c\mathbf{T})$
2a) Combined TC-clause	$m(M \cap \mathbf{T}) + m(cP \cap c\mathbf{T}) \leq m(M^* \cap \mathbf{T}) + m(cP^* \cap c\mathbf{T})$	
&	&	
0b) Basic: FC-clauses: $FC^* \subseteq FC$	$M^* - \mathbf{T} \subseteq M - \mathbf{T}$	$cP^* - c\mathbf{T} \subseteq cP - c\mathbf{T} \equiv$ $P \cap \mathbf{T} \subseteq P^* \cap \mathbf{T}$
1b) FC-clauses: $mFC^* \leq mFC$	$m(M^* - \mathbf{T}) \leq m(M - \mathbf{T})$	$m(cP^* - c\mathbf{T}) \leq m(cP - c\mathbf{T})$
Conditional	$m(M^* \mid c\mathbf{T}) \leq m(M \mid c\mathbf{T})$	$m(cP^* \mid \mathbf{T}) \leq m(cP \mid \mathbf{T}) \equiv$ $m(P \mid \mathbf{T}) \leq m(P^* \mid \mathbf{T})$
2b) Combined FC-clause	$m(M^* - \mathbf{T}) + m(cP^* - c\mathbf{T}) \leq m(M - \mathbf{T}) + m(cP - c\mathbf{T})$	
TC-/FC-clauses united, quantitative: $Ver \leq Ver^*$		
3)Sidewise	$m(M \cap \mathbf{T}) - m(M - \mathbf{T})$ $\leq m(M^* \cap \mathbf{T}) - [m(M^* - \mathbf{T})$	$m(cP \cap c\mathbf{T}) - m(cP - c\mathbf{T})$ $\leq m(cP^* \cap c\mathbf{T}) - m(cP^* - c\mathbf{T})$
4) Combined (= purely quantitative):	$[m(M \cap \mathbf{T}) + m(cP \cap c\mathbf{T})] - [m(M - \mathbf{T}) + m(cP - c\mathbf{T})]$ $\leq [m(M^* \cap \mathbf{T}) + m(cP^* \cap c\mathbf{T})] - [m(M^* - \mathbf{T}) + m(cP^* - c\mathbf{T})]$	

[a] $m(A \mid B) =_{def} m(A \cap B)/m(B)$

and FC-clauses are united, comparing of course the corresponding degrees of verisimilitude. Finally, in all cases we get 'q-closer to the truth' by assuming at least one strict inequality.

It is not difficult to prove that the following entailment relations hold: (0) ⇒ (1) ⇒ (2), (3) ⇒ (4). Note that (2) and (3) may or may not go together.

As an aside it is interesting to simplify, in Table 5.5, the last part of Table 5.4, uniting the TC- and FC-clause, by reformulating the comparisons in terms of symmetric differences,[2] starting with the sidewise basic case, left out in Table 5.4, and subsequently focusing on maximal theories.

This enables us to give a simple propositional example of a theory that is not basically but q-closer the truth. In Table 5.6 it is made clear by truth-tables that $M^* = Mod((p_1 \underline{v} p_2) \Leftrightarrow p_3)$[3] is q-closer to $\mathbf{T} = Mod((p_1 v p_2) \Leftrightarrow p_3)$ than theory $M = Mod((p_1 \& p_2) \Leftrightarrow p_3)$. For we get: $Mod(M) = \{s_1, s_6, s_7, s_8\}$, $Mod(M^*) = \{s_2, s_3, s_4, s_8\}$ and $Mod(\mathbf{T}) = \{s_1, s_2, s_3, s_8\}$. Hence $|M^* \Delta \mathbf{T}| = |\{s_1, s_4\}| = 2 < |M \Delta \mathbf{T}| = |\{s_2, s_3, s_6, s_7\}| = 4$. Note that M* is not basically closer to **T** than M, for s_1 and s_4 make this impossible.

Next, partially quantitative definitions of the second intended comparative judgment, i.e., "<M*, P*> is at least as q-successful as <M, P>", are given in Table 5.7,

Table 5.5 The basic and quantitative TC/FC-clauses united, in terms of symmetric differences: general and specifically for maximal theories

TC/FC-clauses united	M-side	P-side
Sidewise basic	$M^* \Delta \mathbf{T} \subseteq M \Delta \mathbf{T}$	$cP^* \Delta c\mathbf{T} \subseteq cP \Delta c\mathbf{T}$
Quantitative Ver ≤ Ver*		
3) Sidewise	$m(M^* \Delta \mathbf{T}) \leq m(M \Delta \mathbf{T})$	$m(cP^* \Delta c\mathbf{T}) \leq m(cP \Delta c\mathbf{T})$
4) Combined (purely quantitative):	$m(M^* \Delta \mathbf{T}) + m(cP^* \Delta c\mathbf{T}) \leq m(M \Delta \mathbf{T}) + m(cP \Delta c\mathbf{T})$	
If the theories are maximal (Kuipers 2000), i.e., M = P = X M* = P* = X*		
Basic	$X^* \Delta \mathbf{T} \subseteq X \Delta \mathbf{T}$	
Quantitative	$m(X^* \Delta \mathbf{T}) \leq m(X \Delta \mathbf{T})$	

Table 5.6 A propositional example: $M^* = Mod((p_1 \underline{v} p_2) \Leftrightarrow p_3)$ is q-closer to $T = Mod((p_1 v p_2) \Leftrightarrow p_3)$ than $P = Mod((p_1 \& p_2) \Leftrightarrow p_3)$

Models/states	p_1	p_2	p_3	M: $(p_1 \& p_2) \Leftrightarrow p_3$	M*: $(p_1 \underline{v} p_2) \Leftrightarrow p_3$	**T**: $(p_1 v p_2) \Leftrightarrow p_3$
$s_1 = p_1 \& p_2 \& p_3$	1	1	1	1	0	1
$s_2 = \neg p_1 \& p_2 \& p_3$	0	1	1	0	1	1
$s_3 = p_1 \& \neg p_2 \& p_3$	1	0	1	0	1	1
$s_4 = p_1 \& p_2 \& \neg p_3$	1	1	0	0	1	0
$s_5 = \neg p_1 \& \neg p_2 \& p_3$	0	0	1	0	0	0
$s_6 = \neg p1 \& p_2 \& \neg p_3$	0	1	0	1	0	0
$s_7 = p_1 \& \neg p_2 \& \neg p_3$	1	0	0	1	0	0
$s_8 = \neg p_1 \& \neg p_2 \& \neg p_3$	0	0	0	1	1	1

[2] $A \Delta B =_{def} (A-B) \cup (B-A)$.

[3] '$p_1 \underline{v} p_2$' is short for $(p_1 v p_2) \& \neg(p_1 \& p_2)$, so-called exclusive 'or'.

Table 5.7 <M*, P*> is at least as q-successful as <M, P> (relative to <R, S>), 4 versions, see text. Sidewise definitions (0^s, 1^s, 3^s) should again be read as conjunctions of the sidewise specified clauses

	M-side	P-side
<M*, P* > is at least as q-successful as <M, P>, relative to <R, S>, iff		
$0a^s$) Basic: AC-clauses $AC \subseteq AC^*$	$M \cap R \subseteq M^* \cap R$	$cP \cap cS \subseteq cP^* \cap cS$
$1a^s$) AC-clauses: $mAC \leq mAC^*$	$m(M \cap R) \leq m(M^* \cap R)$	$m(cP \cap cS) \leq m(cP^* \cap cS)$
Conditional	$m(M \mid R) \leq m(M^* \mid R)$	$m(cP \mid cS) \leq m(cP^* \mid cS) \equiv m(P^* \mid cS) \leq m(P \mid cS)$
$2a^s$) Combined AC-clause	$m(M \cap R) + m(cP \cap cS) \leq m(M^* \cap R) + m(cP^* \cap cS)$	
&	&	
$0b^s$) Basic: RC-clauses $RC^* \subseteq RC$	$M^* - S \subseteq M - S$	$cP^* - cR \subseteq cP - cR$
$1b^s$) RC-clauses: $mRC^* \leq mRC$	$m(M^* - S) \leq m(M - S)$	$m(cP^* - cR) \leq m(cP - cR)$
Conditional	$m(M^* \mid cS) \leq m(M \mid cS)$	$m(cP^* \mid R) \leq m(cP \mid R) \equiv m(P \mid R) \leq m(P^* \mid R)$
$2b^s$) Combined RC-clause	$m(M^* - S) + m(cP^* - cR) \leq m(M - S) + m(cP - cR)$	
AC/RC-clauses united, quantitative: $Suc \leq Suc^*$		
3^s) Sidewise:	$m(M \cap R) - m(M - S) \leq m(M^* \cap R) - m(M^* - S)$	$m(cP \cap cS) - m(cP - cR) \leq m(cP^* \cap cS) - m(cP^* - cR)$
4^s) Combined (= purely quantitative):	$[m(M \cap R) - m(M - S)] + [m(cP \cap cS) - m(cP - cR)] \leq [m(M^* \cap R) - m(M^* - S)] + [m(cP^* \cap cS) - m(cP^* - cR)]$	

now first in terms of the relevant AC- and RC-clause, first basic ($0^s = 0a^s \& 0b^s$) again both sidewise ($1^s = 1a^s \& 1b^s$) and combined ($2^s = 2a^s \& 2b^s$), and in the sidewise case an *un*conditional and a conditional, but equivalent, variant.[4] Again, all definitions are typically partial or hybrid. Table 5.7 concludes with 'united' sidewise (3^s) and combined (or purely) quantitative (4^s) definitions of 'at least as q-successful', relative to <R, S>, where the AC- and RC-clauses are united, comparing of course the corresponding degrees of empirical success. Again, in all cases we get 'q-more successful' by assuming at least one strict inequality.

Again, it is not difficult to prove that the following entailment relations hold: (0^s) ⇒ (1^s) ⇒ (2^s), (3^s) ⇒ (4^s). Note also that (2^s) and (3^s) may or may not go together. Unfortunately, in the present case there is no simplifying formulation of the last part of Table 5.7 in terms of symmetric differences.

Compared to the basic, purely qualitative, cases, the hybrid definitions of 'at least as q-close to the truth' and 'at least as q-successful' leave much more room for cases in which these comparative judgments apply. Moreover, the combined clauses

[4] Note that superscript 's' refers to a success condition, not to sidewise!

give still more possibilities than the conjunction of the sidewise clauses. However, only when the two relevant clauses are united the quantitative definitions lead to a sidewise or combined complete ordering of theories with respect to verisimilitude and empirical success, i.e., (3) and (4), and (3^s) and $4^s)$, respectively.

5.2.3 Expected Success Theorem

Now we turn to the relation between verisimilitude and empirical success comparisons. The main question is of course to what extent a 'success theorem' can be proven. In the basic case, the easy to prove success theorem states: 'at least as close to the truth' entails 'at least as successful'. Surely, it is now not possible to prove the 'q-version' of this claim, let alone that 'q-closer to the truth' entails 'q-more successful', in whatever of the defined partial or overall senses. However, it is possible to prove under certain probabilistic conditions that 'q-closer to the truth' entails that 'q-more successfulness' is to be expected, in terms of mathematical expectations. This 'expected success theorem' is restricted to the 'numerical measure function' on a finite U, according to which $m(X) = |X| / |U|$ for subset X of U, where |X| indicates the number of elements in X. It is illuminating because it expresses very well what we might intuitively expect.

Recall that the expectation value $Ef(\underline{u})$ of real-valued function $f(\underline{u})$ of random variable $\underline{u}$, governed by probability function p, is defined as $\sum f(u)p(\underline{u} = u)$, with summation over all the possible values of $\underline{u}$. Suppose that the set R, at a certain time, but now more important, with size r, results from random selection out of the set of all subsets of **T** of size r, denoted by $\mathbf{T}_r$. Hence, $\underline{R}$ is a random variable, with probability $p(\underline{R} = R) = 1/|\mathbf{T}_r|$. Similarly, suppose that S, with size s, results from random selection out of the set of all supersets of **T** of size s, denoted by $\mathbf{T}_s$. Hence, $\underline{S}$ is a random variable over the subsets of c**T** united with **T**, with probability $p(\underline{S}=S) = 1/|\mathbf{T}_s|$.

Expected Success Theorem:

if U is finite and q-closer is based on the numerical *m*, if <M*, P* > is q-closer to the truth than <M, P > in version (1), see Table 5.4, and if evidence <R, S > is based on random sampling in the described way, and no mistakes have been made ($R \subseteq \mathbf{T} \subseteq S$), then the expected value of the amount of (rightly) accepted-content on both sides of <M*, P* > is higher than that of the corresponding sides of <M, P>, whereas the expected value of the amount of (rightly) rejected-content on both sides of <M*, P* > is lower than of the corresponding sides of <M, P>. In sum, it is to be expected that <M*, P* > is more q-successful than <M, P > in version (1^s).

We prove one of the four claims involved; the proof of the other three claims is similar. Here is the proof for the M-side TC-clause (1a, for numerical *m*): if $|M \cap \mathbf{T}| \leq |M^* \cap \mathbf{T}|$ then $E\,|M \cap \underline{R}| \leq E\,|M^* \cap \underline{R}|$, i.e., the 'expected' version of $(1a^s$, for numerical *m*).

Recall, that $\mathbf{T}_r$ indicates the set of subsets of $\mathbf{T}$ of size r. Our point of departure is:

$$E|M \cap \underline{R}| = \Sigma_{\underline{R}=R \in \mathbf{T}r} |M \cap R|.p(\underline{R} = R)$$

Given $p(\underline{R} = R) = 1/|\mathbf{T}_r|$, and denoting 'truth value' by tv (being 1 or 0), we get that $E\,|M \cap \underline{R}|$ is equal to:

$$(1/|\mathbf{T}_r|)\Sigma_{\underline{R}=R \in \mathbf{T}r} |M \cap R| = (1/|\mathbf{T}_r|)\Sigma_{\underline{R}=R \in \mathbf{T}r} \Sigma_{x \in M} \, tv(\text{``}x \in R\text{''})$$

$$=$$

$$(1/|\mathbf{T}_r|)\Sigma_{x \in M} \Sigma_{\underline{R}=R \in \mathbf{T}r} \, tv(\text{``}x \in R\text{''})$$

Note now that the last double summation is proportional to $|M \cap \mathbf{T}|$, for every x in $\mathbf{T}$ is equally many times a member of R. Combining this with the similar proportionality of $E\,|M^* \cap \underline{R}|$ to $|M^* \cap \mathbf{T}|$, the claim of the theorem directly follows from the relevant defining condition of q-closer to, viz. $|M \cap \mathbf{T}| \leq |M^* \cap \mathbf{T}|$. Q.e.d.[5]

Of course, variants of the theorem can be proven for the comparative claims based on the other quantitative definitions, and it may be possible to liberate some of the other assumptions to some extent (e.g., the numerical measure, the finiteness assumption). However this may be, the corresponding expected success theorem will remain to have a limited and artificial, random sampling, character. But this does not exclude the general message of the theorem: if a theory remains to be more q-successful than another after prolonged testing, this is a good reason to explain this, at least by default, by supposing that it is q-closer to the truth.

Let us conclude this section by formulating the remaining methodological steps to complete the quantitative theory of nomic truth approximation: assuming that a new theory is at a certain moment more q-successful than the old one, propose and test the '*empirical progress hypothesis*': the new theory (is and) remains more q-successful than the old one. Assuming that after 'sufficient confirmation' this 'empirical progress hypothesis' is accepted (for the time being), argue *on the basis of a reconstruction of the way in which experiments have been done* and the most relevant expected success theorem, that the 'default explanation' for this case of empirical progress is the hypothesis that the new theory is q-closer to the truth than the old one, *i.e., that this is a case of quantitative truth approximation*. Finally, *conclude abductively* (for the time being) that the new theory is q-closer to the truth than the old one, *i.e., that quantitative truth approximation has been achieved*. Of course, we might strengthen this argumentation by assuming relevant thresholds for the conclusion that the new theory is q-more successful than the old one.

[5] Note that the proof of the claim (Sect. 5.2.2) that a theory which is closer to the truth than another in the basic sense will become more successful in the long run is a very special case of the present proof.

5.3 Generalized Purely Quantitative Definitions

In the preceding section we have already given purely quantitative measures of verisimilitude and success. It is of course possible to generalize them by introducing parameters and base comparisons on them. In order to take different absolute weights of truth- and falsity-content into account, as well as relative weights of the two kinds of truth-content and falsity-content, respectively, we introduce three real-valued parameters, viz. φ, γ, and δ, all positive, and define.

Definition:
The degree of verisimilitude of <M,P>:

$$\mathrm{Ver}_{\varphi}^{\gamma,\delta}(\langle \mathrm{M}, \mathrm{P}\rangle) =_{\mathrm{def}} [m(\mathrm{M} \cap \mathbf{T}) + \gamma\, m(\mathrm{cP} \cap \mathrm{c}\mathbf{T})] - \varphi\, [m(\mathrm{M} - \mathbf{T}) + \delta\, m(\mathrm{cP} - \mathrm{c}\mathbf{T})]$$

The degree of success of <M, P> relative to <R, S>:

$$\mathrm{Suc}_{\varphi}^{\gamma,\delta}(\langle \mathrm{M}, \mathrm{P}\rangle_{\langle \mathrm{R}, \mathrm{S}\rangle}) =_{\mathrm{def}} [m(\mathrm{M} \cap \mathrm{R}) + \gamma\, m(\mathrm{cP} \cap \mathrm{cS})] - \varphi\, [m(\mathrm{M} - \mathrm{S}) + \delta\, m(\mathrm{cP} - \mathrm{cR})]$$

Note first that the purely quantitative clauses in Table 5.4, (4), and Table 5.7, (4^s), are based on putting φ=γ=δ=1.

Of course, the definition of the quantitative measures would be inadequate when it would not cohere with the basic definitions. However, it is not difficult to prove the

Coherence Theorem:
if <M*, P*> is closer to the truth (in the basic sense, Table 5.4, (0)) than <M, P> then for all φ, γ, δ > 0 $\mathrm{Ver}_{\varphi}^{\gamma,\delta}(\langle \mathrm{M}, \mathrm{P}\rangle) \leq \mathrm{Ver}_{\varphi}^{\gamma,\delta}(\langle \mathrm{M}^*, \mathrm{P}^*\rangle)$.

if <M*, P*> is more successful (in the basic sense, Table 5.7, (0^s)) than <M, P> then for all φ, γ, δ > 0 $\mathrm{Suc}_{\varphi}^{\gamma,\delta}(\langle \mathrm{M}, \mathrm{P}\rangle) \leq \mathrm{Suc}_{\varphi}^{\gamma,\delta}(\langle \mathrm{M}^*, \mathrm{P}^*\rangle)$.

The crucial point is that the measure of subsets never exceeds the measure of encompassing sets.

In the first definition, φ represents the relative weight of the truth- and falsity-content and γ represents the relative weight of the truth-content of the M- and the P-theory and δ similarly of the falsity-content of them. The parameters γ and δ may or may not depend on $m(\mathbf{T})$ or they may be related in some other way. The reason to claim that they may depend on $m(\mathbf{T})$ is that, for example, if (one may assume that) $m(\mathbf{T})$ is much smaller than $m(\mathrm{c}\mathbf{T})$ it seems reasonable that M-side truth- and falsity-content are valued higher than those of the P-side. For this reason the parameters could even both be set equal to $m(\mathbf{T})/m(\mathrm{c}\mathbf{T})$.[6] More generally, in terms of models and postulates, by taking γ and δ smaller than 1, one might wish to express that providing fully adequate representations is much more impressive than to rightfully

[6] To be sure, by making the weights (strictly) dependent on $m(\mathbf{T})$ and $m(\mathrm{c}\mathbf{T})$, it becomes impossible to calculate the increase of verisimilitude of a revised theory (see below) in the standard situation in which we don't know **T**.

exclude some conceptual possibilities. However, the one exclusion may not be as impressive as the other. Here we enter the possibility of making difference between the value of successes and mistakes in exclusion, and similarly for inclusion, i.e., the idea of the refined approach, see (Kuipers 2000, Chaps. 10 and 7 of this book).

Other kinds of special cases we get by putting $\varphi = 1$ or by putting $\gamma = \delta$ or even further $\gamma=\delta=1$, the latter specification leading to

$$\mathrm{Ver}_{\varphi}^{1,1}(<\mathrm{M, P}>) =_{\mathrm{def}} \mathrm{Ver}(<\mathrm{M, P}>) = [m(\mathrm{M} \cap \mathbf{T}) + m(\mathrm{cP} \cap \mathrm{c}\mathbf{T}) - \varphi[m(\mathrm{M} - \mathbf{T}) + m(\mathrm{cP} - \mathrm{c}\mathbf{T})]$$

This is the definition in Cevolani et al. (2011) for 'conjunctive theories', translated in the two-sided nomic setting, hence our generalized definition is still more general. The authors call $\mathrm{Ver}_{\varphi}^{1,1}$ a 'feature-contrast' measure, because they relate it to a standard statistical practice of feature-contrast representation. Feature-contrast measures of verisimilitude have been introduced (without using this name) by Cevolani and Calandra (2009). Cevolani and Festa (manuscript) give an overview of feature-contrast measures. See also Cevolani and Festa (to appear) as the most recent presentation, including a significant extension to the case of arbitrary theories, notably disjunctive theories.

It is interesting to turn to the most extreme 'empiricist' revision of <M, P> on the basis of true data <R, S>: $<\mathrm{M}^{\mathrm{r}}, \mathrm{P}^{\mathrm{r}}> = <(\mathrm{M} \cap \mathrm{S}) \cup \mathrm{R}, (\mathrm{P} \cup \mathrm{R}) \cap \mathrm{S}>$, which is easily seen to be closer to the truth and (hence) more successful than <M, P>, both in the basic sense.

The coherence theorem automatically implies that the degree of verisimilitude of this revision, viz. $\mathrm{Ver}_{\varphi}^{\gamma,\delta}(<\mathrm{M}^{\mathrm{r}}, \mathrm{P}^{\mathrm{r}}>)$, is higher than (or at least equal to) that of <M, P>, viz. $\mathrm{Ver}_{\varphi}^{\gamma,\delta}(<\mathrm{M, P}>)$. Similarly for the degree of success.

The increase of the verisimilitude of the revision can now not only be expressed but even calculated without knowing **T**. Recall:

$$\mathrm{Ver}_{\varphi}^{\gamma,\delta}(<\mathrm{M, P}>) \quad =_{\mathrm{def}} [m(\mathrm{M} \cap \mathbf{T}) + \gamma m(\mathrm{cP} \cap \mathrm{c}\mathbf{T})] - \varphi[m(\mathrm{M} - \mathbf{T}) + \delta m(\mathrm{cP} - \mathrm{c}\mathbf{T})]$$

Hence, the verisimilitude of the revision, $\mathrm{Ver}_{\varphi}^{\gamma,\delta}(<\mathrm{M}^{\mathrm{r}}, \mathrm{P}^{\mathrm{r}}>)$, becomes:

$$\begin{aligned}&[m(((\mathrm{M} \cap \mathrm{S}) \cup \mathrm{R}) \cap \mathbf{T} + \gamma m(((\mathrm{cP} - \mathrm{R}) \cup \mathrm{cS}) \cap \mathrm{c}\mathbf{T})] \\ &-\varphi[m(((\mathrm{M} \cap \mathrm{S}) \cup \mathrm{R}) - \mathbf{T}) + \delta m(((\mathrm{cP} - \mathrm{R}) \cup \mathrm{cS}) - \mathrm{c}\mathbf{T}]\end{aligned}$$

It is not difficult to sort out the components of the four measure expressions of the revision.

$$\begin{aligned}m\big(((\mathrm{M} \cap \mathrm{S}) \cup \mathrm{R}) \cap \mathbf{T}\big) &= m(\mathrm{M} \cap \mathbf{T}) \quad + \quad m(\mathrm{R} - \mathrm{M}) \\ m\big(((\mathrm{cP} - \mathrm{R}) \cup \mathrm{cS}) \cap \mathrm{c}\mathbf{T}\big) &= m(\mathrm{cP} \cap \mathrm{c}\mathbf{T}) \quad + \quad m(\mathrm{cS} - \mathrm{cP})\end{aligned}$$

$$m\big(((\mathrm{M}\cap\mathrm{S})\cup\mathrm{R})-\mathbf{T}\big) = m(\mathrm{M}-\mathbf{T}) - m(\mathrm{M}-\mathrm{S})$$
$$m\big(((\mathrm{cP}-\mathrm{R})\cup\mathrm{cS})-\mathrm{c}\mathbf{T}\big) = m(\mathrm{cP}-\mathrm{c}\mathbf{T}) - m(\mathrm{cP}-\mathrm{cR})$$

Conceptually, the first two decompositions are easy to see as the sum of the original 'true positives and negatives' and the added ones by the revision, and the second two as the difference between the original 'false positives and negatives' and the removed ones.

Using the decompositions, it is easy to express the increase of verisimilitude:

$$\mathrm{Ver}_{\varphi}^{\gamma,\delta}(<\mathrm{M}^{r},\ \mathrm{P}^{r}>)-\mathrm{Ver}_{\varphi}^{\gamma,\delta}(<\mathrm{M},\ \mathrm{P}>)=[m(\mathrm{R}-\mathrm{M})+\gamma\, m(\mathrm{cS}-\mathrm{cP})]+\varphi[m(\mathrm{M}-\mathrm{S})+\delta\, m(\mathrm{cP}-\mathrm{cR})]$$

Assuming that the parameters do not depend on **T**, we may conclude that the increase of verisimilitude can be calculated on the basis of <M, P> and <R, S> alone.

It will not come as a surprise that this increase of verisimilitude equals the increase of success. Note first that the resulting revision is maximally successful, i.e., it has no rejected-content and only accepted-content, even the maximally possible, R and cS:

$$\begin{aligned}\mathrm{Suc}_{\varphi}^{\gamma,\delta}(<\mathrm{M}^{r},\ \mathrm{P}^{r}>_{<\mathrm{R},\,\mathrm{S}>})=_{\mathrm{def}}\\ &[m((\mathrm{M}\cap\mathrm{S})\cup\mathrm{R})\cap\mathrm{R}+\gamma\, m(\mathrm{c}((\mathrm{P}\cup\mathrm{R})\cap\mathrm{S})\cap\mathrm{cS})]\\ &-\varphi[m((\mathrm{M}\cap\mathrm{S})\cup\mathrm{R})-\mathrm{S})+\delta\, m(\mathrm{c}((\mathrm{P}\cup\mathrm{R})\cap\mathrm{S})-\mathrm{cR})]\\ &=[m(\mathrm{R})+\gamma\, m(\mathrm{cS})]\end{aligned}$$

The original success is:

$$\mathrm{Suc}_{\varphi}^{\gamma,\delta}(<\mathrm{M},\ \mathrm{P}>_{<\mathrm{R},\,\mathrm{S}>})=_{\mathrm{def}}[m(\mathrm{M}\cap\mathrm{R})+\gamma m(\mathrm{cP}\cap\mathrm{cS})]-\varphi[m(\mathrm{M}-\mathrm{S})+\delta\, m(\mathrm{cP}-\mathrm{cR})]$$

Hence, the increase in success, $\mathrm{Suc}_{\varphi}^{\gamma,\delta}(<\mathrm{M}^{r},\mathrm{P}^{r}>_{<\mathrm{R},\,\mathrm{S}>}) - \mathrm{Suc}_{\varphi}^{\gamma,\delta}(<\mathrm{M},\mathrm{P}>_{<\mathrm{R},\,\mathrm{S}>})$, is:

$$[m(\mathrm{R})+\gamma\, m(\mathrm{cS})]-[m(\mathrm{M}\cap\mathrm{R})+\gamma\, m(\mathrm{cP}\cap\mathrm{cS})]+\varphi[m(\mathrm{M}-\mathrm{S})+\delta\, m(\mathrm{cP}-\mathrm{cR})]$$

$$=[m(\mathrm{R})-m(\mathrm{M}\cap\mathrm{R})+\gamma(m(\mathrm{cS})-m(\mathrm{cP}\cap\mathrm{cS}))]+\varphi[m(\mathrm{M}-\mathrm{S})+\delta\, m(\mathrm{cP}-\mathrm{cR})]$$

$$=[m(\mathrm{R}-\mathrm{M})+\gamma\, m(\mathrm{cS}-\mathrm{cP})]+\varphi[m(\mathrm{M}-\mathrm{S})+\delta\, m(\mathrm{cP}-\mathrm{cR})]$$

Which is indeed the same as the increase in verisimilitude.

Turning back to the generalized purely quantitative definitions in general, but applied to the finite numerical case, it is clear that the expected success theorem can be proven, in particular because the two definitions are based on the same three parameter values. This means that the remaining methodological steps to complete the quantitative theory of nomic truth approximation, as indicated at the end of the previous section, remain valid.

5.4 Concluding Remarks

The present chapter is essentially a strong revision of the probabilistic approach to truth approximation (Kuipers 2009) in the light of the new, generalized (two-sided) approach. More specifically, the conditional version of the clause wise and sidewise definitions of 'closer to the truth' and 'more successful' at the P-side relates to that paper. That paper further explained the link between such a conditional approach and the probabilistic version of the hypothetico-deductive method. In line with this paper it will be shown in Chap. 13 that it is well possible to link the hypothetico-probabilistic method to the P-side part of the analysis in the present chapter. However, whether and how this method can be linked to the M-side is a matter for future research.

As stated in the introduction, the present chapter only deals with the first of (at least) three plausible concretizations of the two-sided basic version (in line with such concretizations in Kuipers (2000)). In Chaps. 6 and 7 the other two are presented: a *refined* version, based on an underlying ternary similarity relation, and a *stratified* version, based on a (theory relative) distinction between an observational and an inclusive theoretical level.

The refined version to be presented in Chap. 6 is a refined version of the basic approach. A related refined version of the quantitative approach in this chapter can be obtained by introducing a distance function between the conceptual possibilities. In Kuipers (2000, Chap. 12) this has been done assuming maximal theories. We leave it as a challenge to do so for two-sided theories.

Finally, Chap. 7 presents a stratified version of the basic account. A stratified version of the present quantitative concretization has still to be developed, and hence of a refined stratified quantitative account. However, there is no reason to think that they are impossible.

References

Cevolani, G., & Calandra, F. (2009). Approaching the truth via belief change in propositional languages. In M. Suárez, M. Dorato, & M. Rèdei (Eds.), *Launch of the European philosophy of science association, EPSA epistemology and methodology of science* (Vol. 1, pp. 47–62). Dordrecht: Springer.

Cevolani, G., & Festa, R. Manuscript. *Features of verisimilitude*. Unpublished.

Cevolani, G., & Festa, R. (To appear). *A partial consequence account of truthlikeness*. To appear in a special issue of *Synthese* in honor of Gerhard Schurz.

Cevolani, G., Crupi, V., & Festa, R. (2011). Verisimilitude and belief change for conjunctive theories. *Erkenntnis, 75*(2), 183–202.

Kuipers, T. (2000). *From instrumentalism to constructive realism. On some relations between confirmation, empirical progress, and truth approximation. Synthese library* (Vol. 287). Dordrecht: Kluwer Academic Publishers.

Kuipers, T. (2009). Empirical progress and truth approximation by the 'Hypothetico-probabilistic method'. *Erkenntnis, 70*, 313–330.

Chapter 6
Refined Nomic Truth Approximation

Abstract Assuming that the target of theory oriented empirical science in general and of nomic truth approximation in particular is to characterize the boundary or demarcation between nomic possibilities and nomic impossibilities, I have presented in Chap. 4 ("Models, postulates, and generalized nomic truth approximation") the 'basic' version of generalized nomic truth approximation, starting from 'two-sided' theories. Its main claim is that nomic truth approximation can perfectly be achieved by combining two prima facie opposing views on theories: (1) the traditional (Popperian) view: theories are (sets of models of) postulates that exclude certain possibilities from being realizable, enabling explanation and prediction and (2) the model view: theories are sets of models that claim to (approximately) represent certain realizable possibilities. Nomic truth approximation, i.e., increasing truth-content and decreasing falsity-content, becomes in this way revising theories by revising their models and/or their postulates in the face of increasing evidence.

The basic version of generalized nomic truth approximation is in many respects as simple as possible. Among other things, it does not take into account that one conceptual possibility may be more similar (or closer) to another than a third one (is to that other). However, for example, one theory may include a possibility that is more similar to a wrongly not included possibility than another theory can offer. Similarly, for wrongly not excluded possibilities. In this chapter it will be shown that such 'refined' considerations can be taken into account by adapted clauses based on a ternary similarity relation between possibilities (structures). This allows again abductive conclusions about refined truth approximation if a theory is persistently more successful in the refined sense than another.

It will also be indicated and illustrated that this refined approach enables a specification to the effect that refined truth approximation can be obtained by the method of idealization and subsequent concretization. Finally, the basic and the refined approach will be evaluated with regard to some general principles and objections that have been discussed in the literature.

An *adapted version*, entitled *Refined nomic truth approximation by revising models and postulates*, will appear in a special issue of *Synthese* in honor of Gerhard Schurz. *Acknowledgements*: I like to thank Ilkka Niiniluoto and Gerhard Schurz as well as two anonymous referees for their challenging questions and remarks.

T. A. F. Kuipers, *Nomic Truth Approximation Revisited*, Synthese Library 399,
https://doi.org/10.1007/978-3-319-98388-2_6

Keywords Refined closer to the truth · Refined more successful · Two-sided theories · Structurelikeness · Likeness clauses · Usefulness clauses · Refined success theorem · Idealization and concretization · Ideal gas model · Van der Waals model · Validity research · Child's play objection

6.1 Introduction

In (Kuipers 2000, Chap. 10) we have introduced refined nomic truth approximation for maximal theories in order to compensate for the following problem of basic truth approximation: any mistake is as worse as any other. For that purpose we introduced the *ternary similarity relation s(x,y,z) between structures in U*: y is at least as similar to z as x.[1] Structures x and y are called comparable or *related*, r(x, z), iff there is a y such that s(x,y,z). The basic idea behind r(x, z) is not the existence of a *proper* intermediate, i.e., a y different from x and z, but only that x and z have at least so much in common that it makes sense to talk about (proper and improper) intermediates.

The relation s is supposed to satisfy the following 'minimal' s-conditions: centered, centering and conditionally left and right reflexive. For all x, y, z, s is *centered* iff s(x,x,x) and *centering* iff s(x,y,x) implies x = y. Moreover, s is *conditionally left/right reflexive* if s(x,y,z) implies all kinds of left and right reflexivity, i.e., s(x,x,y), s(x,x,z), s(y,y,z) and s(x,y,y), s(x,z,z), s(y,z,z), respectively. Note that the conditional form, by assuming s(x,y,z), leaves room for incomparable structures; without this assumption, s(x,x,y) for example would entail r(x,y), for all x and y. The relation s is *trivial* if s(x,y,z) iff x = y = z.

A plausible example of such a s-relation we get in the case the conceptual possibilities consist of the propositional constituents generated by a finite number n of elementary propositions $p_1, p_2, \ldots p_n$: $(\pm)p_1\&(\pm)p_2\&\ldots(\pm)p_n$, where each (±) is either a negation sign or no sign at all. It is plausible to say that e.g. $\neg p_1\&p_2$ is more similar to $p_1\&p_2$ than $\neg p_1\&\neg p_2$. In general, for propositional constituents x, y, z, y is more similar to z than x when y agrees with more (non-)negation signs of z than x. This qualitative definition, called the *basic s-relation*, transforms into a quantitative version by just comparing the number of agreements.[2] We get an interpretation of the r-relation by enlarging U to all (partial) constituents generated by a selection of elementary propositions out of a given set: r(x,y) then amounts to 'x and y are generated by the same selection'. Hence, $p_1\&p_2$ and $\neg p_1\&p_2$ are comparable, but $p_1\&p_2$ and $\neg p_1$ are not.

In this chapter we will adapt and revise the definitions in (Kuipers 2000, Chap. 10) of 'refined closer to' and 'refined more successful' between two two-sided theories. Among other things, the refined success theorem will now hold unconditionally, as in the basic case, roughly, 'refined closer to' entails 'refined more successful'.

[1] The ternary relation may be based on Northcott (2013), who focuses on comparisons of the (causally specified, quantitative) verisimilitude of specific models.

[2] This measure is known as the Clifford measure among constituents (Niiniluoto 1987).

We will present the main definitions in two rounds. In Sect. 6.2 we will introduce the definitions in such a way that they are conceptually highly plausible and lead to the refined success theorem. In Sect. 6.3 we will summarize the main notions in a couple of handsome tables. In Sect. 6.4 we will focus on the special cases where refinement concerns the transition of an idealization to a concretization. We conclude (Sect. 6.5) with the evaluation of the refined approach with regard to some general principles and objections that have been discussed in the literature.

6.2 The Main Definitions and the Refined Success Theorem

To take account of structurelikeness, we start with introducing some new general terminology, again suitable for both sides. In this section, MP will be short for <M, P>, and hence **TT** = <**T**, **T** > and RS for <R, S>. The important new terms with respect to theory MP need all some reflection to understand the underlying idea. They are the following ones: a theory's *missing content*, viz. M − **T** and cP − c**T**, respectively; its *maximally possible falsity- (or false) content*, viz. M − R and cP − cS, respectively; and its *maximally possible missing content*, viz. S − M and cR − cP, respectively. Table 6.1 summarizes all relevant previously introduced (Chap. 4) and new notions, including side-independent symbols for them. The sides are indicated by m and p, respectively, with variable k.

The new notions enable conceptually plausible refined definitions of 'at least as close to the truth' and 'at least as successful'. The idea behind 'MP* is refined at least as close to the truth as MP' is that MP* includes and excludes items at least as adequately as MP. More precisely, first, MP* at least matches up to MP in representing **TT** in the following sense: for each individual claim of MP and each related

Table 6.1 Crucial notions

Theory MP = <M, P>	$k \in \{m,p\}$	m = M-side	p = P-side
Claim of MP	$Cl_k(MP)$	$M \subseteq \mathbf{T}$	$\mathbf{T} \subseteq P$ $(cP \subseteq c\mathbf{T})$
Content of MP	$C_k(MP)$	M	cP
Content of **TT** = <**T**, **T**>	$C_k(\mathbf{TT})$	$\mathbf{T}$	$c\mathbf{T}$
Truth-content of MP	$TC_k(MP)$	$M \cap \mathbf{T}$	$cP \cap c\mathbf{T}$
Falsity- (or false) content of MP	$FC_k(MP)$	$M - \mathbf{T}$	$cP - c\mathbf{T}$ $\mathbf{T} - P$
Missing content of MP	$MC_k(MP)$	$\mathbf{T} - M$	$c\mathbf{T} - cP$
Claim of data theory RS = <R, S>, assumed to be accepted!	$Cl_k(RS)$	$R \subseteq \mathbf{T}$	$\mathbf{T} \subseteq S$ $(cS \subseteq c\mathbf{T})$
Content of RS	$C_k(RS)$	R	cS
Accepted-content of MP	$AC_k(MP)$	$M \cap R$	$cP \cap cS$
Rejected-content of MP	$RC_k(MP)$	$M - S$	$cP - cR$
Maximally possible falsity-(or false) content of MP	$MPFC_k(MP)$	$M - R$	$cP - cS$
Maximally possible missing content of MP	$MPMC_k(MP)$	$S - M$	$cR - cP$

individual claim of **TT** MP* has an individual claim that is at least as good in representing that claim of **TT** as that of MP does, and, second, all extra, but mistaken, claims of MP* are useful in matching up to MP in the indicated way. In still other words, $<M^*, P^*>$ characterizes, by M* and cP*, respectively, what is to be included and excluded, i.e., **T** and c**T**, at least as precisely as $<M, P>$ does by M and cP.

For all clauses in the following definition there will first be given a general verbal formulation, followed by a general formalization, and finally a side-wise decomposition.

Definition 1:
$MP^* = <M^*, P^*>$ is *refined at least as close to the truth* $\mathbf{TT} = <\mathbf{T}, \mathbf{T}>$ as $MP = <M, P>$ iff

(i) *Likeness-clauses* (or refined TC-clauses)

For either side k in {m, p}, for all x in the k-content of MP and for all z in the k-content of **TT** if x and z are comparable then there is y in the k-content of MP* such that y is between x and z, i.e.,

$$\text{(i)}\ \forall_{k\in\{m,p\}} \forall_{x\in Ck(MP)} \forall_{z\in Ck(\mathbf{TT})}\, r(x,z) \rightarrow \exists_{y\in Ck(MP^*)}\, s(x,y,z)$$

$\Leftrightarrow$

$$^{M}\text{(i)}\ \forall_{x\in M} \forall_{z\in \mathbf{T}}\, r(x,z) \rightarrow \exists_{y\in M^*}\, s(x,y,z)$$

&

$$^{P}\text{(i)}\ \forall_{x\in cP} \forall_{z\in c\mathbf{T}}\, r(x,z) \rightarrow \exists_{y\in cP^*}\, s(x,y,z)$$

(ii) *Usefulness-clauses* (or refined FC-clauses)

For either side k in {m, p}, for all y in the *extra* false k-content of MP* (relative to MP) there are x in the false k-content of MP and z in the missing k-content of MP such that y is between x and z, i.e.,

$$\text{(ii)}\ \forall_{k\in\{m,p\}} \forall_{y\in FCk(MP^*) - FCk(MP)} \exists_{x\in FCk(MP)} \exists_{z\in MCk(MP)}\, s(x,y,z)$$

$\Leftrightarrow$

$$^{M}\text{(ii)}\ \forall_{y\in M^* - (M\cup\mathbf{T})} \exists_{x\in M-\mathbf{T}} \exists_{z\in \mathbf{T}-M}\, s(x,y,z)$$

&

$$^{P}\text{(ii)}\ \forall_{y\in cP^* - (cP\cup c\mathbf{T})} \exists_{x\in cP-c\mathbf{T}} \exists_{z\in c\mathbf{T}-cP}\, s(x,y,z)$$

It is easy to check, using the minimal s-conditions, that the Likeness-clauses are stronger than the basic TC-clauses and that they reduce to these clauses when s is trivial: s(x,y,z) iff $x = y = z$. Strictly speaking, we prefer to call the refined TC-clauses *Likeness-clauses*, for the term TC-clause is problematic in the refined case, for the clauses deal not only with, for example, at the M-side $M \cap \mathbf{T}$ and $M^* \cap \mathbf{T}$, in the same way as the basic clauses, but also with $M-\mathbf{T}$ and M* in general. In the next section we will give a table with their decomposition, showing in detail that and how they are stronger than the basic clauses.

Note that, in contrast, the refined FC-clauses, which we also call *Usefulness-clauses*, are weaker than the basic FC-clauses and reduce to these clauses when s is trivial. However, they are not only trivially entailed by them, since $M^* - (M \cup \mathbf{T})$ and $cP^* - (cP \cup c\mathbf{T})$ become empty, but they remain to deal only with the (extra) falsity-content (of MP* relative to MP). Hence, they can rightly be called refined FC-clauses. For this reason, we will continue to also speak of the refined TC- and FC-clauses, despite the somewhat misleading term 'refined TC-clause'.

Turning to the definition of 'MP is refined at least as successful as MP*', the idea is that MP* includes and excludes *accepted* items at least as adequate as MP.

Definition 2:
MP* = <M*, P* > is, relative to RS = <R, S>, refined at least as successful as MP = <M, P > iff

$(i)^{RS}$ *Accepted Likeness-clause* (refined AC-clause)

For either side k in {m, p}, for all x in the k-content of MP and for all z in the k-content of RS if x and z are comparable then there is y in the k-content of MP* such that y is between x and z, i.e.,

$$(i)^{RS}\ \forall_{k\in\{m,\,p\}}\ \forall_{x\in Ck(MP)}\ \forall_{z\in Ck(RS)}\ r(x,z) \rightarrow\ \exists_{y\in Ck(MP^*)}\ s(x,y,z)$$

$\Leftrightarrow$

$${}^{M}(i)^{RS}\ \forall_{x\in M}\ \forall_{z\in R}\ r(x,z) \rightarrow\ \exists_{y\in M^*}\ s(x,y,z)$$

&

$${}^{P}(i)^{RS}\ \forall_{x\in cP}\ \forall_{z\in cS}\ r(x,z) \rightarrow\ \exists_{y\in cP^*}\ s(x,y,z)$$

$(ii)^{RS}$ *Accepted Usefulness-clause* (refined RC-clause)

For either side k in {m, p}, for all y in the *extra* rejected k-content of MP* (relative to MP) there are x in the maximally possible false k-content of MP and z in the maximally possible missing k-content of MP such that y is between x and z, i.e.,

$$(ii)^{RS}\ \forall_{k\in\{m,p\}}\ \forall_{y\in RCk(MP^*)\,-\,RCk(MP)}\ \exists_{x\,\in\,MPFCk(MP)}\ \exists_{z\,\in\,MPMCk(MP)}\ s(x,y,z)$$

$\Leftrightarrow$

$${}^{M}(ii)^{RS}\ \forall_{y\in M^*\,-\,(M\cup S)}\ \exists_{x\,\in\,M\,-\,R}\ \exists_{z\,\in\,S\,-\,M}\ s(x,y,z)$$

&

$${}^{P}(ii)^{RS}\ \forall_{y\in cP^*\,-\,(cP\cup cR)}\ \exists_{x\,\in\,cP\,-\,cS}\ \exists_{z\,\in\,cR\,-\,cP}\ s(x,y,z)$$

Now we can summarize clause $(i)^{RS}$ by: MP* at least matches up to MP in representing RS, and clause $(ii)^{RS}$ by: all extra, but mistaken, items of MP* are useful in matching up to MP in representing RS.

Again, it is easy to check, using the minimal s-conditions, that the Accepted Likeness-clauses are stronger than the basic AC-clauses, enabling again a plausible decomposition, and that they reduce to these clauses when s is trivial. Similarly, it is easy to check that the Accepted Usefulness-clauses are weaker than the basic

RC-clauses and that they reduce to these clauses when s is trivial. Terminologically, similar remarks apply: a preference to speak of Accepted Likeness-clauses instead of refined AC-clauses, but no preference between speaking of Accepted Usefulness-clauses and refined RC-clauses.

As announced, the new, more conceptual, approach suggested the above presented revised definition of 'refined at least as successful', such that we can now (easily) prove that the two notions are ideally related, the first entails the second unconditionally.

Theorem 1 Refined Success Theorem If MP* = <M*, P* > is refined at least as close to **TT** as MP = <M, P > and assuming correct data RS = <R, S > then MP* is, relative to RS, refined at least as successful as MP*.

Let us, like in the quantitative case (Chap. 5), conclude this section by formulating the remaining methodological steps to complete the refined theory of nomic truth approximation[3]: assuming that a new theory is at a certain moment refined more successful than the old one, propose and test the '*refined empirical progress hypothesis*': the new theory (is and) remains refined more successful than the old one. Assuming that after 'sufficient confirmation' this 'refined empirical progress hypothesis' is accepted (for the time being), argue *on the basis of a reconstruction of the way in which experiments have been done* and the Refined Success Theorem, that the 'default explanation' for this case of refined empirical progress is the hypothesis that the new theory is refined closer to the truth than the old one, *i.e., that this is a case of refined truth approximation*. Finally, *conclude abductively* (for the time being) that the new theory is refined closer to the truth than the old one, *i.e., that refined truth approximation has been achieved.*

It should be noted that, different from the basic and even more from the quantitative case, the comparative relations of 'closer to' and 'more successful' in the refined sense presuppose very special conditions and hence will seldom hold in full strength. However, in the context of 'idealization and concretization' such relations may well hold as we will argue in Sect. 6.4.[4] But first we will summarize the above in a number of tables.

6.3 A Summary in Tables

In this section, I will summarize the main points of the previous section in some handsome tables, leaving out the side-independent conceptually motivated formulations, and hence we can return to the <M, P>- and <R, S>-representation of theories

---

[3]The main lines are the same as those for the basic account.

[4]A formal way to increase the applicability of the refined approach is by setting up a quantitative refined version by introducing a distance function between the conceptual possibilities. In Kuipers (2000, Chap. 12) this has been done assuming maximal theories. As already remarked in Sect. 6.5.4, we leave it as a challenge to do so for two-sided theories.

and evidence. Table 6.2 summarizes the claims that <M*, P*> is basically and refined at least as close to the truth as <M, P>.

Again, as in the basic and quantitative cases, it is interesting to combine, in Table 6.3, the refined TC- and FC-clauses, in particular for maximal theories.

Table 6.4 summarizes the claims that <M*, P*> is basically and refined at least as successful as <M, P>, relative to <R, S>.

Finally, Table 6.5 shows how the (Accepted) Likeness-clauses can be decomposed in the corresponding basic clauses and extra clauses dealing with the (maximally possible) falsity-content.

6.4 Idealization and Concretization

In this section[5] it will be pointed out that ‘idealization and concretization’ is a special kind of (potential) refined truth approximation. This is illustrated by Van der Waals’s theory of gases. Moreover, it is indicated how idealization and concretization can function as a strategy in validity research around ‘interesting theorems’. The present section is restricted to inclusion theories (‘pure theories of models’), that is, theories that only specify classes of models, hence theories of the form <M, ∅>, which notation will be reduced to M.

6.4.1 Truth Approximation by (Idealization and) Concretization

Concretization or factualization, as it has been presented by the Polish philosophers Władysław Krajewski (1977) and Leszek Nowak (1980), is basically a relation between real-valued functions. Hence, let us assume that the conceptual possibilities (structures) to be considered contain one or more real-valued functions, with or without one or more real constants. Structure y is called a *concretization* of x and x an *idealization* of y, indicated by con(x, y), if y transforms, directly or by a limit procedure, into x when one or more constants occurring in y assume the value 0. It is easy to see that it is a necessary condition for con(x, y) that x and y have the same domain-sets. Moreover, it is easy to check that con is reflexive, antisymmetric and transitive. In a subsection to follow, the example of a Van der Waals gas model will be presented as a concretization of an ideal gas model.

Concretization is primarily a binary relation, but for our purposes, we need the plausible ternary version leading to a *concretization triple*: ct(x, y, z) if and only if con(x, y) and con(y, z). We will assume ct as the underlying notion of structurelikeness. The relation of relatedness based on ct is easily seen to be equivalent to con.

[5] This section is heavily based on Sect. 10.4 of (Kuipers 2000).

Table 6.2 <M*, P*> is (basically and) refined at least as close to the truth as <M, P>

Theory <M, P>	M-side	P-side
M- / P-claim	$M \subseteq \mathbf{T}$	$\mathbf{T} \subseteq P$ $(cP \subseteq c\mathbf{T})$
TC-clauses		
Basic: $TC \subseteq TC^*$	$M \cap \mathbf{T} \subseteq M^* \cap \mathbf{T}$ $\equiv \mathbf{T} - M^* \subseteq \mathbf{T} - M$	$cP \cap c\mathbf{T} \subseteq cP^* \cap c\mathbf{T}$ $\equiv P^* - \mathbf{T} \subseteq P - \mathbf{T} \equiv P^* \cup \mathbf{T} \subseteq P \cup \mathbf{T}$
	Refined TC-clauses entail the basic ones and reduce to them when s() is trivial: s(x,y,z) iff x = y = z	
Refined = **Likeness-clauses**	^{M}(i) $\forall_{x \in M} \forall_{z \in T}\, r(x,z) \rightarrow \exists_{y \in M^*}\, s(x,y,z)$ M* at least matches up to M in representing **T**	^{P}(i) $\forall_{x \in cP} \forall_{z \in cT}\, r(x,z) \rightarrow \exists_{y \in cP^*}\, s(x,y,z)$ cP* at least matches up to cP in representing c**T**
FC-clauses		
Basic: $FC^* \subseteq FC$	$M^* - \mathbf{T} \subseteq M - \mathbf{T}$	$cP^* - c\mathbf{T} \subseteq cP - c\mathbf{T}$ $\equiv \mathbf{T} - P^* \subseteq \mathbf{T} - P \equiv P \cap \mathbf{T} \subseteq P^* \cap \mathbf{T}$
	Refined FC-clauses are entailed by the basic ones and reduce to them when s() is trivial: s(x,y,z) iff x = y = z	
Refined = **Usefulness-clauses**	^{M}(ii) $\forall_{y \in M^* - (M \cup T)} \exists_{x \in M - T} \exists_{z \in T - M}\, s(x,y,z)$ all extra, but mistaken, items of M* are useful in matching up to M	^{P}(ii) $\forall_{y \in cP^* - (cP \cup cT)} \exists_{x \in cP - cT} \exists_{z \in cT - cP}\, s(x,y,z)$ all extra, but mistaken, items of cP* are useful in matching up to cP

Table 6.3 Combined basic and refined TC/FC-clauses: general and specifically for maximal theories

Combined TC/FC-clauses	M-side	P-side
Basic	$M^* \Delta \mathbf{T} \subseteq M \Delta \mathbf{T}$	$cP^* \Delta c\mathbf{T} \subseteq cP \Delta c\mathbf{T}$
Refined	$\forall_{x \in M} \forall_{z \in T} r(x,z) \rightarrow \exists_{y \in M^*} s(x,y,z)$ & $\forall_{y \in M^* - (M \cup T)} \exists_{x \in M - T} \exists_{z \in T - M} s(x,y,z)$	$\forall_{x \in cP} \forall_{z \in cT} r(x,z) \rightarrow \exists_{y \in cP^*} s(x,y,z)$ & $\forall_{y \in cP^* - (cP \cup cT)} \exists_{x \in cP - cT} \exists_{z \in cT - cP} s(x,y,z)$
If the theories are maximal (Kuipers 2000), i.e., M = P = X M* = P* = X*, this results in		
Basic	$X^* \Delta \mathbf{T} \subseteq X \Delta \mathbf{T}$	
Refined[a]	$\forall_{x \in X} \forall_{z \in T} r(x, z) \rightarrow \exists_{y \in X^*} s(x,y,z)$ & $\forall_{x \in cX} \forall_{z \in cT} r(x, z) \rightarrow \exists_{y \in cX^*} s(x,y,z)$	

[a]This corresponds to the refined notion for maximal theories in (Kuipers 1987). On both sides, the second of the refined clauses becomes empty for maximal theories, due to the first clause on the other side. I disliked this, in retrospect, for the M-side, for which reason I introduced in (Kuipers 2000), for maximal theories, a definition which amounts to both M-side clauses with $M^{(*)} = X^{(*)}$. Now it is clear that maximality is the background cause of the problem. For non-maximal theories, both second clauses make perfect sense.

Table 6.4 <M*, P*> is basically and refined at least as successful as <M, P>, relative to <R, S>

Theory <M, P>	M-side	P-side
M- / P-claim	$M \subseteq \mathbf{T}$	$\mathbf{T} \subseteq P$ ($cP \subseteq c\mathbf{T}$)
AC-clauses		
Basic: $AC \subseteq AC^*$	$M \cap R \subseteq M^* \cap R$ $\equiv R - M^* \subseteq R - M$	$cP \cap cS \subseteq cP^* \cap cS$ $\equiv P^* - S \subseteq P - S$
Refined AC-clauses entail the basic ones and reduce to them when s() is trivial: s(x,y,z) iff x = y = z		
Refined = **Accepted Likeness-clauses**	${}^{M}(i)^{RS}\ \forall_{x \in M} \forall_{z \in R}\ r(x,z) \rightarrow \exists_{y \in M^*}\ s(x,y,z)$ M* at least matches up to M in representing R	${}^{P}(i)^{RS}\ \forall_{x \in cP} \forall_{z \in cS}\ r(x, z) \rightarrow \exists_{y \in cP^*}\ s(x,y,z)$ cP* at least matches up to cP in representing cS
RC-clauses		
Basic: $RC^* \subseteq RC$	$M^* - S \subseteq M - S$	$cP^* - cR \subseteq cP - cR$ $\equiv R - P^* \subseteq R - P$
Refined RC-clauses are entailed by the basic ones and reduce to them when s() is trivial: s(x,y,z) iff x = y = z		
Refined = **Accepted Usefulness-clauses**	${}^{M}(ii)^{RS}\ \forall_{y \in M^* - (M \cup S)}\ \exists_{x \in M - R}\ \exists_{z \in S - M}\ s(x,y,z)$ all extra, but given S mistaken, items of M* are useful in matching up to M	${}^{P}(i)^{RS}\ \forall_{y \in cP^* - (cP \cup cR)}\ \exists_{x \in cP - cS}\ \exists_{z \in cR - cP}\ s(x,y,z)$ all extra, but given R mistaken, items of cP* are useful in matching up to cP

Table 6.5 Decomposition of the (accepted) likeness-clauses

Likeness-clauses: decomposition in basic TC-clause & extra clause	$\forall_{x \in M} \forall_{z \in \mathbf{T}} r(x,z) \rightarrow \exists_{y \in M^*} s(x,y,z)$ $\equiv$ $M \cap \mathbf{T} \subseteq M^* \cap \mathbf{T}$ & $\forall_{x \in M - \mathbf{T}} \forall_{z \in \mathbf{T}} r(x,z) \rightarrow \exists_{y \in M^*} s(x,y,z)$	$\forall_{x \in cP} \forall_{z \in c\mathbf{T}} r(x,z) \rightarrow \exists_{y \in cP^*} s(x,y,z)$ $\equiv$ $cP \cap c\mathbf{T} \subseteq cP^* \cap c\mathbf{T}$ & $\forall_{x \in cP - c\mathbf{T}} \forall_{z \in c\mathbf{T}} r(x,z) \rightarrow \exists_{y \in cP^*} s(x,y,z)$
Accepted Likeness-clauses: decomposition in basic AC-clause & extra clause	$\forall_{x \in M} \forall_{z \in R} r(x,z) \rightarrow \exists_{y \in M^*} s(x,y,z)$ $\equiv$ $M \cap R \subseteq M^* \cap R$ & $\forall_{x \in M - R} \forall_{z \in R} r(x,z) \rightarrow \exists_{y \in M^*} s(x,y,z)$	$\forall_{x \in cP} \forall_{z \in cS} r(x,z) \rightarrow \exists_{y \in cP^*} s(x,y,z)$ $\equiv$ $cP \cap cS \subseteq cP^* \cap cS$ & $\forall_{x \in cP - cS} \forall_{z \in cS} r(x,z) \rightarrow \exists_{y \in cP^*} s(x,y,z)$

Note that we have here a clear example in which relatedness is not symmetric, but directed. Note also that ct is trivially decomposable. It is easy to check that ct satisfies the minimal s-conditions of being centered, centering and conditionally left and right reflexive. Moreover, it is antisymmetric (central, left and right) and it satisfies all conceivable kinds of transitivity, e.g., left: if ct(w, x, z) and ct(x, y, z,) then ct(w, y, z). The relation of more truth-(theory-) likeness, M* is closer to **T** (M**) than M, based on this ternary relation, will be indicated by MTLct(M, M*, **T**) and MTLct(M, M*, M**), respectively.

Our next task is to define the binary relation of concretization between theories. Again we will do this as weakly as possible: M* is a *concretization* of M and M an *idealization* of M*, indicated by CON (M, M*), if and only if all members of M have a concretization in M* and all members of M* have an idealization in M. At first sight, one might think that the second clause should be strengthened to: and all members of M* have a *unique* idealization in M. However, this would exclude e.g., 'inclusive' concretization triples <M, M*, M**> with M as a subset of M* and M* of M** and CON (M, M*) and CON(M*, M**).

It is trivial that CON is reflexive and transitive. However, contrary to what one might expect, it need not be antisymmetric. But sufficient for antisymmetry of CON (M, M*) is that M and M* are convex (i.e., closed for intermediates). Here convexity of M amounts to: if con(x, y) and con(y, z), i.e., ct(x, y, z), and x and z are in M, then y is in M.

The ternary relation of concretization of theories we define again as weakly as possible: CT(M, M*, M**) if and only if CON(M, M*) and CON(M*, M**). It is easy to check that CT has the properties of being centered, centering for convex sets, and conditional left and right reflexivity, antisymmetry (central, left, right) for convex sets and all conceivable forms of transitivity. As a consequence, CT(M, M*, M**) is for fixed M** a partial ordering as far as convex theories are concerned.

The main question is whether or under what conditions CT(M, M*, M**) implies MTLct(M, M*, M**). It turns out that some conditions have to be added to guarantee this implication, but there are some alternative possibilities. We are, of course, primarily interested in conditions on M and/or M* or their combination, for in the crucial case we do not dispose of M**, i.e., **T**. One sufficient combination of conditions is the following: M* should be convex as well as *mediating,* the latter condition being defined as: if z is a concretization of x and if x has a concretization in M* and z an idealization in M*, then M* also provides an intermediate for x and z; or, more formally, if con(x, z) and if there are y and y′ in M* such that con(x, y) and con(y′, z), then there is y″ in M* such that con (x, y″) and con(y″, z), i.e., ct(x, y″, z).

Note that both conditions only concern M*. Although being mediating is a more specific property than convexity, it is not a very restrictive condition in the present context. Note also that it follows that any M can be an idealized starting point for successive concretization. However, the starting point M will usually even be *closed for idealizations* in the sense that if x in M and con(x′, x), then x′ is in M. It is easy to check that this trivially implies that M is convex and mediating.

Let us formally state the main claim: it is (easily) possible to prove the following *Double Concretization (DC-)Theorem*: if CT(M, M*, M**) and if M* is convex and mediating, then MTLct(M, M*, M**). In words: the intermediate theory of a concretization triple is closer to the third than the first, assuming that it is convex and mediating.

We may define stronger versions of concretization triples such as $CT^{\#}$(M, M*, M**) = CT(M, M*, M**) and M* convex and mediating or even $CT^{\#\#}$(M, M*, M**) = $CT^{\#}$(M, M*, M**) and M and M** also convex. According to the DC-Theorem, both are special kinds of more theorylikeness. Moreover, it has already been mentioned that CT(M, M*, M**) is antisymmetric (in the central sense) as soon as the three sets are convex; hence $CT^{\#\#}$ is an antisymmetric special type of theorylikenes.

A direct consequence of the DC-Theorem is that, if theory M* is a concretization of theory M, if M* is convex and mediating, and if the true set of nomic possibilities **T** is a concretization of M*, then M* is closer to the truth than M. This may be called the *Truth Approximation by Double Concretization (TADC-)Corollary*– a major goal of this section – viz., to show that and in what sense concretization may be a form of truth approximation. All conditions for truth approximation can be checked, except, of course, the crucial heuristic hypothesis that **T** is a concretization of M*.

To obtain 'good reasons' to assume that the required heuristic hypothesis that **T** is a concretization of M* is true, it is important that the concretization has some type of (necessarily insufficient) justification, of a theoretical or empirical nature, suggesting that the account of the new factor is in the proper direction. In this respect, it is plausible to speak of theoretical and/or empirical concretization. The famous case of Van der Waals to be presented in the next subsection evidently is a case of theoretical concretization, followed by empirical support. The same holds true for Sommerfeld's concretization of the 'old quantum theory', presented in (Kuipers 2000, Chap. 11).

6.4.2 Application to Gas Models

The transition from the theory of ideal gases to Van der Waals's theory of gases has frequently been presented as a paradigmatic case of concretization. The challenge of any sophisticated theory of truthlikeness hence is to show that this transition can be a case of truth approximation.

The Law of Van der Waals takes, informally speaking, two factors into account that are neglected in the ideal gas law: the mutual attraction of the molecules and the volume of the molecules. The idea is that these two concretizations are successive steps in the direction of the truth. To specify this in some formal detail, we start by formulating the relevant models in elementary structuralist terms. (S, n, P, V, T) is a *potential gas model* (PGM) if and only if S represents a set of thermal states of n moles of a gas and P, V and T are real-valued functions defined on S and representing pressure, volume and (empirical absolute) temperature, respectively.

Specific gas models are PGM's satisfying an additional condition. The *ideal gas models* (IGM) satisfy in addition P(s)V(s) = nRT(s) for all s in S, or simply PV = nRT, where R is the so-called ideal gas constant. For *gas models with mutual attraction* (GMa) there is a non-negative real (number) constant a, within a certain fixed interval, such that $(P + (n^2a/V^2))V = nRT$. For *gas models with non-zero volume of molecules* (GMb) there is a non-negative real constant b, within a certain fixed interval, such that P(V− nb) = nRT. Finally, in the case of *Van der Waals gas models* (WGM) there are non-negative real constants a and b, within the previously mentioned two intervals, such that $(P + (n^2a/V^2))(V - nb) = nRT$.[6]

Note first that it is a necessary condition for con(x, y) (x and y in PGM) that x and y have the same set of thermal states (Sx = Sy). Note also that IGM/GMa/GMb/WGM have been defined such that they are all convex and mediating.

It is easy to check that IGM, GMa and WGM as well as IGM, GMa and WGM constitute a concretization triple: an element of WGM transforms into an element of GMa/GMb by substituting the value 0 for b and a, respectively. The resulting elements of GMa and GMb transform into elements of IGM by substituting 0 for a and b, respectively.

Due to the DC-Theorem, it follows that GMa and GMb are both refined closer to WGM than IGM.[7] As a consequence, if WGM represented the true set of nomically possible gases, GMa and GMb would be refined closer to the truth than IGM. Finally, and most importantly, the TADC-Corollary guarantees that WGM is refined closer to the truth than IGM, assuming the heuristic hypothesis that the true set of nomically possible gases is on its turn a concretization of WGM.

6.4.3 Validity Research

Scientific research is not always directed at describing the actual world or characterizing the set of nomically possible worlds. It may also primarily aim at proving interesting theorems for certain conceptual possibilities. Hamminga (1983) showed this for neo-classical economics, in particular the theory of international trade.

Let a certain domain of research and a vocabulary have been chosen, and hence a universe of conceptual possibilities U, let **T** indicate the (unknown) subset of nomic possibilities, and let R_{mf} indicate the (equally unknown) subset (of **T**) of as a *matter of fact* realized (nomic) possibilities, possibly containing just one element, the actual possibility.

Let IT indicate an 'interesting theorem', that is some insightful claim, of which it is interesting to know whether it is true for the nomic possibilities, or at least the

---

[6] Note that the two factors become negligible when the volume is very large.

[7] Note that GMa and GMb are mutually not comparable relative to IGM, e.g. GMb is not closer to GMa than IGM in the refined sense.

realized possibilities. Let VAL(IT), or simply VAL, indicate the set of conceptual possibilities for which IT can be proven. VAL is called the *domain of (provable) validity* of IT, and it is assumed to be not yet explicitly characterized.

A frequently occurring type of scientific progress is the following. Suppose that it was proven earlier that IT holds for M, i.e., that M is a subset of VAL. The new result is that M*, which includes M, is also, like M, included in VAL. Due to concentricity of the basic and refined theorylikeness notions, it follows in this case that M* is closer to VAL than M both in the basic and the refined sense (MTLb/ MTLr(M, M*, VAL). The ultimate purpose of this type of research was to find out whether **T**, or at least R_{mf}, is a subset of VAL. Of course, the larger VAL has been proven to be, as in the described case, the greater the chance, informally speaking, that R_{mf} or even **T** are subsets of VAL. However, simply enlarging the proven domain of validity does not necessarily go in the direction of R_{mf} and **T**. For this purpose, concretization is the standard strategy.

Let it first have been shown that M is a subset of VAL, and later that a concretization M* of M (CON(M, M*), M need not be a subset of M*) is also a subset of VAL. It then trivially follows that MTLr(M, M∪M*, VAL). If, moreover, M* is convex and mediating, it follows from the heuristic hypothesis that M* is a concretization of **T** (CON(M*, **T**)), using the DC-Theorem, that MTLr(M, M*, **T**). Hence, we have proven IT for a set M* which is more similar to **T** than M, which increases the chance that IT holds for **T**, ipso facto for R_{mf}.

A complex form of validity research concerns the case that IT is not fixed, but that realistic factors are successively accounted for. Formally, this is also a form of concretization. IT_2 is called a concretization of IT_1 if $VAL(IT_2) = VAL_2$ is a concretization of $VAL(IT_1) = VAL_1$.

Now suppose that IT_1 is proven for M. The relevant heuristic strategy is to look for a concretization M* of M and a concretization IT_2 of IT_1 such that IT_2 can be proven for M*. The heuristic hypotheses are that **T** is a concretization of M* and that there is a concretization IT* of IT_2 such that IT* holds for **T** and hence for R_{mf}. This makes sense because, if M* and IT_2 are convex and mediating, it not only follows that M* is closer to **T** than M, but also that VAL_2 is closer to VAL(IT*) than VAL_1. Hence, in this case we are not only on the way to **T** but also to IT*.

The concretization of the theory and corresponding theorem of Modigliani and Miller concerning the capital structure of firms by Kraus and Litzenberg turns out to be a perfect example of this kind of approximation of a provable interesting truth. It is presented in Kuipers (2000, Chap. 11).

6.5 Some Evaluation

Although the refined definitions are in our view rather plausible in themselves, we like to conclude with the evaluation of the refined approach with regard to some general principles and objections that have been discussed in the literature, notably by Graham Oddie. In this section we will freely speak of the sets M and P and the theories M and P.

6.5.1 The Value of Content for Truths

Let us start with the highly desirable principle of *the (positive) value of content for truths* (Oddie 2016), which amounts in terms of Ilkka Niiniluoto (1987, p. 230) to: among true theories truthlikeness should covary with logical strength. So, let $\langle M^*, P^*\rangle$ be true and stronger than $\langle M, P\rangle$, hence also true. That is, we assume: $M \subset M^* \subseteq \mathbf{T}$ and $\mathbf{T} \subseteq P^* \subset P$. It is easy to check that the basic definition satisfies it: $M^* \Delta \mathbf{T} \subset M \Delta \mathbf{T}$ and $P^* \Delta \mathbf{T} \subset P \Delta \mathbf{T}$. It is also easy to check that the usefulness-clauses (ii) of the refined definition (see Definition 1, Sect. 6.2) are empty (e.g. $M^*-(M \cup \mathbf{T}) = \emptyset$), and hence that they are trivially satisfied by true theories, and even in both directions. Moreover, $\langle M^*, P^*\rangle$ satisfies the likeness-clauses (i) relative to $\langle M, P\rangle$. E.g. for x in M, and hence in M*, and z in **T**, if r(x,z) then s(x,x,z) holds because structurelikeness (s(x,y,z)) is supposed to be conditionally left reflexive. Hence $\langle M^*, P^*\rangle$ is also refined at least as close to the truth as $\langle M, P\rangle$. Finally, a simple Venn-diagrammatic representation shows that when s(x,y,z) is based on Euclidean distances the likeness-clauses will not be satisfied in the other direction, in which case $\langle M^*, P^*\rangle$ is definitely closer to the truth than $\langle M, P\rangle$ in the refined sense. It is even difficult to see how s(x,y,z) could be defined such that this is not the case. In sum, both the basic and the refined definition satisfy the (positive) value of content for truths.

6.5.2 The (Dis-)Value of Content for Falsehoods

Regarding false theories and strength, let me first quote Oddie (2016):

> [H]ow might strength determine verisimilitude [truthlikeness] amongst false theories? There seem to be just two plausible candidates: that verisimilitude increases with increasing strength (the principle of the value of content for falsehoods) or that it decreases with increasing strength (the principle of the disvalue of content for falsehoods). Both proposals are at odds with attractive judgements and principles.

It is indeed also very interesting to elaborate the way in which the refined approach works out with respect to these two possible principles. Let us first note that in particular the first principle, *the value of content for falsehoods*, is also known as the so-called child's play objection of Pavel Tichý (1974, p. 157, fn. 2) and Oddie (Oddie 1981), for it seems to enable truth approximation by just adding new claims to false theories. It might seem that this objection can be made at the P-side of the basic case. An analysis in terms of truth- and falsity-content focused on our particular setup of nomic theories will be very illuminating. Let theory P be false, hence with a non-empty falsity-content $cP-c\mathbf{T} = \mathbf{T}-P$, with the extreme case $\mathbf{T}-P = \mathbf{T}$, i.e., **T** and P do not overlap. Now it is easy to check that if we strengthen P to P* ($P^* \subset P$) such that the falsity-content remains the same (the FC-constant condition), which is guaranteed in the extreme case, then the truth-content increases: $cP \cap c\mathbf{T} \subset cP^* \cap c\mathbf{T}$. Hence, under the FC-constant condition, it is a child's play to come basically closer to the truth and is 'the value of content' highly plausible for false theories. Of course, not knowing **T**, we can't guarantee the FC-constant condition. In other words, basic truth approximation is only a child's play when we know the nomic truth. But this is difficult to see as an objection. It will be no surprise that the refined definition keeps this property. Both refined clauses are trivially satisfied, the likeness-clause by the assumption that s(x,y,z) is conditionally left symmetric, the usefulness-clause by being empty.

At the M-side, we have of course a similar situation. When we strengthen (the claim of) M to M* ($M \subset M^*$) such that the non-empty falsity-content remains constant ($\varnothing \neq M-\mathbf{T} = M^*-\mathbf{T}$) we increase the truth-content and hence come basically and refined closer to the truth. Again, in practice we have no control over the condition and so the child's play is idle.

There is another type of child's play that is even more interesting, in particular at the M-side. When M is false ($M-\mathbf{T} \neq \varnothing$) and we weaken (the claim of) M to $M^{\#}$ ($M^{\#} \subset M$) such that the truth-content remains constant (TC-constant condition) ($M \cap \mathbf{T} = M^{\#} \cap \mathbf{T}$) the falsity-content decreases ($M^{\#}-\mathbf{T} \subset M-\mathbf{T}$) and hence may even become empty. According to the basic definition such weakening is straightforward truth approximation; again a child's play assuming the non-realistic TC-constant condition. Of course, the reverse side of this is that strengthening (the claim) of M, under constant truth-content, has a negative effect on truth approximation. Hence, it illustrates *the principle of the (negative value or) disvalue of content for falsehoods* when TC is kept constant. At the P-side we obtain the parallel situation by assuming P false ($cP-c\mathbf{T} = \mathbf{T}-P \neq \varnothing$) and weakening it ($P \subset P^{\#}$) such that its truth-content remains constant ($cP^{\#} \cap c\mathbf{T} = cP \cap c\mathbf{T}$). The falsity-content will then decrease ($\mathbf{T}-P^{\#} \subset \mathbf{T}-P$) and hence basic truth approximation is obtained. Again, strengthening under this condition brings us further from the truth.

It is important to note that we may already conclude that Oddie's suggestion in the quotation that either the principle of the value of content for falsehoods or the principle of the disvalue of content for falsehoods will hold in general is mistaken. For in the basic case we have seen that the first principle holds when the falsity-content is kept constant and the second when the truth-content is kept constant.

The interesting point regarding the second principle is that this child's play is now blocked by the refined definition, and for good reasons. This is best illustrated by an example at the M-side. We might give up a wrongly included possibility, which is however more similar to a non-included nomic possibility than a remaining wrongly included possibility.

Here is an example. Assume three elementary propositions p_1, p_2, p_3 and let

$$\mathbf{T} = \mathrm{Mod}\left(\left(p_1 v p_2\right) \Rightarrow p_3\right) = \left\{s_1, s_2, s_3, s_5, s_8\right\}$$

$$M = \mathrm{Mod}\left(\left(p_1 v p_2\right) \& \neg p_3\right) = \left\{s_4, s_6, s_7\right\}$$

$$M^{\#} = \mathrm{Mod}\left(\left(p_1 \Leftrightarrow \neg p_2\right) \& \neg p_3\right) = \left\{s_6, s_7\right\}$$

Table 6.6 $M^{\#} = Mod((p_1 \Leftrightarrow \neg p_2)\&\neg p_3)$ is basically closer to **T** = $Mod(((p_1 \vee p_2) \Rightarrow p_3)$ than $M = Mod((p_1 \vee p_2)\&\neg p_3)$, but M is refined closer to **T** than $M^{\#}$

Propositional constituents/ models	p_1	p_2	p_3	M: models of $(p_1 \vee p_2)\&\neg p_3$	$M^{\#}$: models of $(p_1 \Leftrightarrow \neg p_2)\&\neg p_3$	**T**: models of $(p_1 \vee p_2) \Rightarrow p_3$
$s_1 = p_1\&p_2\&p_3$	1	1	1	0	0	1
$s_2 = \neg p_1\&p_2\&p_3$	0	1	1	0	0	1
$s_3 = p_1\&\neg p_2\&p_3$	1	0	1	0	0	1
$s_4 = p_1\&p_2\&\neg p_3$	1	1	0	1	0	0
$s_5 = \neg p_1\&\neg p_2\&p_3$	0	0	1	0	0	1
$s_6 = \neg p_1\&p_2\&\neg p_3$	0	1	0	1	1	0
$s_7 = p_1\&\neg p_2\&\neg p_3$	1	0	0	1	1	0
$s_8 = \neg p_1\&\neg p_2\&\neg p_3$	0	0	0	0	0	1

where the right side, in terms of propositional constituents, can be read of from Table 6.6. We will assume the basic s-relation among constituents, see Sect. 6.1.

It is easy to check that $M^{\#}$ and M are false, even such that neither $M^{\#}$ nor M overlaps with **T**, and that $M^{\#}$ is a proper subset of M. Hence $M^{\#}$ is *basically* closer to **T** than M.

According to the refined definition (Definition 1, Sect. 6.2) we get however the reverse. The claim that M is refined at least as close to **T** as $M^{\#}$ amounts to (simplified by neglecting relatedness, for all propositional constituents of the same elementary propositions, i.c. p, q and r, are related):

Likeness clause: ^{M}(i) $\forall_{x \in M^{\#}} \forall_{z \in \mathbf{T}} \exists_{y \in M}\, s(x,y,z)$
Usefulness clause: ^{M}(ii) $\forall_{y \in M-(M^{\#} \cup \mathbf{T})} \exists_{x \in M^{\#}-\mathbf{T}} \exists_{z \in \mathbf{T}-M^{\#}}\, s(x,y,z)$

It is easy to check that this holds in the case at hand: ^{M}(i) is almost trivial due to the simplified minimal s-condition of left symmetry, s(x,x,z), and $M^{\#}$ being a subset of M; for ^{M}(ii) it is crucial, in view of the fact that $M-(M^{\#} \cup \mathbf{T}) = \{p_1\&p_2\&\neg p_3\} = \{s_4\}$, that e.g. $s(s_6, s_4, s_1) = s(\neg p_1\&p_2\&\neg p_3, p_1\&p_2\&\neg p_3, p_1\&p_2\&p_3)$ holds, where s_6 is in $M^{\#}-\mathbf{T}$ and s_1 in $\mathbf{T}-M^{\#}$.

M is refined even closer to **T** than $M^{\#}$, for the reversed first clause

$${}^{M}(i)\text{rev} \quad \forall_{y \in M} \forall_{z \in \mathbf{T}} \exists_{x \in M^{\#}} s(x,y,z)$$

does not hold, for e.g. neither $s(s_4, s_6, s_1)$ nor $s(s_4, s_7, s_1)$ holds, whereas at least one of them would be required.

Of course, at the P-side, a similar case could illustrate that weakening of a false P such that the truth-content remains the same results in basic truth approximation, but in general not in refined truth approximation, on the contrary. Instead of spelling this also out for nomic truth approximation, we like to turn to *actual* truth approximation.

6.5.3 Application to Actual Truth Approximation

It is interesting to apply the above findings to actual truth approximation, i.e., the case that **T** = {**t**} where '**t**' represents the actual truth. Actual truth approximation is in the first place a matter of comparing structures (see Kuipers 2000, Sect. 7.1), e.g. in terms of s(x,y,**t**). But at the level of comparing sets of structures it is also possible.

Regarding the positive value of content for truths at the P-side the situation is simple. The truth- and strength condition together ($\mathbf{T} \subseteq P^* \subset P$) entail for **T** = {**t**} that $\mathbf{t} \in P^* \subset P$, from which it is easy to see that the truth-content increases and the falsity-content remains empty, leading to basic and refined actual truth approximation. At the M-side, the condition $M \subset M^* \subseteq \mathbf{T}$ entails for **T** = {**t**} the unique case $\varnothing = M$, $M^* = \{\mathbf{t}\}$, which is a trivial case of basic and refined actual truth approximation.

Let us now turn to the child's play of strengthening a false theory while keeping the falsity-content constant. At the P-side, P being false amounts to: $\mathbf{t} \notin P$. By strengthening P to P* we keep this falsity-content automatically the same, viz. {**t**}, ($cP - c\{\mathbf{t}\} = \{\mathbf{t}\} - P = \{\mathbf{t}\} = \{\mathbf{t}\} - P^* = cP^* - c\{\mathbf{t}\}$), but increase the truth-content ($cP \cap c\{\mathbf{t}\} = cP - \{\mathbf{t}\} \subset cP^* \cap c\{\mathbf{t}\} = cP^* - \{\mathbf{t}\}$). Hence, basic and refined truth approximation result. At the M-side, the strengthening while keeping the falsity-content constant can only be done by just including **t** (i.e., $M^* = M \cup \{\mathbf{t}\}$), again resulting in basic and refined truth approximation. Hence, not surprisingly, also basic and refined *actual* truth approximation respect the principle of the value of content for falsehoods in the present setup, assuming constant falsity-content, hence again only useful on condition of knowing the actual truth.

Remains to study the child's play of actual truth-approximation by weakening while keeping the truth-content constant. To be sure, on the M-side the case of actual truth approximation may seem at first sight somewhat weird in general as soon as M is not a singleton set {m}, because the claim of M amounts to $M \subseteq \{\mathbf{t}\}$, that is, in terms of 'individual claims', for all $m \in M$ m = **t**. Hence, formally speaking, the claim is inconsistent, assuming the structures in M are really different.[8] But it is nevertheless interesting to see what happens when we weaken a false theory ($M - \{\mathbf{t}\} \neq \varnothing$, $M^{\#} \subset M$) with constant truth-content ($M \cap \{\mathbf{t}\} = M^{\#} \cap \{\mathbf{t}\}$), hence **t** is either in both M and $M^{\#}$ or in neither of the two. Of course we get basic actual truth approximation. Again by way of a propositional example we will show that this is not necessarily so in the refined approach.[9] Assuming two elementary propositions, p_1 and p_2, let

$$\mathbf{T} = \{p_1 \,\&\, p_2\} = \{\mathbf{t}\}$$
$$M = \mathrm{Mod}(\neg p_1 \vee \neg p_2) = \{\neg p_1 \,\&\, p_2, \neg p_1 \,\&\, \neg p_2, p_1 \,\&\, \neg p_2\} = M^{\#} \cup \{p_1 \,\&\, \neg p_2\}$$
$$M^{\#} = \mathrm{Mod}(\neg p_1) = \{\neg p_1 \,\&\, p_2, \neg p_1 \,\&\, \neg p_2\}$$

[8] Note that this is not so in the nomic case, because then we do not know the size of **T**.

[9] The example was provided to me by Gerhard Schurz and initiated in fact this whole section.

It is easy to check that (the claim of) $M^{\#}$ is a weakening of (that of) M, with both an empty truth content, and hence that $\mathbf{T}\Delta M^{\#}\subset\mathbf{T}\Delta M$, that is, $M^{\#}$ is basically closer to **T** than M.

According to the refined definition (again neglecting relatedness, for all propositional constituents of the same elementary propositions, i.c. p_1 and p_2, are related) we get however the reverse. The claim that M is refined at least as close to **T** as $M^{\#}$ amounts in the case of actual truth approximation to:

Likeness clause:

$$^{M}(i)\ \forall_{x\in M^{\#}}\ \forall_{z\in \mathbf{T}}\ \exists_{y\in M}\ s(x,y,z) \equiv \forall_{x\in M^{\#}}\ \exists_{y\in M}\ s(x,y,\mathbf{t})$$

Usefulness clause:

$$^{M}(ii)\ \forall_{y\in M-(M^{\#}\cup \mathbf{T})}\ \exists_{x\in M^{\#}-\mathbf{T}}\ \exists_{z\in \mathbf{T}-M^{\#}} s(x,y,z) \equiv \forall_{y\in M-M^{\#}}\ \exists_{x\in M^{\#}}\ s(x,y,\mathbf{t})$$

It is easy to check that this holds:

Ad ^{M}(i): the two possible cases follow from the minimal s-conditions, notably s(x,x,z),

Ad ^{M}(ii): for the only case $y = p_1\&\neg p_2$ we take $x=\neg p_1\&\neg p_2$, for, of course, $s(\neg p_1\&\neg p_2, p_1\&\neg p_2, \mathbf{t})$.

M is refined even closer to **T** than $M^{\#}$, for the reversed first clause

$$^{M}(i)\text{rev}\ \forall_{y\in M}\ \forall_{z\in \mathbf{T}}\ \exists_{x\in M^{\#}}\ s(x,y,z) \equiv \forall_{y\in M}\ \exists_{x\in M^{\#}}\ s(y,x,\mathbf{t})$$

does not hold for $y = p_1\&\neg p_2$.

Hence, the example shows that the refined account does not respect the principle of the disvalue of content for falsehoods. At the M-side, it is blocked as a general principle, for just dropping a mistaken structure is not a step forward when that structure is relatively close to the true structure compared to some other structures.

At the P-side the relevant conditions are not applicable in the case of actual truth approximation. P false amounts to $\{\mathbf{t}\}-P^{\#}\neq\emptyset$, weakening to $P\subset P^{\#}$ and constant truth-content to $cP\cap c\{\mathbf{t}\} = cP^{\#}\cap c\{\mathbf{t}\}$, i.e., $cP-\{\mathbf{t}\} = cP^{\#}-\{\mathbf{t}\}$. Since **t** is not in $P^{\#}$, hence not in P, because $P\subseteq P^{\#}$, the last condition can only be satisfied by $P=P^{\#}$, hence not by $P\subset P^{\#}$.

In sum, it is possible that the refined account of nomic and actual truth approximation does not only refine the basic account in the sense that it decides cases of incomparability; it may also disagree with the basic account.[10] This is plausible since, as we have seen (Sect. 6.2), the first clause is a strengthening and the second a weakening of the corresponding basic clauses. Hence, the question arises, which

[10] Recall that the refined definition reduces to the basic one for trivial similarity (i.e., s(x,y,z) iff $x = y = z$).

Table 6.7 Survey of the scores of the basic and refined account regarding Oddie's principle of the (positive) value of content for truths and the principle of the (positive) value of content for falsehoods versus the principle of the (negative or) disvalue of content for falsehoods

The value of content / strengthening	N.B. actual: $\mathbf{T} = \{\mathbf{t}\}$	truth–/falsehood-and strength-condition	extra condition	effect for truth approximation: basic	effect for truth approximation: refined
for truths	nomic	$\mathbf{T} \subseteq P^* \subset P$		positive	positive
	actual	$\mathbf{t} \in P^* \subset P$		positive	positive
	nomic	$M \subset M^* \subseteq \mathbf{T}$		positive	positive
	actual	$\varnothing = M,\ M^* = \{\mathbf{t}\}$		positive	positive
for falsehoods, with constant falsity-content	nomic	$\mathbf{T} - P \neq \varnothing,\ P^* \subset P$	$\mathbf{T} - P = \mathbf{T} - P^{*a}$	positive	positive
	actual	$\mathbf{t} \notin P,\ P^* \subset P\ (\Rightarrow \mathbf{t} \notin P^*)$	no extra condition needed	positive	positive
	nomic	$M - \mathbf{T} \neq \varnothing,\ M \subset M^*$	$M - \mathbf{T} = M^* - \mathbf{T}$	positive	positive
	actual	$M - \{\mathbf{t}\} \neq \varnothing,\ M \subset M^*$	$M - \{\mathbf{t}\} = M^* - \{\mathbf{t}\}^{b}$	positive	positive
for falsehoods with constant truth-content	nomic	$\mathbf{T} - P \neq \varnothing,\ P^* \subset P$	$cP^* \cap c\mathbf{T} = cP \cap c\mathbf{T}$	negative	blocked
	actual	$\{\mathbf{t}\} - P \neq \varnothing,\ P^* \subset P$	$cP^* \cap c\{\mathbf{t}\} = cP \cap c\{\mathbf{t}\}^{c}$	n.a.	n.a.
	nomic	$M - \mathbf{T} \neq \varnothing,\ M \subset M^*$	$M \cap \mathbf{T} = M^* \cap \mathbf{T}$	negative	blocked
	actual	$M - \{\mathbf{t}\} \neq \varnothing,\ M \subset M^*$	$M \cap \{\mathbf{t}\} = M^* \cap \{\mathbf{t}\}$	negative	blocked

Remarks: [a]guaranteed when P and **T** do not overlap, [b]together, the conditions can only be satisfied when **t** does not belong to M and $M^* = M \cup \{\mathbf{t}\}$, [c]since **t** is not in P, hence not in P* if $P^* \subseteq P$, this condition can only be satisfied by $P^* = P$

account should be preferred in a case of conflict? In my view the refined version should in general be preferred, because from the 'scientific common sense' point of view the basic account is rather naïve and coarse. However, there may be special, e.g. propositional, cases where the basic approach is to be preferred.

Table 6.7 summarizes the results of Sects. 6.5.1, 6.5.2 and 6.5.3. It is important to note that, in order to get strengthening as the uniform action in the table, in the last main row (dealing with strengthening of falsehoods with constant truth-content) P* and M* are the stronger versions of P and M, respectively, whereas in the text $P^{\#}$ and $M^{\#}$ were the weaker versions.

6.5.4 Some Other Principles

Ilkka Niiniluoto rightly conjectured (personal communication) that the refined definition satisfies the qualitative version of Graham Oddie's (2016) 'uniform distance principle', that is, assuming propositional constituents x and y: "if x and y are equally distant from **t**, then also their disjunction x v y is at the same distance."

Qualitatively it amounts to the claim that s(x,y,**t**) and s(y,x,**t**) together imply (using the minimal s-condition s(x,x,z)) that, e.g.{x,y} is refined equally close to {**t**} as {x} and {y}, respectively. Assuming general relatedness,[11] it also holds in general, and hence in nomic cases, for e.g. on the M-side: if M and M* are refined equally close to **T**, then M∪M* is as well. Again, s(x,x,z) and s(x,z,z) are crucial. For this kind of reasons I have proposed in debate with Thomas Mormann (see Kuipers 2000, p. 256) the strong boundary condition for 'refined closer to' that the size of M* should be between that of M and **T**. This is never satisfied by M∪M* when **T** = {**t**}. I have also considered there to liberate the condition to: the size of M* should be between that of M∩**T** and M∪**T**. However, this would not prevent the first case: M = {x}, M* = {y}, **T** = {**t**}, as is easy to check. Instead of seeing this as a problem, one might as well agree with Oddie's principle, with the consequence that no further condition is needed.

It is interesting to check whether the definition of 'refined closer to' satisfies the other two principles proposed by Oddie, at least if we assume general relatedness. *The 'Pareto principle' reads (in adapted symbols)*: if y is at least as close to the actual world (**t**) as x is, i.e., s(x,y,**t**), then the result of substitution of y for x in M, i.e., (M∪{y})−{x} $=_{def}$ $M^{x/y}$, is at least as close to the truth as M is. It is easy to check, using s(x,x,**t**), that this principle also holds in general, and hence in nomic cases: if y is refined at least as close to all z in **T** as x is, i.e., s(x,y,z), then $M^{x/y}$ is at least as close to **T** as M is. To be sure, the condition is in a qualitative setting rather strong, but quantitatively (in terms of distances) not at all.

Finally, Oddie submits the 'difference principle', which is typically quantitative (again in adapted symbols): the difference in closeness to the truth {**t**} of M and $M^{x/y}$ is some function or other of at most three factors: the distance of x from the actual world **t**, the distance of y from the actual world, and the size of M. In a qualitative reading it amounts to the claim that the question whether $M^{x/y}$ is at least as close to {**t**} as M, and/or the other way around, or whether they are incomparable, only depends on whether s(x,y,**t**) or s(y,x,**t**) holds, respectively. It is easy to check that this holds as well. The general version, hence including nomic cases, also holds. It amounts to: the question whether $M^{x/y}$ is refined at least as close to **T** as M, and/or the other way around, or whether they are incomparable, only depends on whether either s(x,y,z) or s(y,x,z) holds for all z in **T**.

Table 6.8 summarizes the principles dealt with in the present subsection. That they trivially hold in the basic case is easy to check in view of the fact that the refined definitions reduce to the basic ones when s(x,y,z,) is trivial (s(x,y,z) iff x = y = z).

It is just a matter of tedious checking that all three general principles also hold if relatedness is taken into account. Regarding the P-side, in view of the nature of the principles and the 'mirror-symmetric' correspondence between the two sides, they hold here as well.

---

[11] For all x and z in U r(x, z), that is, for all x and z in U there is y in U such that s(x, y, z). Hence, the in Sect. 6.5.3 given general versions of $M_{(i)}$ and $M_{(ii)}$ apply.

Table 6.8 Survey of the principles: uniform distance, Pareto, and difference

Principle		condition	consequence	basic	refined
uniform distance	actual	s(x,y,**t**) & s(y,x,**t**)	{x,y} ect {**t**} as {x}/{y}	trivial	yes
	nomic	M ect **T** as M*	M∪M* ect **T** as M/M*	trivial	yes
Pareto	actual	s(x,y,**t**)	$M^{x/y}$ lct {**t**} as M	trivial	yes
	nomic	for all z∈**T** s(x,y,z)	$M^{x/y}$ lct **T** as M	trivial	yes
difference	actual	s(x,y,**t**) and/or s(y,x,**t**)	iff $M^{x/y}$ lct {**t**} as M and/or M lct {**t**} as $M^{x/y}$	trivial	yes
	nomic	for all z∈**T** s(x,y,z) and/or for all z∈**T** s(y,x,z)	iff $M^{x/y}$ lct **T** as M and/or M lct **T** as $M^{x/y}$	trivial	yes

Abbreviations: *ect* equally close to, *lct* at least as close to. The used minimal s-conditions are suppressed.

An intriguing question remains the following. Oddie (2013) proved that the three principles implied for actual truth approximation that the so-called average distance measure d_{av} is the only quantitative measure satisfying them. Let $d_{av}(M, t) =_{df} \sum_{x\in M} d(x,t) / |M|$. The above finding that the refined definition satisfies qualitative versions of these principles, generates the question whether the refined definition of "M is at least as close to **t** as M#" entails $d_{av}(M^{\#}, t) \geq d_{av}(M, t)$, assuming s(x,y,t) is specified as $d(x,t) \geq d(y,t)$. Attempts to prove this starting from the specification of the likeness and the usefulness clause for actual truth approximation, i.e. ^{M}(i) and ^{M}(ii) in the previous subsection, failed. On second thoughts this is plausible. Although the refined clauses are certainly compatible with the average measure, they are too weak to imply the corresponding quantitative claim.[12]

In sum, on the basis of this section we may conclude that the refined approach deals in a sophisticated way with a number of principles and objections that have been discussed in the literature.

References

Hamminga, B. (1983). *Neoclassical theory structure and theory development*. Berlin: Springer.

Krajewski, W. (1977). *Correspondence principle and growth of science*. Dordrecht: Reidel.

Kuipers, T. (1987). A structuralist approach to truthlikeness. In T. Kuipers (Ed.), *What is closer-to-the-truth? A parade of approaches to truthlikeness. Poznan studies in the philosophy of the sciences and the humanities* (Vol. 10, pp. 79–99). Amsterdam: Rodopi.

Kuipers, T. (2000). *From instrumentalism to constructive realism. On some relations between confirmation, empirical progress, and truth approximation. Synthese library* (Vol. 287). Dordrecht: Kluwer Academic Publishers.

Niiniluoto, I. (1987). *Truthlikeness. Synthese library* (Vol. 185). Dordrecht: Reidel.

Northcott, R. (2013). Verisimilitude: A causal approach. *Synthese, 190*(9), 1471–1488.

[12] Note that the refined clauses do entail the non-increase of the so-called min-max-average distance measure: $(d_{min}(M, t) + d_{max}(M,t))/2$. However, as Oddie (2013, 2016) shows, this distance measure is in itself problematic as a likeness distance measure, which is of course not an objection to the refined clauses.

Nowak, L. (1980). *The structure of idealization. Synthese library* (Vol. 139). Dordrecht: Reidel.

Oddie, G. (1981). Verisimilitude reviewed. *The British Journal for the Philosophy of Science, 32*, 237–265.

Oddie, G. (2013). The content, consequence and likeness approaches to verisimilitude: compatibility, trivialization, and underdetermination. *Synthese, 190*, 1647–1687.

Oddie, G. (2016). Truthlikeness. In *The Stanford encyclopedia of philosophy* (Winter 2016 Edition, E. N. Zalta, Ed.). https://plato.stanford.edu/archives/win2016/entries/truthlikeness/

Tichý, P. (1974). On Popper's definition of verisimilitude. *The British Journal for the Philosophy of Science, 25*, 155–160.

Chapter 7
Stratified Nomic Truth Approximation

Abstract Assuming that the target of theory oriented empirical science in general and of nomic truth approximation in particular is to characterize the boundary or demarcation between nomic possibilities and nomic impossibilities, I have presented in Chap. 4 ("Models, postulates, and generalized nomic truth approximation") the 'basic' version of generalized nomic truth approximation, starting from 'two-sided' theories. Its main claim is that nomic truth approximation can perfectly be achieved by combining two prima facie opposing views on theories: (1) the traditional (Popperian) view: theories are (sets of models of) postulates that exclude certain possibilities from being realizable, enabling explanation and prediction and (2) the model view: theories are sets of models that claim to (approximately) represent certain realizable possibilities. Nomic truth approximation, i.e., increasing truth-content and decreasing falsity-content, becomes in this way revising theories by revising their models and/or their postulates in the face of increasing evidence.

The basic version of generalized nomic truth approximation is in many respects as simple as possible. Among other things, the basic version does not make a (theory-relative) distinction between an observational and a theoretical level. This raises the question to what extent in a stratified set-up (nomic) truth approximation on a theoretical level entails truth approximation on the observational level. In other words, the question whether 'closer to the truth' can be projected from the theoretical to the observational level. Of course, 'empirical success' is now explicated in observational terms, and 'closer to the observational truth' and 'more successfulness' are now related in the same way as in the unstratified case.

The chapter closes with *quantitative* and *refined* versions of stratified nomic truth approximation in line with Chaps. 5 and 6, respectively. (Referential truth approximation is dealt with in Chap. 11)

Keywords Theoretically closer to the truth · Observationally closer to the truth · Truth projection · Truthlikeness projection · Theoretical realization principle · Stratified success theorem · Two-sided theories · Quantitative stratified truth approximation · Refined stratified truth approximation

T. A. F. Kuipers, *Nomic Truth Approximation Revisited*, Synthese Library 399,
https://doi.org/10.1007/978-3-319-98388-2_7

7.1 Introduction

Typical for many philosophers of science is the presupposition of a (theory-relative) distinction between an observational and a theoretical level,[1] even such that e.g. Lakatosian hard core postulates alone may have no bite on the observational level: hence, stratification is needed (see Kuipers 2000, Chap. 9). In Sect. 7.2 we will deal in the first place with truth approximation on the theoretical level and its projection on the observational level, leading to the stratified success theorem. Then we will discuss the strength of one of the assumptions needed for projection, the so-called Theoretical Realization (TR-)principle, in particular in relation to the (non-)reference of theoretical terms. In Sect. 7.3 the main lines of *quantitative* and *refined* versions of stratified nomic truth approximation are presented in line with Chaps. 5 and 6, respectively.

7.2 Basic Truth Approximation on the Theoretical Level, and Its Observational Consequences.

7.2.1 *The Stratified Success Theorem*

In view of the previous chapters it is plausible to study truth approximation on the theoretical level in terms of two-sided theories, that is, theories with a M(odel)- and a P(ostulate)-side. Stratification of the P-side of theories has already been dealt with to some extent in (Kuipers 2014) and was taken over in Chap. 2. It is easy to see that the same procedure can be used for the M-side, and hence for two-sided theories. However, we will present here weaker conditions under which the corresponding (i.e., stratified) success theorems hold, than the one for the P-side presented in Chap. 2. They will strengthen the point that there may be substantial reason for the abductive conclusion, for the time being, that observational and even theoretical truth approximation has been achieved.

The subscript or superscript ‘t’ will refer to the theoretical level and ‘o’ to the observational level. Let V_t indicate the total vocabulary used on the t-level. It includes the observational vocabulary V_o. U_t and U_o indicate the set of theoretical and observational (conceptual) possibilities generated by V_t and V_o, respectively. Applying the Nomic Principle on both levels, let $\mathbf{T}$ and $\mathbf{T_o}$, being subsets of U_t and U_o, indicate the set of nomic possibilities on the t-level and the o-level, respectively. Let π indicate the projection of U_t on U_o, that is, $\pi(x)$ strips of all elements of x in U_t using t-terms. Let πA simply indicate $\pi(A)$ for a subset A of U_t.

[1] The electric circuit example (Sect. 2.2) becomes a (toy) example by assuming that some or even all connections between bulbs and switches are hidden.

In setting up a theory of truth approximation for the stratified case, assuming the general steps I and III (see Chap. 3), it is useful to present in one go the specific steps II, IV, and V:

II Specific logico-semantic steps: Define '*closer to the truth*'
IV Specific epistemological steps: Define '*more successful*'
V Theoretical step: From verisimilitude to success

A theory X = <M, P> as defined in the previous chapters is now of course primarily conceived on the t-level, i.e., M and P are subsets of U_t. On the other hand, the data D = <R, S>, see Sect. 3.3, are of course on the o-level, hence R and S are subsets of U_o. The observational counterpart of theory X = <M, P> is πX = <πM, πP>. It is formally not evident that the claims of X are automatically projected on the o-level such that they are equivalent to the claims of πX, that is, "$M \subseteq \mathbf{T}$" and "$\mathbf{T} \subseteq P$" do not unconditionally entail "$\pi M \subseteq \mathbf{T}_o$" and "$\mathbf{T}_o \subseteq \pi P$", respectively. However, this is trivially the case if we may assume:

$\mathbf{T}_o = \pi\mathbf{T}$ ($c\mathbf{T}_o = c\pi\mathbf{T}$) Truth Projection Principle (TPP or TP-principle)

One inclusion, $\pi\mathbf{T} \subseteq \mathbf{T}_o$, is a simple semantic fact, for if x in U_t is nomically possible on the t-level its projection, i.e., π(x), must be nomically possible on the o-level:

$\pi\mathbf{T} \subseteq \mathbf{T}_o$ ($c\mathbf{T}_o \subseteq c\pi\mathbf{T}$) projective semantic fact (psf)

The other inclusion, $\mathbf{T}_o \subseteq \pi\mathbf{T}$, is not trivial, it requires that every observational nomic possibility is 'theoretically realizable', given V_t, that is, V_t must be rich enough to represent any item in $\mathbf{T}_o$ in $\mathbf{T}$. We will make its use explicit, i.e., when we use:

$\mathbf{T}_o \subseteq \pi\mathbf{T}$ ($c\pi\mathbf{T} \subseteq c\mathbf{T}_o$) Theoretical Realization Principle (TRP or TR-principle)

That is, to every observational nomic possibility corresponds at least one theoretical nomic possibility that is projected on it or that realizes it on the theoretical level.[2]

Expressions on the o-level in terms of $\pi\mathbf{T}$, the projected truth, will be called π-expressions. Expressions on the o-level not in terms of $\pi\mathbf{T}$, but of $\mathbf{T}_o$, will be called o-expressions.

In Table 7.1 we present for a stratified two-sided theory its claims, and their relation, and the definitions of content and of truth-, falsity- and missing content[3] on both levels and for both sides, including a generic symbolic expression for both sides, some of which will be used later.

---

[2] Below we will discuss the strength of this postulate.

[3] The missing content is for later purposes.

Table 7.1 The claims of theory X = <M, P>, their relation, and the definitions of content and of truth, falsity and missing content on both levels and for both sides

Theory X = <M, P>	s∈{m,p}	m = M-side	p = P-side
Claim of X	$Cl_s(X)$	$M \subseteq \mathbf{T}$	$\mathbf{T} \subseteq P$ ($\equiv cP \subseteq c\mathbf{T}$)
		$\Downarrow$	$\Downarrow$
π-Claim of X	$^{\pi}CL_s(X)$	$\pi M \subseteq \pi\mathbf{T}$	$\pi\mathbf{T} \subseteq \pi P$ ($\equiv c\pi P \subseteq c\pi\mathbf{T}$)
		$\Downarrow$psf	$\Downarrow$TRP
o-Claim of X	$^{o}Cl_s(X)$	$\pi M \subseteq \mathbf{T_o}$	$\mathbf{T_o} \subseteq \pi P$ ($\equiv c\pi P \subseteq c\mathbf{T_o}$)
Content of X	$C_s(X)$	M	cP
o-Content of X	$^{o}C_s(X)$	πM	cπP
Truth-content of X	$TC_s(X)$	$M \cap \mathbf{T}$	$cP \cap c\mathbf{T}$
π-Truth-content of X	$^{\pi}TC_s(X)$	$\pi M \cap \pi\mathbf{T}$	$c\pi P \cap c\pi\mathbf{T}$
o-Truth-content of X	$^{o}TC_s(X)$	$\pi M \cap \mathbf{T_o}$	$c\pi P \cap c\mathbf{T_o}$
Falsity-content of X	$FC_s(X)$	$M - \mathbf{T}$	$cP - c\mathbf{T}$ ($\equiv \mathbf{T} - P$)
π-Falsity-content of X	$^{\pi}FC_s(X)$	$\pi M - \pi\mathbf{T}$	$c\pi P - c\pi\mathbf{T}$ ($\equiv \pi\mathbf{T} - \pi P$)
o-Falsity-content of X	$^{o}FC_s(X)$	$\pi M - \mathbf{T_o}$	$c\pi P - c\mathbf{T_o}$ ($\equiv \mathbf{T_o} - \pi P$)
Missing content of X	$MC_s(X)$	$\mathbf{T} - M$	$c\mathbf{T} - cP$
π-Missing content of X	$^{\pi}MC_s(X)$	$\pi\mathbf{T} - \pi M$	$c\pi\mathbf{T} - c\pi P$
o-Missing content of X	$^{o}MC_s(X)$	$\mathbf{T_o} - \pi M$	$c\mathbf{T_o} - c\pi P$

Table 7.2 The correct data hypothesis and the definitions of accepted- and rejected-content on both levels

Theory X = <M, P>	s∈{m,p}	m = M-side	p = P-side
Correct data hypothesis: CD(D) = CD(R)&CD(S) based on data theory D = <R, S>		CD(R) = $R \subseteq \mathbf{T_o}$	CD(S) = $\mathbf{T_o} \subseteq S$ ($\equiv cS \subseteq c\mathbf{T_o}$)
(o-)Content of D	$^{o}C_s(D)$	R	S
Accepted o-content of X	$^{o}AC_s(X)$	$\pi M \cap R$	$c\pi P \cap cS$
Accepted t-content of X	$^{t}AC_s(X)$	$\pi^{-1}(\pi M \cap R)$	$\pi^{-1}(c\pi P \cap cS)$
Rejected o-content of X	$^{o}RC_s(X)$	$\pi M - S$	$c\pi P - cR$
Rejected t-content of X	$^{t}RC_s(X)$	$\pi^{-1}(\pi M - S)$	$\pi^{-1}(c\pi P - cR)$

Table 7.2 gives, in line with Chap. 4, some relevant notions dealing with the data, notably the correct data hypothesis and the accepted- and rejected-content on the o-level and their 'reproduction' on the t-level. The 'reproduction' $\pi^{-1}(A)$ of a subset A of U_o on the t-level is defined as $\{x \in U_t | \pi(x) \in A\}$.

We now turn to the comparative notions and their relations. Crucial are four comparative notions, which we will specify later on in this section in their basic sense, that is, similarly to the specifications in Chap. 4, but now each on its 'own' level:

X* = <M*, P*>is at least as close to the (*theoretical*) truth (**T**) as X = <M, P>

X* = <M*, P*> is observationally at least as close to the *projected* truth ($\pi\mathbf{T}$) as X = <M, P>
X* = <M*, P*> is observationally at least as close to the *observational* truth ($\mathbf{T_o}$) as X = <M, P>
X* = <M*, P*>is at least as successful as X = <M, P>, relative to D = <R, S>

The main question is of course to what extent the basic truth approximation hypothesis is projected from the t-level on the o-level. Let us first focus on the relation between 'closer to the (theoretical) truth' and 'closer to the projected truth'. In Sect. 2.3 we saw already that on the P-side the truth-content (TC-)clause is simply projected. Of course, on the M-side this holds for the falsity-content (FC-)clause, in view of its formal correspondence to the truth-content at the P-side. However, the projection of the FC-clause on the P-side and that of the TC-clause on the M-side do not hold in general. Let us first specifically focus on the P-side. The FC-clause reads $cP^* - c\mathbf{T} \subseteq cP - c\mathbf{T}$, which is equivalent to:

FC-clause $\quad (P \cap \mathbf{T}) - P^* = \varnothing$

The question is to what extent the FC-clause is implied for the projected sets (in terms of $\pi\mathbf{T}$):

$^{\pi}$FC-clause $\quad (\pi P \cap \pi\mathbf{T}) - \pi P^* = \varnothing$

It is easy to check that the latter can be written as a conjunction of two claims:

$$[(\pi P \cap \pi\mathbf{T}) - \pi(P \cap \mathbf{T})] - \pi P^* = \varnothing \qquad \& \qquad \pi(P \cap \mathbf{T}) - \pi P^* = \varnothing$$

Of course, the second conjunct follows trivially from the FC-clause, for in general if $A - B = \varnothing$ then $\pi(A) - \pi(B) = \varnothing$. Hence, the $^{\pi}$FC-clause is implied by the FC-clause if the following condition holds in addition:

NEL(P, P*) $\quad [(\pi P \cap \pi\mathbf{T}) - \pi(P \cap \mathbf{T})] - \pi P^* = \varnothing$

We like to call this the 'no extra luck' (NEL-) assumption, for the following reason. As we in fact saw already in Sect. 2.3 (Fig. 2.8), a potential member of $[(\pi P \cap \pi\mathbf{T}) - \pi(P \cap \mathbf{T})] - \pi P^*$ originates from a theoretical nomic *im*possibility x, that is rightly excluded by P*, but not by P (hence, *missed* by P), however such that there is also a theoretical nomic possibility z, wrongly excluded by both, that has the same observational projection as x. We called this an 'extra lucky miss' of P relative to P*.

By analogy, we may on the M-side exclude 'extra lucky hits' by:

NEL (M, M*) $\quad [(\pi M \cap \pi\mathbf{T}) - \pi(M \cap \mathbf{T})] - \pi M^* = \varnothing$

in order to guarantee that TC(M, M*), i.e., $(M \cap \mathbf{T}) - M^* = \emptyset$, entails $^{\pi}$TC(M, M*), i.e., $(\pi M \cap \pi \mathbf{T}) - \pi M^* = \emptyset$.

In sum, 'no extra luck' (NEL) amounts on the P-side to 'no extra lucky misses' and on the M-side to 'no extra lucky hits'. How strong are these NEL-assumptions? As far as we can see, they are not very strong. The problem seems more to argue for them in a particular case. In this respect it is interesting to pay attention to stronger conditions that do not only seem to be not much stronger but are also better to argue for in specific cases. Already in Sect. 2.3 we introduced the notion of 'relative correctness' in order to guarantee, on the P-side, the projection of the FC-clause from the t-level to the o-level. Let us add first the complementary condition to guarantee on the M-side the projection of the TC-clause. The assumptions that M and P are 'relatively correct' amount to:

RelCor(M) $\quad \pi M \cap \pi \mathbf{T} \subseteq \pi(M \cap \mathbf{T})$ $\quad$ RelCor(P) $\quad \pi P \cap \pi \mathbf{T} \subseteq \pi(P \cap \mathbf{T})$

In view of the fact that the converse versions of the inclusions RelCor(M) and RelCor(P) hold generally, we may also write:

RelCor(M) $\quad \pi(M \cap \mathbf{T}) = \pi M \cap \pi \mathbf{T}$ $\quad$ RelCor(P) $\quad \pi(P \cap \mathbf{T}) = \pi P \cap \pi \mathbf{T}$

It is easy to check that these assumptions entail the corresponding NEL-assumption on the relevant side, hence they are stronger.

Let us begin with the strength and applicability of RelCor(P). Note first that it is solely an assumption about P and not about P*. Moreover, it has been defined such that it is independent of the TR-principle. As explained in Chap. 2, the name of the assumption is based on the fact that it precisely guarantees that (relative to $\pi\mathbf{T}$) correct items of πP (i.e., in the sense of correctly not excluded by the claim associated with πP) can be extended to correct items of P (relative to **T**, in the corresponding sense). In other words, it guarantees that P has theoretically correct reproductions of all its correct items on the o-level relative to $\pi\mathbf{T}$. How strong is RelCor(P)? To be sure, it means something like: P is on the right track, with the important relativization, that it is on the right track *as far as* $V_t - V_o$ *is concerned*, that is, it is relatively correct in this precise sense. But apart from this, it is not a very strong condition, as argued in some detail in Kuipers (2000, p. 213), roughly as follows. First, RelCor(P) has above been defined for the projection of theory P from the t-level on the o-level in terms of $\pi\mathbf{T}$. In the next subsection (7.2.2) it will be argued that its 'intermediate' projection on the level of referring terms is trivially relatively correct in the relevant sense. Second, it is important to note that the set on the left of (the substantial conditional of) RelCor(P), $\pi P \cap \pi \mathbf{T}$, requires items *x* in P and *z* in **T** for which $\pi(x) = \pi(z)$, that is, it requires a perfect fit between an item on the o-level and the relevant projected theoretical possibility on the t-level. The occurrence of such perfectly fitting observational pairs will, in general, be rather exceptional. (Of course, it is even more unlikely to have perfectly fitting items on the t-level.) But if P has no perfectly fitting observational models, and hence $\pi P \cap \pi \mathbf{T}$ is empty, RelCor(P) is not only trivial, but the relevant observational clause, $\pi P \cap \pi \mathbf{T} \subseteq \pi P^* \cap \pi \mathbf{T}$, itself is trivial as well. Third, as soon as we take refined versions of truth approximation into account (see Sect. 7.3.1), perfect fit even is more or less excluded. Hence, RelCor(P) is not a very strong assumption about P, as far as $V_t - V_o$ is concerned. Of course, roughly the same story can be told about RelCor(M), i.e., $\pi M \cap \pi \mathbf{T} \subseteq \pi(M \cap \mathbf{T})$, but now in terms of items that are correctly included by the claim associated with πM. In what follows we will only use the minimal NEL-conditions, but keep in mind that the stronger RelCor-conditions may be at least as plausible in particular cases.

In Table 7.3 we present the relations between the 'positive' clauses of four kinds: the truth-content (TC-)clauses on the t-level, its projection in terms of $\pi\mathbf{T}$, its $\mathbf{T}_o$-version, and the accepted-content (AC-)clauses. On the P-side each clause entails the next, where only in the last case a substantial assumption, viz. the correct data hypothesis concerning S, CD(S), has to be made, i.e., the same as in the case of the unstratified success theorem. However, on the M-side we need for all three transitions a substantial assumption, viz. 'no extra luck' of M relative to M*, (NEL(M, M*)), the theoretical realization postulate (TRP), and, of course, the relevant correct data hypothesis, viz. CD(R). In other words, the precise form of the 'stratified success theorem', as far as positive clauses are concerned, differs substantially between the sides.[4]

In Table 7.4 we present the relations between the 'negative' clauses of four kinds: the falsity-content (FC-)clauses on the t-level, its projection in terms of $\pi\mathbf{T}$, its $\mathbf{T}_o$-version, and the rejected-content (RC-)clauses. Of course, the tables have just switched because of the formal relations between the notions on the two sides. On the M-side, each clause entails the next, where only in the last case a substantial assumption, CD(R), has to be made, i.e., again the same as in the case of the unstratified success theorem. However, on the P-side we need for all three transitions a substantial assumption, viz. 'no extra luck of P relative to P*, (NEL(P, P*)), the theoretical realization postulate (TRP), and, of course, the relevant correct data hypothesis, viz. CD(S). Hence, again, the precise form of the 'stratified success theorem', as far as negative clauses are concerned, differs between the sides as a kind of mirror image relative to that of the positive clauses.

In sum, we have the

Stratified success theorem (two-sided):
The claim that 'closer to the truth on the theoretical level' entails being 'at least as (and in the long run more) successful on the observational level', holds, assuming correct data, on both sides unconditionally for one clause, viz. the positive clause on the P-side and the negative clause on the M-side, and on condition of the 'no extra luck' (NEL-) assumption, or the stronger 'relative correct' (RelCor-) assumption, and the theoretical realization (TR-) principle for the other clause, viz. the negative clause on the P-side and the positive clause on the M-side.

[4] In general, formulas in the same cell on different lines in a table represent equivalent versions.

Table 7.3 The relations between the truth-content (TC-)clauses on the t-level, its projection in terms of $\pi\mathbf{T}$, its $\mathbf{T}_o$-version, and the accepted-content (AC-)clauses

	$s \in \{m,p\}$	m = M-side	p = P-side
^{t}TC-clauses of: X* = <M*, P*>is at least as close to the theoretical truth (**T**) as X = <M, P>	$TC_s(X) \subseteq TC_s(X^*)$	$M \cap \mathbf{T} \subseteq M^* \cap \mathbf{T}$ $\mathbf{T} - M^* \subseteq \mathbf{T} - M$ $(M \cap \mathbf{T}) - M^* = \varnothing$	$cP \cap c\mathbf{T} \subseteq cP^* \cap c\mathbf{T}$ $P^* - \mathbf{T} \subseteq P - \mathbf{T}$ $P^* \cup \mathbf{T} \subseteq P \cup \mathbf{T}$ $P^* - (P \cup \mathbf{T}) = \varnothing$
		$\Downarrow$ NEL(M, M*)	$\Downarrow$
$^{\pi}$TC-clauses of: X* = <M*, P*> is observationally at least as close to the projected truth ($\pi\mathbf{T}$) as X = <M, P>	$^{\pi}TC_s(X) \subseteq {}^{\pi}TC_s(X^*)$	$\pi M \cap \pi\mathbf{T} \subseteq \pi M^* \cap \pi\mathbf{T}$ $\pi\mathbf{T} - \pi M^* \subseteq \pi\mathbf{T} - \pi M$ $(\pi M \cap \pi\mathbf{T}) - \pi M^* = \varnothing$	$c\pi P \cap c\pi\mathbf{T} \subseteq c\pi P^* \cap c\pi\mathbf{T}$ $\pi P^* - \pi\mathbf{T} \subseteq \pi P - \pi\mathbf{T}$ $\pi P^* \cup \pi\mathbf{T} \subseteq \pi P \cup \pi\mathbf{T}$ $\pi P^* - (\pi P \cup \pi\mathbf{T}) = \varnothing$
		$\Downarrow$TRP: $\mathbf{T}_o \subseteq \pi\mathbf{T}$	$\Downarrow$psf: $\pi\mathbf{T} \subseteq \mathbf{T}_o$
^{o}TC-clauses of: X* = <M*, P*>is observationally at least as close to the observational truth ($\mathbf{T}_o$) as X = <M, P>	$^{o}TC_s(X) \subseteq {}^{o}TC_s(X^*)$	$\pi M \cap \mathbf{T}_o \subseteq \pi M^* \cap \mathbf{T}_o$ $\mathbf{T}_o - \pi M^* \subseteq \mathbf{T}_o - \pi M$ $(\pi M \cap \mathbf{T}_o) - \pi M^* = \varnothing$	$c\pi P \cap c\mathbf{T}_o \subseteq c\pi P^* \cap c\mathbf{T}_o$ $\pi P^* - \mathbf{T}_o \subseteq \pi P - \mathbf{T}_o$ $\pi P^* \cup \mathbf{T}_o \subseteq \pi P \cup \mathbf{T}_o$ $\pi P^* - (\pi P \cup \mathbf{T}_o) = \varnothing$
		$\Downarrow$CD(R): $R \subseteq \mathbf{T}_o$	$\Downarrow$CD(S): $\mathbf{T}_o \subseteq S$
^{o}AC-clauses of: X* = <M*, P*> is at least as successful as X = <M, P>, relative to D = <R, S>	$^{o}AC(X) \subseteq {}^{o}AC(X^*)$	$\pi M \cap R \subseteq \pi M^* \cap R$ $R - \pi M^* \subseteq R - \pi M$ $(\pi M \cap R) - \pi M^* = \varnothing$	$c\pi P \cap cS \subseteq c\pi P^* \cap cS$ $\pi P^* - S \subseteq \pi P - S$ $\pi P^* \cup S \subseteq \pi P \cup S$ $\pi P^* - (\pi P \cup S) = \varnothing$

7.2.2 *The Theoretical Realization (TR–)Principle and Reference*

In view of the above relations, and having already discussed the strength of the no-extra luck assumptions, the interesting remaining question is of course the strength of the theoretical realization (TR-)principle. It amounts to the claim that every observational nomic possibility has at least one nomic representative on the theoretical level. Though there is formally room for exceptions, it is difficult to imagine such cases, in particular if all theoretical terms are supposed to refer.[5] In Sect. 11.4 we will define in terms of **T** and the P-side when a term refers, leading to a subset V_r of V_t of all referring terms, after some rephrasing, called *the referential truth*. Of course, we assume that the V_o-terms refer and hence that V_o is a subset of V_r. In sum: $V_o \subseteq V_r \subseteq V_t$. Let U_r indicate the set of conceptual possibilities of V_r and $\mathbf{T}_r$ the set of nomic possibilities on this U_r-level, called *the substantial truth*. Here we restrict the attention to a discussion of the TR-principle (see last note).

[5] What now follows is heavily based on (Kuipers 2000, Sect. 9.1.2, pp. 216–7), where also the topics of referential and of substantial truth approximation are dealt with. As far as referential truth approximation is concerned, Sect. 11.4 in this book reproduces the main lines.

Table 7.4 The relations between the falsity-content (FC-)clauses on the t-level, its projection in terms of $\pi\mathbf{T}$, its $\mathbf{T}_o$-version, and the rejected-content (RC-)clauses

	$s \in \{m,p\}$	m = M-side	p = P-side
^{t}FC-clauses of: $X^* = \langle M^*, P^*\rangle$ is at least as close to theoretical truth ($\mathbf{T}$) as $X = \langle M, P\rangle$	$FC_s(X^*) \subseteq FC_s(X)$	$M^*-\mathbf{T} \subseteq M-\mathbf{T}$ $M^* \cup \mathbf{T} \subseteq M \cup \mathbf{T}$ $M^* - (M \cup \mathbf{T}) = \varnothing$	$cP^*-c\mathbf{T} \subseteq cP-c\mathbf{T}$ $\mathbf{T}-P^* \subseteq \mathbf{T}-P$ $P \cap \mathbf{T} \subseteq P^* \cap \mathbf{T}$ $(P \cap \mathbf{T}) - P^* = \varnothing$
		$\Downarrow$	$\Downarrow$ NEL(P, P*)
$^\pi$FC-clauses of: $X^* = \langle M^*, P^*\rangle$ is observationally at least as close to the projected truth ($\pi\mathbf{T}$) as $X = \langle M, P\rangle$	$^\pi FC_s(X^*) \subseteq {}^\pi FC_s(X)$	$\pi M^*-\pi\mathbf{T} \subseteq \pi M-\pi\mathbf{T}$ $\pi M^* \cup \pi\mathbf{T} \subseteq \pi M \cup \pi\mathbf{T}$ $\pi M^* - (\pi M \cup \pi\mathbf{T}) = \varnothing$	$c\pi P^*-c\pi\mathbf{T} \subseteq c\pi P-c\pi\mathbf{T}$ $\pi\mathbf{T}-\pi P^* \subseteq \pi\mathbf{T}-\pi P$ $\pi P \cap \pi\mathbf{T} \subseteq \pi P^* \cap \pi\mathbf{T}$ $(\pi P \cap \pi\mathbf{T}) - \pi P^* = \varnothing$
		$\Downarrow$psf: $\pi\mathbf{T} \subseteq \mathbf{T}_o$	$\Downarrow$TRP: $\mathbf{T}_o \subseteq \pi\mathbf{T}$
^{o}FC-clauses of: $X^* = \langle M^*, P^*\rangle$ is observationally at least as close to the observational truth ($\mathbf{T}_o$) as $X = \langle M, P\rangle$	$^o FC_s(X^*) \subseteq {}^o FC_s(X)$	$\pi M^*-\mathbf{T}_o \subseteq \pi M-\mathbf{T}_o$ $\pi M^* \cup \mathbf{T}_o \subseteq \pi M \cup \mathbf{T}_o$ $\pi M^* - (\pi M \cup \mathbf{T}_o) = \varnothing$	$c\pi P^*-c\mathbf{T}_o \subseteq c\pi P-c\mathbf{T}_o$ $\mathbf{T}_o-\pi P^* \subseteq \mathbf{T}_o-\pi P$ $\pi P \cap \mathbf{T}_o \subseteq \pi P^* \cap \mathbf{T}_o$ $(\pi P \cap \mathbf{T}_o) - \pi P^* = \varnothing$
		$\Downarrow$CD(S): $\mathbf{T}_o \subseteq S$	$\Downarrow$CD(R): $R \subseteq \mathbf{T}_o$
^{o}RC-clauses of: $X^* = \langle M^*, P^*\rangle$ is at least as successful as $X = \langle M, P\rangle$, relative to $D = \langle R, S\rangle$	$^o RC(X^*) \subseteq {}^o RC(X)$	$\pi M^*-S \subseteq \pi M-S$ $\pi M^* \cup S \subseteq \pi M \cup S$ $\pi M^* - (\pi M \cup S) = \varnothing$	$c\pi P^*-cR \subseteq c\pi P-cR$ $R-\pi P^* \subseteq R-\pi P$ $\pi P \cap R \subseteq \pi P^* \cap R$ $(\pi P \cap R) - \pi P^* = \varnothing$

The TR-principle between the t- and the o-level, indicating the relevant projection function by $\pi_{t\text{-}o}$, reads:

TRP (t, o) $\quad \mathbf{T}_o \subseteq \pi_{t\text{-}o}\mathbf{T}\ (c\pi_{t\text{-}o}\mathbf{T} \subseteq c\mathbf{T}_o) \quad$ (entailing $\mathbf{T}_o = \pi_{t\text{-}o}\mathbf{T}$)

can now be decomposed into two principles:

TRP (r, o) $\quad \mathbf{T}_o \subseteq \pi_{r\text{-}o}\mathbf{T}_r\ (c\pi_{r\text{-}o}\mathbf{T}_r \subseteq c\mathbf{T}_o) \quad$ (entailing $\mathbf{T}_o = \pi_{r\text{-}o}\mathbf{T}_r$)

TRP (t, r) $\quad \mathbf{T}_r \subseteq \pi_{t\text{-}r}\mathbf{T}\ (c\pi_{t\text{-}r}\mathbf{T} \subseteq c\mathbf{T}_r) \quad$ (entailing $\mathbf{T}_r = \pi_{t\text{-}r}\mathbf{T}$)

For they entail $\quad \mathbf{T}_o \subseteq \pi_{r\text{-}o}\pi_{t\text{-}r}\mathbf{T}\ (c\pi_{r\text{-}o}\pi_{t\text{-}r}\mathbf{T} \subseteq c\mathbf{T}_o) \quad$ (entailing $\mathbf{T}_o = \pi_{r\text{-}o}\pi_{t\text{-}r}\mathbf{T}$)

which is equivalent to TRP(t, o) for the projection function is of course transitive: $\pi_{r\text{-}o}\pi_{t\text{-}r} = \pi_{t\text{-}o}$.

A detailed motivation of TRP(t, o) will be given by motivating TRP(t, r) and TRP(r, o) separately. Let us first consider the relation between the r- and the o-level. Hence, let $\pi_{r\text{-}o}$ refer to the projection of U_r onto U_o. As in the case of TRP(t, o), that $\pi_{r\text{-}o}\mathbf{T}_r$ is a subset of $\mathbf{T}_o$ is a semantic fact, for if a conceptual possibility on the r-level is nomically possible, it will remain nomically possible when one skips some components. However, the substantial claim that $\mathbf{T_o}$ is a subset of $\pi_{r\text{-}o}\mathbf{T_r}$ is now also plausible, for how could $\mathbf{T_o}$ leave room for an 'observational' possibility without leaving room for at least one reproduction of it on the r-level? In other words, how could $\mathbf{T_o}$ leave room for more observational possibilities than only those descending from the nomic possibilities on the level of all really referring terms taken into consideration?

Let us now turn to the relation between the r-level and the t-level, and hence the relation between substantial and theoretical truth approximation. Let $\pi_{t\text{-}r}$ refer to the projection of U_t onto U_r. The terms in $V_t - V_r$ are supposed not to '**T**-refer' in the sense that they do not curtail the nomic possibilities on the t-level. Formally, it is assumed that none of these terms satisfies the T-reference clause (see Sect. 11.4 for details). In other words, as far as **T** is concerned, they are fictions. Hence, we may assume that **T** is the V_t-reproduction of $\mathbf{T_r}$, i.e., $\mathbf{T} = \pi_{t\text{-}r}{}^{-1}\mathbf{T_r}$, which may be called the Fiction (F-) Principle (FP(t, r)). This name is somewhat misleading, for the mentioned condition is only a necessary feature of fictions. That $\mathbf{T} = \pi_{t\text{-}r}{}^{-1}\mathbf{T_r}$ is not a sufficient condition for letting $V_t - V_r$ concern fictions is due to the fact that one or more extra terms may not narrow down the number of nomic possibilities in co-production with V_r, but there may be other referring terms, outside V_t, that would do. However this may be, note that V_r could have been defined as the smallest set V' between V_o and V_t such that its truth **T'** satisfies the fiction postulate relative to **T**. Now it is important to note and easy to check that the F-Principle implies the TR-Principle between **T** and $\mathbf{T_r}$, i.e., $\mathbf{T_r} \subseteq \pi_{t\text{-}r}\mathbf{T}$, indicated by TRP(t, r). Hence, in combination with the motivation of TRP(r, o) we may conclude that the motivation for TRP(t, o) has been completed. However this may be, above and below we indicate when the TR-principle is used or presupposed.

7.2.3 *Remaining Methodological Steps*

Let us finally, like in the quantitative and refined cases, conclude this section by formulating the remaining methodological steps to complete the stratified theory of nomic truth approximation: assuming that a new theory in V_t-terms is at a certain moment more successful in the basic sense and in V_o-terms than the old one, propose and test the '*empirical progress hypothesis*': the new theory (is and) remains more successful in the basic sense than the old one. Assuming that after 'sufficient confirmation' this 'empirical progress hypothesis' is accepted (for the time being), argue *on the basis of a reconstruction of the way in which experiments have been done* and the (Basic) Success Theorem, that the 'default explanation' for this case of empirical progress is the hypothesis that the new theory is observationally closer to the truth in the basic sense than the old one, *i.e., that this is a case of basic observational truth approximation*. Argue next that the no extra luck assumption, or even the relative correctness assumption, in the relevant senses and the theoretical realization principle are plausible. Argue on this basis that the 'default explanation' for this case of observational truth approximation is the hypothesis that the new theory is theoretically closer to the truth in the basic sense than the old one, *i.e., that this is a case of basic theoretical truth approximation*. Finally, *conclude abductively* (for the time being) that the new theory is closer to the theoretical truth in the basic sense than the old one, *i.e., that basic theoretical truth approximation has been achieved.*

Finally, it is important to note that failures of the projection of "<M*, P*> is closer to the truth than <M, P>" on the observational level create room for truth approximation on the theoretical level 'against appearances', i.e., when "<M*, P*> is not at least as successful as <M, P>" and hence they create room for debate about how to proceed.

7.3 Quantitative and Refined Stratified Nomic Truth Approximation

Quantified and refined versions of stratified truth approximation are not difficult to imagine, as will be indicated in Sects. 7.3.1 and 7.3.2, respectively. They are mutually independent. We leave open the question of a synthesis, leading to refined quantified stratified nomic truth approximation. We restrict ourselves to the relevant tables to indicate the projection relations, including some open questions about them. In view of the discussion of the TR-principle above we simplify the presentation by assuming that principle throughout. In other words, a $\pi\mathbf{T}$ stage is left out by equating $\pi\mathbf{T}$ with $\mathbf{T}_o$.

7.3.1 Quantitative Stratified Nomic Truth Approximation

In Table 7.5 and 7.6 we repeat first the main clauses and (simplified) relations from the previous section and present then the unconditional quantitative clauses of quantitative stratified nomic truth approximation. But recall first the 'no-extra-luck' (NEL)-conditions:

NEL(M, M*) $[(\pi M \cap \mathbf{T}_o) - \pi(M \cap \mathbf{T})] - \pi M^* = \emptyset$
NEL(P, P*) $[(\pi P \cap \mathbf{T}_o) - \pi(P \cap \mathbf{T})] - \pi P^* = \emptyset$

We will replace them for the quantitative relations by the following NEL-q(quantitative) conditions:

NEL(M, M*)-q $m([(\pi M \cap \mathbf{T}_o) - \pi(M \cap \mathbf{T})] - \pi M^*) \leq m([(\pi M^* \cap \mathbf{T}_o) - \pi(M^* \cap \mathbf{T})] - \pi M)$
$\& \; m(\pi(M \cap \mathbf{T}) - \pi M^*) \leq m(\pi(M^* \cap \mathbf{T}) - \pi M)$

NEL(P, P*)-q $m([(\pi P \cap \mathbf{T}_o) - \pi(P \cap \mathbf{T})] - \pi P^*) \leq m([(\pi P \cap \mathbf{T}_o) - \pi(P \cap \mathbf{T})] - \pi P^*)$
$\& \; m(\pi(P \cap \mathbf{T}) - \pi P^*) \leq m(\pi(P^* \cap \mathbf{T}) - \pi P)$

In both cases, the first conjunct might be paraphrased as expressing 'not-more-extra-luck' of M and P relative to M* and P*, respectively. It will again be difficult to argue for these first conjuncts. After specifying the tables, we will come back to the second conjuncts in relation to the following 'gaps'.

At the 'opposite' sides of the NEL-conditions there arise in the quantitative case substantial gaps: GAP(P, P*)-q and GAP(M, M*)-q. In the tables they are simplified by using ($\mathbf{T}_o = \pi\mathbf{T}$ and) the 'general projective fact' (gpf) that the projection of the union of two sets equals the union of the projections of the two sets. In this light we will return to the gaps (in comparison with the second conjuncts above) after the tables.

Table 7.5 <M*, P*> is basically and quantitatively at least as close to the truth as <M, P>, on both levels, and their relation. *TC-clauses*

TC-clauses	M-side	P-side
(1) basic: $TC \subseteq TC^*$	$M \cap \mathbf{T} \subseteq M^* \cap \mathbf{T}$ $(M \cap \mathbf{T}) - M^* = \emptyset$	$cP \cap c\mathbf{T} \subseteq cP^* \cap c\mathbf{T}$ $P^* - \mathbf{T} \subseteq P - \mathbf{T}$ $P^* - (P \cup \mathbf{T}) = \emptyset$
	$\Downarrow$ NEL(M, M*)	$\Downarrow$
°(1) basic: $^oTC \subseteq {}^oTC^*$	$\pi M \cap \mathbf{T}_o \subseteq \pi M^* \cap \mathbf{T}_o$ $(\pi M \cap \mathbf{T}_o) - \pi M^* = \emptyset$	$c\pi P \cap c\mathbf{T}_o \subseteq c\pi P^* \cap c\mathbf{T}_o$ $\pi P^* - \mathbf{T}_o \subseteq \pi P - \mathbf{T}_o$ $\pi P^* - (\pi P \cup \mathbf{T}_o) = \emptyset$
Both basic TC-clauses entail the corresponding quantitative ones		
(2) quantitative: $m(TC) \leq m(TC^*)$	$m(M \cap \mathbf{T}) \leq m(M^* \cap \mathbf{T})$ $m\{(M \cap \mathbf{T}) - M^*\} \leq m\{(M^* \cap \mathbf{T}) - M\}$	$m(cP \cap c\mathbf{T}) \leq m(cP^* \cap c\mathbf{T})$ $m(P^* - \mathbf{T}) \leq m(P - \mathbf{T})$ $m\{(P^* - (P \cup \mathbf{T})\} \leq m\{P - (P^* \cup \mathbf{T})\}$
	$\Downarrow$ NEL(M, M*)-q	$\Downarrow$ GAP(P, P*)-q
°(2)quantitative: $m(^oTC) \leq m(^oTC^*)$	$m(\pi M \cap \mathbf{T}_o) \leq m(\pi M^* \cap \mathbf{T}_o)$ $m\{(\pi M \cap \mathbf{T}_o) - \pi M^*\} \leq m\{(\pi M^* \cap \mathbf{T}_o) - \pi M\}$	$m(c\pi P \cap c\mathbf{T}_o) \leq m(c\pi P^* \cap c\mathbf{T}_o)$ $m(\pi P^* - \mathbf{T}_o) \leq m(\pi P - \mathbf{T}_o)$ $m\{\pi P^* - (\pi P \cup \mathbf{T}_o)\} \leq m\{(\pi P - (\pi P^* \cup \mathbf{T}_o)\}$ $\Rightarrow_{gpf} m\{\pi P^* - \pi(P \cup \mathbf{T})\} \leq m\{(\pi P - \pi(P^* \cup \mathbf{T})\}$

Table 7.6 <M*, P*> is basically and quantitatively at least as close to the truth as <M, P>, on both levels, and their relation. *FC-clauses*

FC-clauses	M-side	P-side
(1)basic: $FC^* \subseteq FC$	$M^* - \mathbf{T} \subseteq M - \mathbf{T}$ $M^* - (M \cup \mathbf{T}) = \emptyset$	$cP^* - c\mathbf{T} \subseteq cP - c\mathbf{T}$ $P \cap \mathbf{T} \subseteq P^* \cap \mathbf{T}$ $(P \cap \mathbf{T}) - P^* = \emptyset$
	$\Downarrow$	$\Downarrow$ NEL(P, P*)
°(1)basic: $^{o}FC^* \subseteq {}^{o}FC$	$\pi M^* - \mathbf{T}_o \subseteq \pi M - \mathbf{T}_o$ $\pi M^* - (\pi M \cup \mathbf{T}_o) = \emptyset$	$c\pi P^* - c\mathbf{T}_o \subseteq c\pi P - c\mathbf{T}_o$ $\pi P \cap \mathbf{T}_o \subseteq \pi P^* \cap \mathbf{T}_o$ $(\pi P \cap \mathbf{T}_o) - \pi P^* = \emptyset$
Both basic FC-clauses entail the corresponding quantitative ones		
(2)quantitative: $m(FC^*) \leq m(FC)$	$m(M^* - \mathbf{T}) \leq m(M - \mathbf{T})$ $m\{M^* - (M \cup \mathbf{T})\} \leq m\{(M - (M^* \cup \mathbf{T})\}$	$m(cP^* - c\mathbf{T}) \leq m(cP - c\mathbf{T})$ $m(P \cap \mathbf{T}) \leq m(P^* \cap \mathbf{T})$ $m\{(P \cap \mathbf{T}) - P^*\} \leq m\{(P^* \cap \mathbf{T}) - P\}$
	$\Downarrow$ GAP(M, M*)-q	$\Downarrow$ NEL(P, P*)-q
°(2)quantitative: $m(^{o}FC^*) \leq m(^{o}FC)$	$m(\pi M^* - \mathbf{T}_o) \leq m(\pi M - \mathbf{T}_o)$ $m\{\pi M^* - (\pi M \cup \mathbf{T}_o)\} \leq m\{\pi M - (\pi M^* \cup \mathbf{T}_o)\}$ $\Rightarrow_{gpf} m\{\pi M^* - \pi(M \cup \mathbf{T})\} \leq m\{\pi M - \pi(M^* \cup \mathbf{T})\}$	$m(c\pi P^* - c\mathbf{T}_o) \leq m(c\pi P - c\mathbf{T}_o)$ $m(\pi P \cap \mathbf{T}_o) \leq m(\pi P^* \cap \mathbf{T}_o)$ $m\{(\pi P \cap \mathbf{T}_o) - \pi P^*\} \leq m\{(\pi P^* \cap \mathbf{T}_o) - \pi P\}$

Focusing on the P-side, there is a formal parallel between GAP(P, P*)-q, dealing with the TC-clause, that is the transition[6]

from $m(P^* - (P \cup \mathbf{T})) \leq m(P - (P^* \cup \mathbf{T}))$ to $m(\pi P^* - \pi(P \cup \mathbf{T})) \leq m(\pi P - \pi(P^* \cup \mathbf{T}))$

and the transition from the FC-clause on the theoretical level to the second conjunct of NEL(P, P*)-q, that is

from $m((P \cap \mathbf{T}) - P^*) \leq m((P^* \cap \mathbf{T}) - P)$ to $m(\pi(P \cap \mathbf{T}) - \pi P^*) \leq m(\pi(P^* \cap \mathbf{T}) - \pi P)$

Of course, the general situation is similar at the M-side. One could imagine that in certain special cases, e.g. in finite numerical cases, that these transitions on both sides are valid on formal grounds.

However, we leave the further study of these relations, including possible quantitative versions of the stronger 'relatively-correct'- conditions, at this point, and conclude that the projection of quantitative truth approximation from the theoretical to the observational level is far from trivial. However, this does not mean that in specific cases we may not have good reasons to assume the projectibility, and hence may abductively argue from quantitatively more successfulness, via quantitatively closer to the observational truth, to quantitatively closer to the theoretical truth.

[6] For both transitions, see the last listed equivalent versions in the relevant cells in Table 7.5 and Table 7.6, respectively.

Again, it is important to note that failures of the projection of "<M*, P*> is quantitatively closer to the truth than <M, P>" on the observational level create room for quantitative truth approximation on the theoretical level 'against appearances', i.e., when "<M*, P*> is quantitatively not at least as successful as <M, P>", and hence they create room for debate about how to proceed.

7.3.2 Refined Stratified Nomic Truth Approximation

Now we turn to refined stratified nomic truth approximation. Recall that a $\pi\mathbf{T}$ stage will be left out by equating $\pi\mathbf{T}$ with $\mathbf{T_o}$.

In Table 7.7 it will be indicated that for the projection of the likeness-clause on the M-side we need a refined version of NEL(M, M*) and for that on the P-side we need a refined condition called GAP(P,P*)-refined. Unfortunately, there seems no interesting conceptual interpretation of the latter. The same applies to the two strengthenings, see Table 7.8, that are needed of the usefulness-clauses on the t-level to guarantee the projection of the validity of these clauses on the o-level. Of course, the strengthening at the P-side is in some way a refined version of NEL(P, P*).

We start with the table of the TC-/Likeness-clauses.

Both refined Likeness-clauses, at both sides, entail the corresponding basic TC-clauses and reduce to them when s( ) is trivial. NEL(M, M*)-refined reduces to NEL(M,M*) for trivial s. Instead of NEL(M, M*)-refined, the stronger

RelCor(M)-refined $\forall_{x_o \in \pi M} \forall_{z_o \in \mathbf{T_o}} r(x_o, z_o) \rightarrow \exists_{x \in M} \exists_{z \in \mathbf{T}}\, x_o = \pi x$
$\& \; z_o = \pi z \; \& r(x, z)$

would of course do the job as well (and reduce to RelCor(M), $\pi M \cap \mathbf{T_o} \subseteq \pi(M \cap \mathbf{T})$, for trivial s). GAP(P,P*)-refined reduces for trivial s to the relevant basic clause on the o-level, $\pi P^* - \mathbf{T_o} \subseteq \pi P - \mathbf{T_o}$, but that is already entailed by the corresponding clause on the t-level: $P^* - \mathbf{T} \subseteq P - \mathbf{T}$, so the 'resulting reduced GAP' is not needed.

Now we turn to the table for the FC-/Usefulness-clauses, where at both sides, strengthening of the refined theoretical clause is needed to enable the projection on the observational level.

Now both refined usefulness-clauses, including the strengthened ones, are entailed by the corresponding basic FC-clauses and reduce to them when s( ) is trivial. However, to get back NEL(P, P*), it will be necessary to decompose the strengthened clause at the P-side in the theoretical one and an extra clause, NEL(P, P*)-refined, reducing to NEL(P, P*) for trivial s. Similarly, it will be possible to find a still stronger extra clause, RelCor(P)-refined that reduces to RelCor(P) for trivial s.

The above tables at least show that there may be specific cases in which we may have good reasons to assume the projectibility, and hence may abductively argue from 'refined more successfulness', via 'refined closer to the observational truth', to 'refined closer to the theoretical truth'. Of course, similar as in the previous cases, it is important to note that failures of the projection of "<M*, P*> as refined closer to

Table 7.7 <M*, P*> is basically and refined at least as close to the truth as <M, P>, on both levels, and their relation. *TC-/Likeness-clauses*

TC-clauses	M-side	P-side
(1) basic: $TC \subseteq TC^*$	$M \cap \mathbf{T} \subseteq M^* \cap \mathbf{T}$	$cP \cap c\mathbf{T} \subseteq cP^* \cap c\mathbf{T}$ $P^* - \mathbf{T} \subseteq P - \mathbf{T}$
	$\Downarrow$ NEL(M, M*): $[(\pi M \cap \mathbf{T}_o) - \pi(M \cap \mathbf{T})] - \pi M^* = \varnothing$	$\Downarrow$
°(1) basic: $^oTC \subseteq {}^oTC^*$	$\pi M \cap \mathbf{T}_o \subseteq \pi M^* \cap \mathbf{T}_o$	$c\pi P \cap c\mathbf{T}_o \subseteq c\pi P^* \cap c\mathbf{T}_o$ $\pi P^* - \mathbf{T}_o \subseteq \pi P - \mathbf{T}_o$
Both refined TC-clauses entail the corresponding basic ones and reduce to them when s() is trivial[a]		
(3) refined: Likeness-clauses	$\forall_{x \in M} \forall_{z \in \mathbf{T}} r(x,z) \rightarrow \exists_{y \in M^*} s(x,y,z)$ $\equiv^b \forall_{x \in M-M^*} \forall_{z \in \mathbf{T}-M^*} r(x,z) \rightarrow \exists_{y \in M^*} s(x,y,z)$ M* at least matches up to M in representing **T**	$\forall_{x \in cP} \forall_{z \in c\mathbf{T}} r(x,z) \rightarrow \exists_{y \in cP^*} s(x,y,z)$ $\equiv^b \forall_{x \in cP-cP^*} \forall_{z \in c\mathbf{T}-cP^*} r(x,z) \rightarrow \exists_{y \in cP^*} s(x,y,z)$ $\equiv \forall_{x \in P^*-P} \forall_{z \in P^*-\mathbf{T}} r(x,z) \rightarrow \exists_{y \in cP^*} s(x,y,z)$ cP* at least matches up to cP in representing c**T**
	$\Downarrow$ NEL(M,M*)-refined[c]: $\forall_{xo \in \pi M-\pi M^*} \forall_{zo \in \mathbf{To}-\pi M^*} r(xo,zo) \rightarrow$ $\exists_{x \in M-M^*} \exists_{z \in \mathbf{T}-M^*}$ $r(x,z)\ \&\ xo = \pi x\ \&\ zo = \pi z$	$\Downarrow$ GAP(P,P*)-refined[d]: $\forall_{xo \in c\pi P-c\pi P^*} \forall_{zo \in c\mathbf{To}-c\pi P^*} r(xo,zo) \rightarrow$ $\exists_{x \in cP-cP^*} \exists_{z \in c\mathbf{T}-cP^*} \exists_{y \in cP^*: \pi y \in c\pi P^*}{}^e$ $s(x,y,z)\ \&\ xo = \pi x\ \&\ zo = \pi z$ $\equiv \forall_{xo \in \pi P^*-\pi P} \forall_{zo \in \pi P^*-\mathbf{To}} r(xo,zo) \rightarrow$ $\exists_{x \in P^*-P} \exists_{z \in P^*-\mathbf{T}} \exists_{y \in cP^*: \pi y \in c\pi P^*}$ $s(x,y,z)\ \&\ xo = \pi x\ \&\ zo = \pi z$

(continued)

Table 7.7 (continued)

TC-clauses	M-side	P-side
[o](3) refined: Likeness-clauses	$\forall_{xo\in\pi M}\forall_{zo\in To}\, r(xo,zo) \rightarrow \exists_{yo\in\pi M^*}\, s(xo,yo,zo)$ $\equiv \forall_{xo\in\pi M-\pi M^*}\forall_{zo\in To-\pi M^*}\, r(xo,zo) \rightarrow$ $\exists_{yo\in\pi M^*}\, s(xo,yo,zo)$ πM* at least matches up to πM in representing **T**$_o$	$\forall_{xo\in c\pi P}\forall_{zo\in cTo}\, r(xo,zo) \rightarrow \exists_{yo\in c\pi P^*}\, s(xo,yo,zo)$ $\equiv \forall_{xo\in c\pi P-c\pi P^*}\forall_{zo\in cTo-c\pi P^*}\, r(xo,zo) \rightarrow$ $\exists_{yo\in c\pi P^*}\, s(xo,yo,zo)$ $\equiv \forall_{xo\in\pi P^*-\pi P}\forall_{zo\in\pi P^*-To}\, r(xo,zo) \rightarrow$ $\exists_{yo\in c\pi P^*}\, s(xo,yo,zo)$ cπP* at least matches up to cπP in representing c**T**$_o$

[a]s(x,y,z) iff x = y = z

[b]This equivalence holds due to the s-conditions: use: if x or z are in M* then s(x,x,z), respectively s(x,z,z). Similarly for the o-level clause and for the P-side clauses

[c]Using the semantic fact: s(x,y,z) entails s(πx, πy, πz)

[d]Again using the semantic fact: s(x,y,z) entails s(πx, πy, πz)

[e]Instead of $\exists_{y\in cP^*:\pi y\in c\pi P^*}$ we cannot simply put r(x, z), for then it need not be the case that the resulting y in cP* is such that πy is in cπP*

Table 7.8 <M*, P*> is basically and refined at least as close to the truth as <M, P>, on both levels, and their relation. *FC-/Usefulness-clauses*

FC-clauses	M-side	P-side
(1)basic: $FC^* \subseteq FC$	$M^* - \mathbf{T} \subseteq M - \mathbf{T}$ $M^* - (M \cup \mathbf{T}) = \varnothing$	$cP^* - c\mathbf{T} \subseteq cP - c\mathbf{T}$ $(P \cap \mathbf{T}) - P^* = \varnothing$
	$\Downarrow$	$\Downarrow$ NEL(P, P*): $[(\pi P \cap \mathbf{T}_o) - \pi(P \cap \mathbf{T})] - \pi P^* = \varnothing$
°(1)basic: $^oFC^* \subseteq {}^oFC$	$\pi M^* - \mathbf{T}_o \subseteq \pi M - \mathbf{T}_o$ $\pi M^* - (\pi M \cup \mathbf{T}_o) = \varnothing$	$c\pi P^* - c\mathbf{T}_o \subseteq c\pi P - c\mathbf{T}_o$ $(\pi P \cap \mathbf{T}_o) - \pi P^* = \varnothing$
Both refined FC-clauses are entailed by the corresponding basic ones and reduce to them when s( ) is trivial		
(3) refined: Usefulness-clauses	$\forall_{y \in M^* - (M \cup \mathbf{T})} \exists_{x \in M - \mathbf{T}} \exists_{z \in \mathbf{T} - M}\, s(x,y,z)$ all extra, but mistaken, items of M* are useful in matching up to M	$\forall_{y \in cP^* - (cP \cup c\mathbf{T})} \exists_{x \in cP - c\mathbf{T}} \exists_{z \in c\mathbf{T} - cP}\, s(x,y,z)$ $\equiv \forall_{y \in (P \cap \mathbf{T}) - P^*} \exists_{x \in \mathbf{T} - P} \exists_{z \in P - \mathbf{T}}\, s(x,y,z)$ all extra, but mistaken, items of cP* are useful in matching up to cP
	$\Downarrow$If (3) can be strengthened to $\forall_{y \in M^* - (M \cup \mathbf{T})} \exists_{x \in M - \mathbf{T}: \pi x \in \pi M - \mathbf{To}} \exists_{z \in \mathbf{T} - M: \pi z \in \mathbf{To} - \pi M}\, s(x,y,z)$ $\equiv^a \forall_{y \in M^* - (M \cup \mathbf{T})} \exists_{x: \pi x \in \pi M - \mathbf{To}} \exists_{z: \pi z \in \mathbf{To} - \pi M}\, s(x,y,z)$	$\Downarrow$If (3) can be strengthened to $\forall_{y \in (P \cap \mathbf{T}) - P^*} \exists_{x \in \mathbf{T} - P: \pi x \in \mathbf{To} - \pi P} \exists_{z \in P - \mathbf{T}: \pi z \in \pi P - \mathbf{To}}\, s(x,y,z)$ $\equiv^a \forall_{y \in (P \cap \mathbf{T}) - P^*} \exists_{x: \pi x \in \mathbf{To} - \pi P} \exists_{z: \pi z \in \pi P - \mathbf{To}}\, s(x,y,z)$
°(3) refined: Usefulness-clauses	$\forall_{yo \in \pi M^* - (\pi M \cup \mathbf{To})} \exists_{xo \in \pi M - \mathbf{To}} \exists_{zo \in \mathbf{To} - \pi M}\, s(xo,yo,zo)$ all extra, but mistaken, items of πM* are useful in matching up to πM	$\forall_{yo \in c\pi P^* - (c\pi P \cup c\mathbf{To})} \exists_{xo \in c\pi P - c\mathbf{To}} \exists_{zo \in c\mathbf{To} - c\pi P}\, s(xo,yo,zo)$ $\equiv \forall_{yo \in (\pi P \cap \mathbf{To}) - \pi P^*} \exists_{xo \in \mathbf{To} - \pi P} \exists_{zo \in \pi P - \mathbf{To}}\, s(xo,yo,zo)$ all extra, but mistaken, items of cπP* are useful in matching up to cπP

[a]Use the semantic fact: s(x,y,z) entails s(πx, πy, πz) and e.g. that $\pi x \in \pi M - \mathbf{T}_o$ entails $x \in M - \mathbf{T}$

the truth than <M, P>" on the observational level create room for refined truth approximation on the theoretical level 'against appearances', i.e., when "<M*, P*> is refined not at least as successful as <M, P>" and hence they create room for debate about how to proceed.

We closed Chap. 4 with the following remark about research programs:

> Lakatos's distinction between hardcore and auxiliary postulates can easily be made on the P-side, on the M-side we can make the distinction between prototypical and tentative models. Whereas this leads on the P-side to a playground for revising the auxiliary postulates it leads on the M-side to a similar playground of revising (the set of) tentative models. Hence, Lakatosian empirical progress and truth approximation can then be achieved when such revisions respect the prototypical models and the hardcore postulates.

Taking now refinement and stratification into account, and neglecting quantification, and granting the importance of Lakatos's distinction, the most representative type of major theories in the process of truth approximation in the natural sciences seems to be the following:

> Two-sided theories, consistent and stratified (in a theory-relative way), the P-side in refined terms, with a distinction between hardcore and auxiliary postulates, and the M-side in refined terms, with a distinction between prototypical and tentative models.

With this conjecture we conclude the chapters dealing with three concretizations of basic nomic truth approximation.

References

Kuipers, T. (2000). *From instrumentalism to constructive realism. On some relations between confirmation, empirical progress, and truth approximation, Synthese library 287*. Dordrecht: Kluwer Academic Publishers.

Kuipers, T. (2014). Empirical progress and truth approximation revisited. Proceedings Tilburg conference on progress in science (April, 2012). *Studies in History and Philosophy of Science, special section Progress in Science, 46*, 64–72.

Part III
Extensions

Chapter 8
Theories Looking for Domains. Nomic Truth Approximation by Domain Revision

Abstract The theory of nomic truth approximation essentially deals primarily with truth approximation by theory revision for a fixed domain. However, variable domains and a fixed theory can also be taken into account, where the main changes concern domain extensions and restrictions. In this chapter I will present a coherent set of definitions of 'more truthlikeness', 'empirical progress' and 'truth approximation' due to a revision of the 'domain of intended applications'. This set of definitions seems to be the natural counterpart of the basic definitions of similar notions as far as theory revision is concerned.
The resulting formal aspects of theory and domain revision strongly suggest an analogy between truth approximation and design research, for example, drug research: a new drug (theory) may be better for a certain disease (domain) than an old one. However, it also occurs that a certain drug may be better for another disease than for the original target disease, a phenomenon which was nicely captured by the title of a study by Rein Vos (1991): *Drugs Looking for Diseases*. Similarly, truth approximation may not only take the shape of theory revision but also of domain revision, naturally suggesting the phenomenon of 'Theories looking for domains'. However, whereas Vos documented his title with a number of examples, so far, apart from plausible cases of 'truth accumulation by domain extension' and the conjecture that cases of so-called Kuhn-loss can be reconstructed as cases of 'truth approximation by domain restriction', I did not find clear-cut but more complex empirical instantiations of the analogy, only some very interesting, non-empirical examples.

Keywords Domain revision · Domain of intended applications · Truth approximation by domain revision · Domain restriction · Analogy with design research · Kuhn-loss · Truth accumulation

---

Revised version of: "Theories looking for domains. Fact or fiction? Reversing Structuralist Truth Approximation" In: L. Magnani (ed.), *Model-Based Reasoning in Science and Engineering*, Studies in Logic, Vol. 2, College Publications, King's College, London, 2006, pp. 33–50.
Acknowledgements: This paper was first presented on the Workshop 'Structures and Dynamics of Science', in Paris in 2004, organized by Ulises Moulines. I am grateful for all comments that I received in Groningen, in Paris, and, last but not least, in Pavia. Special thanks go to David Atkinson, Gordana Dodig-Crnkovic, Nico Krijn, Andrés Rivadulla, Jan-Willem Romeijn and Rein Vos for their useful comments and suggestions on an earlier draft.

T. A. F. Kuipers, *Nomic Truth Approximation Revisited*, Synthese Library 399,
https://doi.org/10.1007/978-3-319-98388-2_8

8.1 Introduction

This chapter elaborates a point that was already made explicit by Sjoerd Zwart (1998). In (Kuipers 2000, p. 207) I wrote, in concluding the second chapter on truth approximation (Chap. 8):

> Finally, variable domains can also be taken into account, where the main changes concern extensions and restrictions. We will not study this issue, but see (Zwart 1998, Chap. 2, Chap. 4), for some illuminating elaborations in this connection, among other things, the way in which strengthening/weakening of a theory and extending/reducing its domain interact.

More specifically, in this chapter I will present a coherent set of definitions of 'more truthlikeness', 'empirical progress' and 'truth approximation' due to a revision of the 'domain of intended applications'. This set of definitions seems to be the natural counterpart of the basic definitions of similar notions in (Kuipers 2000) and Chap. 4 as far as theory revision is concerned.

The formal aspects of theory and domain revision strongly suggest an analogy between truth approximation and design research, for example drug research. Whereas a new drug may be better for a certain disease than an old one, a certain drug may be better for another disease than the original target disease, a phenomenon which was nicely captured by the title of a study by Rein Vos (1991): *Drugs Looking for Diseases*. Similarly, truth approximation may not only take the shape of theory revision but also of domain revision, naturally suggesting the phenomenon of 'Theories looking for domains'. However, whereas Vos documented his title with a number of examples, so far, apart from plausible cases of 'truth accumulation by domain extension' and the conjecture that cases of so-called Kuhn-loss can be reconstructed as cases of 'truth approximation by domain restriction', I did not find clear-cut but more complex empirical instantiations of the analogy, only some very interesting, non-empirical examples.

Section 8.2 briefly states the nomic theory of truth approximation by theory revision for a fixed domain and concludes with the formal analogy with design research. Suggested by this analogy, the corresponding theory of 'truth approximation by domain revision for a fixed theory' is presented in Sect. 8.3, which ends with a synthesis of truth approximation by theory and/or domain revision. As a kind of meta- and self-application of the 'Theories looking for domains' theme, Sect. 8.4 is devoted to the search for historical cases exemplifying this theory. I conclude, Sect. 8.5, with arguing that it is plausible to expect that truth approximation by domain revision in the empirical sciences frequently occurs, interestingly enough, in particular as far as the instrumentalist method is practiced.

8.2 Truth Approximation by Theory Revision

My qualitative theory of (two-sided) nomic truth approximation will here explicitly be represented in the terms of the structuralist theory of theory representation. Surprisingly enough, and if I am not mistaken, the latter theory of representation has not been presented in Model Based Reasoning (MBR-) conferences before and after 2004 as a pre-eminent case of model-based reasoning in the sense that not the sentences but the models of a theory form the primary domain of discourse.[1]

Starting from a fixed vocabulary and a suitable similarity type of set-theoretic structures, usually provided by a research program, let U indicate the set of structures of that type, also called the *conceptual possibilities* or *potential models* of the theory. In (Kuipers 2000, Chap. 13) I have already indicated that it is possible to introduce a domain vocabulary, as an observational sub-vocabulary of the vocabulary constituting U, such that it is possible to define the so-called *domain of intended applications* explicitly in that vocabulary, leaving the further nature of the applications as the unknown to be specified. The advantage of this is that domains become comparable sets, that is, it enables us to say that an instance of one domain is or is not an instance of another. This will be an important, though implicit, assumption in dealing with the main topic of this paper.

Finally, let us assume that our target is the set of physically or, more broadly, nomically possible phenomena within that domain on the level of (the vocabulary constituting) U. Let **T** indicate this domain 'as seen through U', hence a subset of U, and be called the set of (intended) nomic possibilities or the truth about the domain of intended applications. Note that we do not yet suppose to dispose of a general characterization of **T** as a subset of U. What we only know is that each intended application can be represented as a potential model in U.[2] In the present context, a general characterization of **T** is 'the great unknown' where theories, as represented by their models and postulates, are looking for.

[1] With the kind help of Lorenzo Magnani, I only found a quasi-relevant reference by Ronald Giere (1999) in his MBR98-contribution, entitled "Using models to represent reality". Giere pays some attention to the pioneering work of Patrick Suppes in this respect. However, Suppes' equally pioneering work with respect to "Using models to represent theories" is not touched upon. This work of Suppes was the starting point of Joseph Sneed and later followed by Stegmüller, Balzer, Moulines, and others, and called the structuralist representation of theories, i.e., theories as sets of structures, forming their models.

[2] We may assume that **T** is a subset of the set of all nomic possibilities that correspond to all possible domains, the overall domain, that can be characterized by the domain vocabulary. In the previous chapters the domain is fixed and may be this overall domain or a specific sub-domain.

8.2.1 The Main Lines of Truth Approximation by Theory Revision

Let the subsets M and P of U indicate, respectively, the set of specified (intended and interpreted) models of a theory and the set of models of the postulates of the theory. The theory is now represented by the tuple <M, P>, together with the claims that **T** is a superset of M and a subset of P ($M \subseteq \mathbf{T} \subseteq P$). It is called maximal when M = P. A theory is said to be true (false) when both claims are true (at least one is false). It is easy to check that (a general characterization of) the maximal theory for which $M = \mathbf{T} = P$ is the theory with the strongest true claims. That is why **T** may well be called 'the truth' in this context, that is, the truth about the given domain in the given vocabulary.

Now it is plausible to define what it means, for fixed U and **T**, and hence theories of the same research program, that one theory <M*, P* > is closer (or more similar) to the truth than another theory <M, P>. Intuitively, when <M*, P* > is moving from <M, P > in the direction of **T**. Formally, in terms of symmetric differences: $M^* \Delta \mathbf{T} \subseteq M \Delta \mathbf{T}$ and $P^* \Delta \mathbf{T} \subseteq P \Delta \mathbf{T}$, and at least once a proper subset. In Fig. 8.1, the two sides have to be combined,[3] shaded areas are empty and at least one of the four #-areas has to be empty.

Knowing what it means that one theory is closer to the truth than another is one thing, in view of the fact that we don't know the truth, judging that this is in fact the case, in the light of our evidence, is another. Although there is no theorem guaranteeing that more-successfulness entails closer-to-the-truth, there is a theorem, the Success Theorem, guaranteeing almost the reverse entailment, viz., closer-to-the-truth entails at-least-as-successfulness.

Let R, after a certain time, represent the set of (fully specified) realized applications belonging to the domain, that is, the experimentally or otherwise realized nomic possibilities at that time. Let S indicate the strongest law about the domain induced on the basis of R. Of course, R will partly be the result of testing hypothetical

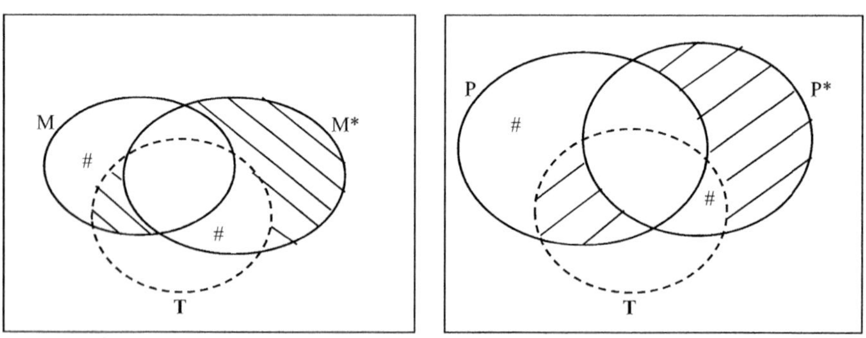

Fig. 8.1 <M*, P*> is closer to the truth **T** than <M, P>

[3] To combine both figures it is important to note that M (M*) is a subset of P(P*).

laws. If no mistakes have been made in representing the realized possibilities, R is not only a subset of U, but even of **T**, for, by definition, nomic *im*possibilities within the domain can't be realized. Moreover, in the structuralist representation S is also a subset of U, such that it has been concluded, at least provisionally, that conceptual possibilities outside S, that is, in $U - S$ are representations of nomically impossible intended applications. Finally, if our inductive jump from R to S is correct, S has to be a superset of **T**. To sum up: $R \subseteq \mathbf{T} \subseteq S \subseteq U$. <R, S> is called the data set or data theory. We refer to Table 4.4 for the precise definition of the claim that, relative to <R, S>, <M*, P*> is more successful than <M, P>.

The Success Theorem states, assuming that <R, S> is correct, that is, $R \subseteq \mathbf{T} \subseteq S$, that <M*, P*> will be at least as successful as <M, P>, and will become more successful in the long run, if <M*, P*> is closer to the truth than <M, P>. This theorem is such that *persistently* being-more-successful is functional or instrumental for truth approximation, though not guaranteeing it, in the sense that it is very difficult/special for <M*, P*>to be at least as successful as, let alone more successful than, <M, P>, when <M*, P*> is not closer to the truth than <M, P>.

More in detail we may argue as follows. Let <M*, P*> be more successful than <M, P> relative to <R, S>. This suggests the Empirical Progress Hypothesis, according to which it is hypothesized that this will remain the case, whatever experiments we design and perform. Testing of this hypothesis may result in the conclusion, at least for the time being, that this is in fact the case: <M*, P*> persistently remains more successful than <M, P>. This is the paradigm situation for speaking of empirical progress, and for applying the 'instrumentalist rule of success', viz. replace, for the time being, <M, P> in favor of <M*, P*>. Moreover, under this condition it is very hard to imagine that <M*, P*> is, despite appearances, not closer to the truth than <M< P>. To be sure, if this would nevertheless be the case, we have not been creative enough in designing new experiments. In other words, the situation even allows the tentative conclusion that <M*, P*> is in fact closer to the truth. Note that it is an abductive conclusion, based on the Success Theorem.

So far I have presented the basic theory of nomic truth approximation, in structuralist terms. It is basic in the sense that several idealizations have been made, requiring concretization.[4] Let us indicate some of the main concretizations as they were indicated in Chap. 3 and have been elaborated in (Kuipers 2000) and Chaps. 5, 6, and 7.

Above we did not make a distinction between (relatively) observational and theoretical terms. In Chap. 7 I have spelled out how this distinction works out, with the main conclusion that empirical progress, of course, in observational terms, remains functional for truth approximation on the theoretical level, though with some greater risk of being wrong in this.

Another important idealization was that we implicitly assumed that any established counterexample of a theory, that is, a realized possibility not being a model of the theory, is as bad for one theory as for any other. However, in Chap. 6 I have introduced the idea that one structure may be more similar to another than a third.

[4] In Kuipers (2007) I have extensively illustrated the (philosophical) method of concept explication by idealization and concretization by the example of truth approximation.

In this way, the possibility arises, for example, that a counterexample is less dramatic for one theory than for another, because the former has a model that is more similar to the 'countermodel' (i.e., the potential model representing the counterexample), than any model of the other theory. In this way refined definitions of more-successfulness and closer-to-the-truth lead again to the conclusion that empirical progress is functional for truth approximation. As a matter of fact we only know of real life scientific examples of (potential) truth approximation, e.g. the Law of Van der Waals as successor of the ideal gas law, when this concretization is introduced. Without this, we only know of toy examples.

Finally, in the beginning of this section I have already introduced an important concretization relative to the previous chapters, that is, for our domain revision purposes it is crucial to assume a domain vocabulary as an observational sub-vocabulary of the given vocabulary, such that it is possible to define the domain explicitly in that vocabulary, leaving the further nature of the applications as the unknown to be specified. But first I will introduce the analogy with design research in order to make domain revision a plausible subject of theoretical and (meta-)empirical research.

8.2.2 *Analogy with Design Research*

The formal pattern of (the development of) design research, as presented in (Kuipers 2001, Chap. 10), is to some extent analogous to that of (descriptive or explanatory) 'nomological' research. Starting from a set of relevant characteristics, let us call the subset of characteristics of the intended product (or process) its intended profile. As soon as we have a prototype we can delineate the subset of its actual characteristics, the prototype profile. A second prototype amounts to genuine progress when the symmetric difference between its profile and the intended profile is a proper subset of that between the profile of the first and the intended profile. In this way, for example, improving a (medical) drug for a given disease becomes formally analogous with improving a theory for a specific domain. See (Forbus and Gentner 1997) for another illustration of progress in design research, viz. designing 'qualitative simulators' to be evaluated in terms of useful and problematic properties derived from qualitative reasoning. The latter example makes clear that there are, as a rule, two (non-overlapping) profiles involved: that of desired characteristics and that of undesired ones, together constituting the set of relevant characteristics. In this way, the analogy between the demarcation of the nomic possibilities from the nomic impossibilities and the demarcation between desired and undesired characteristics is clear. However, also the difference between design and nomic research will be clear, for in the design case the demarcation is a point of departure and in the nomic case it is the great unknown. Our interest concerns of course the analogy as far as it applies.

Rein Vos (1991), who elaborated the formal description of progress in design research for the case of drugs, entitled his book *Drugs Looking for Diseases*, because by historical research he came to the conclusion that pharmaceutical research often goes in the other direction. That is, there may be an interesting chemical substance that does not do what it is supposed to do for a given disease, even after some manipulation. However, instead of discarding the substance for this reason, one starts to look for other diseases, whether or not related to the first one, for which the substance may have curing effects. In this way, for example, beta blockers have been discovered as a drug for certain heart diseases different from the original targeted heart disease.

Hence the plausible question arises whether in the context of nomological research something similar, viz., 'theories looking for domains' or, equivalently, 'models looking for applications', does also exist. One type of example is well known, viz., the case that a theory which is or is not successful for a certain domain turns out to be (also) successful for a totally unrelated domain. In this case we have a purely formal analogy. A famous example is Keynes' transfer of hydraulic modeling from fluid mechanics to macroeconomics. Our interest in this paper concerns however cases in which the new and old domains are to some extent related, that is, genuine domain revision.

8.3 Truth Approximation by Domain Revision

Let us assume that specific domains are specified in some subvocabulary of the vocabulary of the theories we consider and that the truth about such domains amounts to the set of nomic possibilities of that domain on the level of the encompassing vocabulary.

8.3.1 Closer to the Truth About One Domain than About another

It is plausible to try to define what it means that, for fixed theory <M, P>, the truth **T*** about domain D* is closer (or more similar) to <M, P > than the truth **T** about domain D by assuming that **T*** is moving from **T** such that it comes more between M and P. In terms of the claims of theory <M, P > this amounts to assuming, informally, that the inclusion and the exclusion claims of <M, P > make less mistakes w.r.t. **T*** than w.r.t. **T**. Formally, when $M-\mathbf{T}^* \subseteq M-\mathbf{T}$ ($\varnothing_1$-area empty in Fig. 8.2) and $cP-c\mathbf{T}^* \subseteq cP-c\mathbf{T}$ ($\equiv \mathbf{T}^* - P \subseteq \mathbf{T} - P$) ($\varnothing_2$-area empty) and at least once it should be a proper subset relation (at least one #-area non-empty).[5] Or, equivalently, when the inclusion and the exclusion claims of <M, P> have more matches w.r.t. **T*** than w.r.t. **T**. Formally, when $M \cap \mathbf{T} \subseteq M \cap \mathbf{T}^*$ ($\varnothing_1$-area empty in Fig. 8.2) and $cP \cap c\mathbf{T} \subseteq cP \cap c\mathbf{T}^*$ ($\varnothing_2$-area empty) and at least once it should be a proper subset relation (at

[5] Note that we do not speak of truth (and falsity)content for this might be misleading because the truth is now changing.

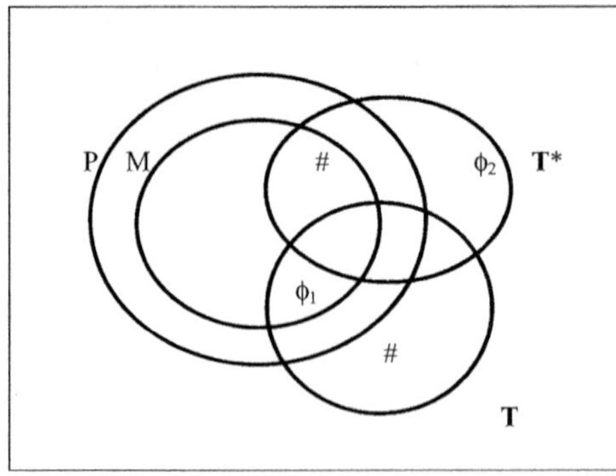

Fig. 8.2 The truth **T*** about domain D* is closer to theory <M, P > than the truth **T** about domain D

least one #-area non-empty). Note that the #-areas are differently located than those in the case of theory revision, since **T*** is now moving from **T** such that it comes more between M and P, whereas <M*, P*> was moving from <M, P> such that **T** became more in between <M*, P*> than in between <M, P>.[6] Now it is not difficult to argue that this definition amounts to: <M, P> has more correct models and more true consequences relative to **T*** than relative to **T**. Formally, $M \cap \mathbf{T}$ is a proper subset of $M \cap \mathbf{T}^*$ and the set of supersets of $P \cup \mathbf{T}$ is a proper subset of that of $P \cup \mathbf{T}^*$, respectively.

8.3.2 *More Successful Relative to One Domain than to Another*

Empirical progress by domain revision becomes a bit more complicated because we have to compare two data sets, relative to **T** (<R, S>) and to **T*** (<R*, S*>), assuming that we can distinguish domain specific realized possibilities and induced laws.

<M, P > is at least as successful relative to <R*, S*> as to <R, S> if and only if $M \cap R \subseteq M \cap R^*$ and $S^* \cup P \subseteq S \cup P$. The first condition now amounts to the claim that M does not loose realized examples by going from R to R*; and the second condition amounts to the claim that P does not loose induced examples and hence no explained induced laws by going from S to S*.

Assuming again correct data, that is, $R \subseteq \mathbf{T} \subseteq S$ and $R^* \subseteq \mathbf{T}^* \subseteq S^*$, it is now easy to derive the Success Theorem for Domain Revision: "<M, P> is closer to **T*** than to **T**" entails "M is at least as successful relative to <R*, S*> as to <R, S>".[7]

[6] Note also that the definition reduces to that in (Kuipers 2006, p. 39) for the extreme special case that M = P, in which case it can be restated in terms of symmetric differences (and M): $M \Delta \mathbf{T}^*$ should be a proper subset of $M \Delta \mathbf{T}$.

[7] Starting from one set of realized possibilities R and one strongest established empirical law S, that is, when they are not yet specified relative to the two domains, we may also define these specifications, replacing **T** by **T1** and **T*** by **T2**, for i = 1, 2 as follows: $Ri =_{def} R \cap \mathbf{Ti}$ and $Si =_{def} S \cup \mathbf{Ti}$, which even guarantees the condition $Ri \subseteq \mathbf{Ti} \subseteq Si$.

The present Success Theorem is again such that persistently being-more-successful relative to one domain than to another is functional or instrumental for truth approximation. Let <M, P > be more successful relative to <R*, S*> than to <R, S>. This suggests the (Empirical) Progress Hypothesis, according to which it is hypothesized that this will remain the case, whatever experiments we design and perform. Testing of this hypothesis may result in the inductive conclusion, at least for the time being, that this in fact is the case: <M, P> persistently remains more successful relative to the resulting evidence relative to **T*** than to **T**. This certainly is a kind of empirical progress: empirical progress by domain revision. Moreover, in the indicated situation it is very hard to imagine that **T*** is, despite appearances, not closer to <M, P> than **T**, and hence it is tempting to abductively conclude that it is closer to, for then the empirical progress becomes 'a matter of course'. Again, this certainly is a kind of truth approximation: truth approximation by domain revision.

We have restricted ourselves to the basic theory of truth approximation by domain revision, containing several idealizations. Similar concretizations as in the case of theory revision can be made.

8.3.3 A Synthesis of Revisions

There is a plausible formal synthesis of basic truth approximation by theory and domain revision, being a generalization of both.

The basic definitions of truth approximation by theory revision and domain revision, respectively, amount to comparisons at both sides of the set of successes and failures relative to the corresponding claim. Both can be generalized qualitatively, leading to the same result:

Definition: <M*, P*> and **T*** fit better than <M, P> and **T** if and only if

$M \cap \mathbf{T} \subseteq M^* \cap \mathbf{T}^*$ and $M^* - \mathbf{T}^* \subseteq M - \mathbf{T}$

and

$cP \cap c\mathbf{T} \subseteq cP^* \cap c\mathbf{T}^*$ and $cP^* - c\mathbf{T}^* \subseteq cP - c\mathbf{T}$

It is easy to check that for equal domains (D = D*) and hence **T** = **T*** this definition reduces to the basic definition of "theory <M*, P*> is closer to the truth **T** about domain D than theory <M, P>", and that for M = M* and P=P* it reduces to the basic definition of "the truth **T*** about domain D* is closer (or more similar) to theory <M, P> than the truth **T** about domain D".

Recall that in the special case of theory revision (**T** = **T***) the conditions amount to

$M \Delta \mathbf{T} \subset M^* \Delta \mathbf{T}$ and $cP \Delta c\mathbf{T} \subset cP^* \Delta c\mathbf{T}$ ($\equiv P \Delta \mathbf{T} \subset P^* \Delta \mathbf{T}$)

In the special case of domain revision ($M = M^*$, $P=P^*$) the conditions amount for the M-side to:

$M \cap \mathbf{T} \subseteq M \cap \mathbf{T}^*$ and $M - \mathbf{T}^* \subseteq M - \mathbf{T}$

and on the P-side to:

$cP \cap c\mathbf{T} \subseteq cP \cap c\mathbf{T}^*$ and $cP - c\mathbf{T}^* \subseteq cP - c\mathbf{T}$
or, equivalently,
$P \cup \mathbf{T}^* \subseteq P \cup \mathbf{T}$ and $\mathbf{T}^* - P \subseteq \mathbf{T} - P$

Here, for both sides, the resulting left condition is just equivalent to the one at the right.

In general, special cases of trying to obtain better fit than between $<M, P>$ and **T** are the following. First, for fixed $<M, P>$, by domain *restriction*, and hence $\mathbf{T}^* \subseteq \mathbf{T}$, leading to a stronger inclusion claim and a weaker exclusion claim; or domain *extension*, and hence $\mathbf{T} \subseteq \mathbf{T}^*$, leading to a weaker inclusion claim and a stronger exclusion claim. Second, for a fixed domain and hence fixed **T**, a better fit can be reached by theory *weakening*, $M^* \subseteq M$ and $P \subseteq P^*$ and at least once proper, leading to a weaker inclusion and/or exclusion claim, by theory *strengthening*, $M \subseteq M^*$ and $P^* \subseteq P$ and at least once proper, leading to a stronger inclusion and/or exclusion claim, by theory *generalization*, $M \subseteq M^*$ and $P \subseteq P^*$ and at least once proper, leading to a stronger inclusion and/or weaker exclusion claim, and by theory *specialization*, $M^* \subseteq M$ and $P^* \subseteq P$ and at least once proper, leading to a weaker inclusion claim and/or stronger exclusion claim. Of course, such special forms of theory and domain revision can be combined in successive steps.

8.4 In Search of Applications of 'Truth approximation by Domain Revision'

So far we have shown that it is perfectly possible to explicate the idea of 'Truth approximation by domain revision', and hence of the possibility of 'Theories looking for domains'. Now the question arises, as a kind of meta- and self-application of the 'Theories looking for domains' theme, whether there are historical cases exemplifying this possibility.

Apart from a plausible type of cases the 'predicted' phenomenon may seem rare in the empirical sciences. Of course, there are well-known examples of theories of which the formal core, i.e., $<M, P>$, finds a totally new area of applications. It may be that the new type of applications behave in relevant respects analogously to the originally intended ones, in which case we have a kind of domain extension, say, domain extension by analogy. An example is the Law of Coulomb about attraction (and repulsion) between static electrically charged particles which is formally analogous to Newton's law of gravitation.

However, and here I depart from the 'Drugs Looking for Diseases' perspective, in which it doesn't really matter whether the diseases are or are not related, my main interest is in truth approximation by genuine domain revision in the sense of domain manipulation within an area of, in some sense, related empirical phenomena, that is, the vocabulary constituting U should have the same kind of interpretation. Given this assumption, we may distinguish between basic (or set-theoretic) kinds of domain revision and refined kinds.

8.4.1 Domain Restriction and Extension

Of the basic kinds we can distinguish domain restriction, domain extension, and their combinations: domain restriction followed by domain extension or vice versa. Note that speaking of 'truth *approximation* by domain extension (leading to the move) from **T** to **T***', in accordance with our definition, viz., **T** is a proper subset of **T***, such that M−**T*** ⊆ M−**T** and **T***−P ⊆ **T**−P, may sound somewhat strange in the case that <M, P> is already true relative to **T**, i.e., M ⊆ **T** ⊆ P, and hence that it must be true relative to **T*** as well, i.e., M ⊆ **T*** ⊆ P (which amounts to a weaker inclusion but a stronger exclusion claim). For this amounts to the case that we correctly extend the domain such that the theory remains true, without having to correct the original domain. Let us speak in this case of 'truth *accumulation* by domain extension'.[8] Historical cases of this type are well known: successful extensions of the domain with respect to which the theory was already (assumed to be) successfully applied. The successful search for new domains of application of Newton's theory, by Newton and his followers belongs to this type. Examples are, starting from falling objects and planetary curves on to ballistic curves, pendulae and so forth, all still within the vocabulary of particle mechanics.[9] It is interesting to note that from the two-sided perspective these particle mechanics examples can not only be described as cases of basic domain extension, but also as cases of introducing or, more precisely, uncovering new (sets of) exemplars within the fixed domain of all particle mechanical systems. Formally, this amounts to extending M, by fixed P and **T**, and hence to strengthening of the theory.

Other extensions of Newton's theory need also enrichment of the vocabulary, notably, rigid body mechanics, introducing notions like angular momentum and torque forces, and fluid mechanics. Still other extensions bring us from the macro-world to the micro-world, notably the kinetic theory of gases and more encompass-

[8] Note that there is a counterpart in the sense of 'truth accumulation by theory strengthening'.

[9] Of course, starting from an uninteresting (purported) truth, domain extension will be equally uninteresting. In other words, it is perfectly possible to let the search for domain revision be guided by 'interesting (purported) truths', which reflects, apart from the terminology, a typically instrumentalist perspective. The case of domain extension guided by the law of conservation of energy, below, is a case in point. However, in other cases this may well lead to domain restriction, rather than to domain extension.

ing statistical mechanics, with explanatory links to macro gas laws, such as the ideal gas law and the law of Van der Waals,[10] and more generally (phenomenological) thermodynamics.

It is not always unambiguous whether we have a case of pure domain extension or a case of domain extension by analogy. For example, from the point of view of Newton's theory of gravitation the Rutherford-Bohr planetary model of the atom is a kind of domain extension by analogy with the planetary system, since the Coulomb forces behave analogously to the gravitational forces. However, looked from Newton's general theory of motion we have a proper case of pure domain extension within particle mechanics.

Of course, later findings may relativize the correctness of the original and/or later claims. In this case one option is theory revision, as in the case of the discovered deviations from Newton's predictions, leading to Einstein's special and general theory of relativity. The special theory is, relative to Newton's theory, a typical case of 'truth approximation by (idealization followed by a, to be sure, fundamental) concretization of the theory'. In order to define this type of truth approximation by theory revision I have developed in Part III of (Kuipers 2000) a general theory of 'refined truth approximation by theory revision', which was indicated in Sect. 8.2, and adapted for two-sided theories in Chap. 6. As a side remark, in contrast to Andrés Rivadulla's general claim (Rivadulla 2007), for refined truth approximation it is not at all relevant that the theories concerned are incompatible, e.g., those of Newton and Einstein, of Copernicus and Kepler, and of Boyle/Gay-Lussac (ideal gas law) and Van der Waals. Unfortunately, it needs some preparation to explain this possibility of refined truth approximation in detail. The reader is referred to Chap. 6, where the gas example is presented in some detail.

Besides concretization of the theory, when later findings relativize the correctness of the original and/or later claims related to 'truth accumulation by domain extension', another option is to concretize the (specification of the) domain. More generally, it is plausible that also a general form of 'refined truth approximation by domain revision' can be defined with 'truth approximation by (idealization followed by) concretization of the (specification of the) domain' as a special subtype.

8.4.2 *Other Kinds of Domain Revision*

Hence, the remaining question is whether there are historical examples of truth approximation by basic or refined kinds of domain revision, apart from cases of (putative) 'truth accumulation by domain extension'. As far as the basic kind is concerned it may be not surprising if there are none. Also in the case of 'truth approximation by theory revision' basic kinds are either toy examples or examples of 'truth approximation by theory strengthening', that is, the counterpart of 'truth

[10] These links enabled e.g. the study of the domain of validity of the ideal gas law in terms of the 'critical point' and its corresponding type of microstates.

accumulation by domain extension'. However, so far, I also did not find convincing refined cases, but this may well be a matter of insufficient knowledge of the history of science. One reason is that from my primarily structuralist perspective it is tempting to focus on empirical theories of a mathematical nature, while it may be easier to find cases in areas where this is less typically the case, e.g., biology and psychology, because the target phenomenon of domain revision may occur in these areas at least as frequently.

It is easier to find related types of non-empirical cases. In what follows we may or may not assume that (constant) M is empty, but P should be substantial, i.e., a *proper* subset of U. Let us begin with cases of 'truth accumulation by domain extension'. Mathematics provides evident cases. For example, every extension of a set of natural numbers for which Goldbach's conjecture has been proven belongs to it. Similarly, even before the full proof of the positive answer of the four-color problem, every new type of map for which the four or, for that matter, the five color problem was positively solved, provides another illustration. Mathematical economics also provides beautiful illustrations. As Bert Hamminga (1983) has documented, in the development of the theory of international trade new results frequently amounted to an extension of the proven domain of validity of a certain 'interesting theorem', such as the 'factor prize equalization theorem'; for a brief further indication see Chap. 6. Recall that this type of research, though not empirical as such, can be motivated in terms of truth approximation with respect to the actual world, because any extension of the proven domain of validity increases the chance that the theorem also holds in the actual world.

I do not yet know of basic cases of 'truth approximation by domain revision' in mathematical economics other than of the cumulative type. However, (mixed) refined cases have been studied. In particular, 'truth approximation by (idealization followed by) concretization of the domain and the theory', where the theory is an interesting theorem in the sense indicated above. In Cools et al. (1994), see also (Kuipers 2000, Section 11.2), important developments in the theory of the capital structure of firms have been reconstructed along these lines, more specifically, the transition from the theory of Modigliani and Miller to that of Kraus and Litzenberger.

8.4.3 Domain Revision by Definition Revision

I only know of one, beautiful, case study of (presumably[11]) basic truth approximation by domain revision in mathematics, viz., Lakatos' (1976) reconstruction of the history of Euler's theorem about polyhedra. As is well-known, Lakatos shows convincingly, using a lively class room setting, with historical notes, that the history of Euler's claim that the polyhedra satisfy $V - E + F = 2$ (V the number of vertices, E that of edges, and F that of faces) is a history of domain revision. More specifically,

[11] To be honest, I did not (yet) check whether there is an episode that satisfies the basic definition perfectly, without being merely a case of domain extension, but this seems very likely.

it is domain revision by all kinds of revision of the definition of a polyhedron. Hence, the case also illustrates that basic truth approximation by domain revision may well take the form of 'truth approximation by definition revision'.

David Atkinson provided me with a clear case in physics of 'truth approximation by domain extension' in the form of definition revision, more precisely, in the form of successive definition extensions: (putative) truth accumulation of the law of conservation of energy by extending the definition of 'energy'. With his kind permission, I quote his description of the case in full.

> In the elastic collision of balls, there is a conserved scalar quantity, the kinetic energy: if we know the mass of each ball, and measure its speed at a given time, we can calculate from these numbers the total kinetic energy of the system, and this does not change in time, at least if the effects of inelasticity and friction are negligible. If the balls are not free, however, but are attached to one another by an array of springs, the total kinetic energy of the balls does not remain constant in time, i.e., the law of conservation of energy breaks down. Or does it? By extension of the definition of energy to include the potential energy that is stored in an extended or a compressed spring – which can be calculated if one knows the (Hooke) spring constant –, one finds that the law of conservation of energy holds, on condition that one defines the energy as the sum of the total kinetic energy of all the balls, and the sum of the potential energies stored in all the springs. For the motion of cannon balls in the gravitational field of the earth, one finds that the same law is valid, where now the potential energy is stored in the gravitational field, and is proportional to the height of the ball at any instant.
>
> A proviso in the above is that friction should be negligible. If it is not – or if one makes measurements with sufficiently great accuracy, or waits long enough – one finds that energy, as defined above, does not seem after all to be conserved, but appears to diminish slowly with time. The law is reinstated by a further extension of the notion of energy to include heat. A precise equivalence was found in the XIX century (Joule) between heat and work, which is a form of energy. If one adds the heat produced by friction, and that produced by any inelasticity in collisions, to the kinetic and the potential energy to produce the total energy, one finds that this is indeed conserved.
>
> The conservation of kinetic energy + potential energy + heat was thought to be exact until, at the end of the XIX century, it was found that, in the decay of radioactive substances, there is a small mismatch in energy before and after decay. The law was reinstated by Einstein, who showed in his theory of special relativity that mass and energy are equivalent. In radioactive decay the sum of the masses of the constituents before and after decay was not the same, and if one extends the notion of energy once more to include mass, m, and adds mc^2 to the kinetic energy + potential energy + heat, one finds that the law of conservation of energy is reinstated.

Since it is a case of successive domain extension by successive definition extension, it is, however, not the type of varied domain revision as in the case of polyhedra. Moreover, it is easy to see that the successive domain extension can be described as cases of domain concretization, keeping the theory fixed, at least the exclusion side (P). At first, only kinetic energy is taken into account, neglecting the role of potential energy, heat and reactive decay. They are successively taken into account.

Gordana Dodig-Crnkovic provided me with an example that may be less straightforwardly cumulative with respect to the (definition of the) domain, viz., the study of magnetism. With her kind permission, I quote her indication of this case as well.

> First, a number of observations (starting in ancient Greece) are made about what magnets are and how they behave. Later on the concept of magnetic field is introduced. Then the fact that the moving electric charge induces magnetic field was established. Earlier concepts about magnetism were transplanted to the new environment. Maxwell predicted the existence of electromagnetic waves and identified light as an electromagnetic phenomenon. One has noticed that even subatomic particles possess magnetic properties, so the concept and the theory have shifted the physical domain again. Evidently, the domains of the first proto-theories of magnetism were very different from the domains of the theory of electromagnetism of today, that is, electromagnetism and quantum electrodynamics.

Only a full reconstruction could show whether this is a case of not purely cumulative domain revision. However, apart from that we have to face the fact that there is no fixed theory, that is, there is no fixed theory of magnetism originating in ancient Greece.

8.5 Concluding Remarks

The question remains whether there are convincing cases in the empirical sciences of basic or refined truth approximation by domain revision, other than pure domain extension, whether or not in the form of definition revision, other than successive definition extension. One reason to expect many explicit or hidden cases of this kind is the frequently confessed instrumentalism; for domain revision in practice evidently is a plausible form of instrumentalist practice. Happily enough for epistemological realists, it is possible to show (Kuipers 2000) in general that the instrumentalist methodology with respect to theory revision is in theory, and hence in practice, functional for truth approximation and, even more surprisingly, more efficient than the falsificationist methodology. It is rather easy to see that a variant of the argumentation for this claim can be used for domain revision. The first claim is argued by realizing that, starting from a falsified theory, it may well be that the shortest route to the truth will go via intermediate theories that are also false. The falsificationist method instead forces one to start over again with a new theory as soon as the next theory again gets falsified. Similarly, given a theory, starting from a 'non-fitting' domain, that is, a domain with respect to which the (in general, two-sided) claim is false, it may well be that the shortest route to a fitting domain, that is, a domain with respect to which the double claim of the theory is true, hence a true theory for that domain, will go via intermediate non-fitting domains. Again, the plausibly adapted version of the falsificationist method to domain revision, if that is acceptable from the falsificationist perspective at all, forces one instead to start over again with a new domain as soon as the next domain again turns out not to fit the given theory. Of course, the same holds even more strongly when we search for a fitting domain for a (fixed) maximal theory (M = P).

8.5.1 Dialectical Interaction

Of course, in practice, neither the domain nor the theory will be fixed stubbornly, while the other is revised. As I wrote at the end of (Kuipers 2000, p. 332), 'science in the making', that is, truth approximation in practice, is a matter of dialectical interaction between vocabulary, domain and theory. The above mentioned example of magnetism illustrates this process nicely.

Let us consider from this point of view the various attitudes that are possible with regard to a (realized) counterexample of (the P-claim of) a theory, assuming that we at least fix the basic vocabulary. Hence, I leave it to the reader to think about the way in which revision of the vocabulary can be brought into the picture (Kuipers 2000, pp. 129–130). So far we have suppressed in this chapter the plausible Lakatosian distinction, among the postulates, between the 'core theory' and auxiliary hypotheses. Hence, in the spirit of the Duhem-Quine thesis, we may first look for auxiliary hypotheses that might be blamed for the counterexample, which may lead to revision of one or more of the auxiliary hypotheses, leaving the 'core theory' intact. However, as soon as we accept the counterexample as a counterexample of the core theory for the domain, we may, secondly, look for an appropriate theory revision, or, thirdly, for an appropriate domain revision.

A similar story can be told about dealing with explanatory problems of a theory. An explanatory problem amounts to an established empirical law that cannot be derived from (the P-claim of) the theory; hence it is too weak for the domain. Again we may revise an auxiliary hypothesis, the core theory, the domain, the vocabulary, or a combination of these.

Of course, in all cases, we want to avoid ad hoc revision; we prefer revisions that lead to successful new predictions. However, we have not paid attention to 'ad hocness', because from the truth approximation analysis it is clear that ad hoc changes are unproblematic if and only if they survive the severe testing of the relevant 'empirical progress hypothesis' (called 'comparative success hypothesis' in Kuipers 2000, pp. 168–169), and this we have assumed throughout.

In sum, truth approximation by revision of theories, domains and vocabularies is perfectly possible, in particular when choices are guided by the instrumentalist methodology. This is in sharp contrast to one of the main claims of Rivadulla (2007, end of Section 2): empirical success is no indicator of truth, verisimilar or truth approximation.

8.5.2 Kuhn-Loss

As a matter of fact, Rivadulla's illustrations of 'historical failures of the Newtonian gravitational model in astrophysics and cosmology' nicely provide an historical case (Rivadulla 2007, Sect. 4.1) I have been looking for. According to the Kelvin-Helmholtz gravitational collapse hypothesis the luminosity of the stars originates from the fact that a star is the result of a collision of smaller bodies, due to mutual gravitation. This collision produces, according to the theory of Joule, an amount of heat equivalent to the kinetic energy lost in the collision. As Eddington noted, this explanation leads to bizarre conclusions about the age of the sun, reason for which 'the source of stellar energy', was no longer considered to belong to the domain of the theory of gravitation: hence, here we have a case of domain restriction. In the literature this type of restriction is known as 'Kuhn-loss'.

It is important to note that the conclusion was not that the theory should be revised. As is well known, and reported by Rivadulla in Sect. 4.2, this was essentially done in a rather drastic way by Einstein's general theory of relativity, in particular, in light of the classical failure of Newton's theory to explain the advance of Mercury's perihelion. This was soon followed by the successful prediction of light deflection by the Sun and, much later, by the successful explanation of the phenomenon of red shift. Of course, these phenomena disappeared out of the domain of Newton's theory, if they were ever considered as such at all, which is only the case for Mercury's behavior. Moreover, one might say that the domain of Newton's theory became essentially restricted to phenomena with small velocities and weak gravitational fields, that is, velocities and fields for which the theory is approximately true. Relative to Newton's theory these restrictions might also be called cases of Kuhn-loss.

For completeness, I add that Rivadulla (Sect. 3.1) also shows that Newton's theory can be successfully used to calculate the (relative) masses of binary stars, illustrating that the orbiting of binary stars, as might be expected, belongs to its domain. Moreover, he reports (Sect. 3.2) the way in which Newton's theory (or model, as he prefers) can explain the stability of stars, that is, that they do not collapse. A clear case of domain extension.

Finally, the illustration of domain revision by restriction, i.e., the exclusion of the problem of the source of stellar energy from the domain of Newton's theory, suggests that several examples of so-called Kuhn-loss might be reconstructed as cases of domain restriction. Hence, we may conclude that, besides truth accumulation by domain extension, truth approximation by domain revision in the form of domain restriction is fact rather than fiction in the empirical sciences.

References

Cools, K., Hamminga, B., & Kuipers, T. (1994). Truth approximation by concretization in capital structure theory. In B. Hamminga & N. B. de Marchi (Eds.), *Idealization VI: Idealization in economics* (pp. 205–228). Amsterdam: Rodopi.

Forbus, K., & Gentner, D. (1997). *Qualitative mental models: Simulations or memories?* In Proceedings of the eleventh international workshop on qualitative reasoning, (pp. 1–8), Cortona, Italy. http://www.qrg.northwestern.edu/papers/Files/QMM QR97.pdf

Giere, R. (1999). Using models to represent reality. In L. Magnani, N. J. Nersessian, & P. Thagard (Eds.), *Model-based reasoning in scientific discovery* (pp. 41–57). New York: Kluwer Academic Publishers.

Hamminga, B. (1983). *Neoclassical theory structure and theory development*. Berlin: Springer.

Kuipers, T. (2000). *From instrumentalism to constructive realism. On some relations between confirmation, empirical progress, and truth approximation, Synthese library 287*. Dordrecht: Kluwer Academic Publishers.

Kuipers, T. (2001). *Structures in science. Heuristic patterns based on cognitive structures. An advanced textbook in neo-classical philosophy of science, Synthese library 301*. Dordrecht: Kluwer Academic Publishers.

Kuipers, T. (2006). Theories looking for domains. Fact or fiction? Reversing structuralist truth approximation. In L. Magnani (Ed.), *Model-based reasoning in science and engineering* (pp. 33–50). London: College Publications.

Kuipers, T. (2007). Empirical and conceptual idealization and concretization. The case of truth approximation. In J. Brzezinski et al. (Eds.), *The courage of doing philosophy. Essays presented to Leszek Nowak* (pp. 75–101). Amsterdam: Rodopi.

Lakatos, I. (1976). *Proofs and refutations: The logic of mathematical discovery*. Cambridge: Cambridge University Press.

Rivadulla, A. (2007). The role of theoretical models in the methodology of physics. In L. Magnani (Ed.), *Model-based reasoning in science and engineering* (pp. 75–85). London: College Publications.

Vos, R. (1991). *Drugs looking for diseases*. Dordrecht: Kluwer Academic Publishers.

Zwart, S. (1998). *Approach to the truth. Verisimilitude and truthlikeness*. PhD thesis. Amsterdam: ILLC, 1998. Revised version: Zwart, Sjoerd. 2001. *Refined verisimilitude*. Synthese library 307. Dordrecht: Kluwer Academic Publishers.

Chapter 9
Beauty, a Road to the Truth?

> *We may find totally opposite things beautiful: a simple mathematical principle as well as a series of unrepeatable complex contingencies. It is a matter of psychology.*
> (Stephen Jay Gould, translated passage from (Kayzer 2000, p. 30)).

Abstract In this chapter I give a naturalistic-cum-formal analysis of the relation between beauty, empirical success, and truth. The analysis is based on the one hand on a hypothetical variant of the so-called 'mere-exposure effect' which has been more or less established in experimental psychology regarding exposure-affect relationships in general and aesthetic appreciation in particular. It was initiated by Robert Zajonc (Attitudinal effects of mere exposure. J Pers Soc Psychol Monogr Suppl 9:1–27, 1968). On the other hand it is based on the formal theory of truthlikeness and truth approximation as presented in my *From instrumentalism to constructive realism* (2000).

The analysis supports the findings of James McAllister in his beautiful *Beauty and revolution in science* (1996), by explaining and justifying them. First, scientists

---

Revised version of: "Beauty, a road to the truth", *Synthese*, ***131***.3, 2002, 291–328. The question mark in the present title got lost in the production process. *Acknowledgements*: This paper is an enlarged and revised version of a translation by Eefke Meijer of my Dutch paper: "Kan schoonheid de weg wijzen naar de waarheid?", *Algemeen Nederlands Tijdschrift voor Wijsbegeerte* (ANTW), **91**.3, 1999, 174–193. James McAllister corrected the English. The paper was read in 1998 at a meeting of the General Dutch Union for Philosophy (ANVW), in Utrecht, in 1999 at the University of Trieste, and in 2000 at a meeting of the Forum for European Culture, in Amsterdam, at the Catholic University of Lublin, at the annual meeting of the British Society for the Philosophy of Science (BSPS) in Sheffield, and at the Gesellschaft für Analytische Philosophie (GAP-4) congress in Bielefeld. I thank David Atkinson, Joop Doorman, Job van Eck, Roberto Festa, Erik Krabbe, Jan Albert van Laar, Anne Ruth Mackor, Jeanne Peijnenburg, Henk de Regt and André de Vries for comments on the Dutch version and Dirk Povel for his information leading to the literature on the mere-exposure effect. I thank David Miller in particular for his extensive and critical comments on the first English translation, which he presented at the BSPS meeting in Sheffield. I also thank Jeffrey Koperski and Elliott Sober for their suggestions on the same occasion. Finally, I like to thank the three anonymous referees who pressed me to clarify a number points.

T. A. F. Kuipers, *Nomic Truth Approximation Revisited*, Synthese Library 399,
https://doi.org/10.1007/978-3-319-98388-2_9

are essentially right in regarding aesthetic criteria useful for empirical progress and even for truth approximation, provided they conceive of them as less hard than empirical criteria. Second, the aesthetic criteria of the time, the 'aesthetic canon', may well be based on 'aesthetic induction' regarding non-empirical features of paradigms of successful theories which scientists have come to appreciate as beautiful. Third, aesthetic criteria can play a crucial, schismatic role in scientific revolutions. Since they may well be wrong, they may, in the hands of aesthetic conservatives, retard empirical progress and hence truth approximation, but this does not happen in the hands of aesthetically flexible, 'revolutionary' scientists.

Keywords Beauty · Road to the truth · Aesthetic features · Truth approximation · Mere-exposure effect · Aesthetic induction · Aesthetic canon · Revolutionary science

9.1 Introduction

Elucidating the concepts of the True, the Good, and the Beautiful, and elucidating their relations, is frequently called the classical task of philosophy. In my opinion, this task should be carried out as much as possible in agreement with scientific insights and findings. Moreover, if possible, it should be demystifying, and, if necessary, even disenchanting. The topic of this chapter is the relation between truth and beauty. Among scientists the intuition is widespread that there is a strong bond between them. Dirac is the classical representative of this intuition. Many other examples of natural scientists expressing this opinion could be given. For example, in an interview series on Dutch television, entitled 'Concerning Beauty and Consolation',[1] the physicist Steven Weinberg and the biologist Stephen Jay Gould emphasized that considerations of beauty play an important role in their appreciation of theories and findings. The best philosopher's response seems to me to be, OK, this intuition of scientists should be taken seriously, but since it is not self-evident, to say the least, an explanation and justification of it should be given. However, to be honest, before reading McAllister (1996), I always thought, like many other philosophers of science, that there must be some kind of misunderstanding among scientists. It not only seemed implausible that aesthetic appreciation has anything to do with empirical success, let alone with truth or truth approximation. But even a relation between non-empirical features frequently mentioned as examples of aesthetic features, such as symmetry and simplicity, on the one hand and truth approximation on the other seemed rather implausible on the basis of my own approach to truth approximation.

[1]The 25 interviews are available on videotape (http://www.vpro.nl/frontend/index.shtml) and have also been published in Dutch under the title *Het boek over de schoonheid en de troost (The book concerning beauty and consolation)* (Kayzer 2000). Unlike many of the other interviews, those with Gould and Weinberg have been transcribed and translated fairly literally.

9.1.1 Some Examples and Preliminary Considerations

So let us be open-minded and try to explain and justify the fact that purported truths are frequently deemed beautiful[2] and that this is why the beauty of an idea, of a potential truth, is considered to be an indication of its soundness. Expressions such as 'the splendor of truth' and 'the simplicity of truth' support this. However, as James McAllister has shown in his very inspiring book *Beauty and Revolution in Science* (1996) (for brief expositions, see also McAllister 1998, 1999), our aesthetic judgements are subject to change. We are not only inclined to find the heliocentric worldview of Copernicus more beautiful, because it is simpler, than the geocentric view of Ptolemy, but we are also inclined to find Kepler's elliptic planetary orbits at least as beautiful as Copernicus's circular orbits. However, ellipses are undoubtedly more complicated than circles, and this is precisely the reason why they were found less beautiful, if not problematically ugly, at the time. Moreover, I would like to add, aesthetic criteria not only change with time within a discipline, but may also differ greatly between disciplines. For example, after expressing in the interview series mentioned above his agreement with Weinberg about the importance of beauty considerations, Gould hastens to stress that his criteria for beauty totally differ from those of Weinberg. Whereas Weinberg mentions 'inevitability' of desired consequences as his dominant aesthetic criterion – as exemplified by Einstein's theory, which, in contrast to Newton's theory, made the inverse square in the law of gravitation inevitable – Gould stresses that, besides diversity, unrepeatable contingencies and irregularities are the sources of his ultimate aesthetic satisfaction.[3] Ironically enough, Weinberg not only refers to Bach's music in general and his two-part Inventionen in particular as showing a similar kind of inevitability, but also mentions the gravedigger scene in Shakespeare's Hamlet as a surprising intermezzo in a logical sequence of events, which, according to Weinberg, illustrates the fact that in the arts there are even higher aesthetic phenomena than in science.[4]

---

[2] As a matter of fact, in science we come across two kinds of beauty. We speak of the beauty of methods of proof and problem solving on the one hand, and of results such as propositions, laws, theories, and truths on the other. The so-called diagonal proof of the non-denumerability of real numbers is an example of a method that strikes almost everyone for its simplicity and inventiveness. In Kuipers (1991) I have collected ten examples of beautiful problem-solving methods for quite mundane problems such as "What is the shortest network of roads between four cities located on the corners of an imaginary square?". However, as suggested, and as in this case, solutions themselves may also be considered as beautiful, like new results in general. Moreover, regarding results themselves, we might distinguish between new results that are found beautiful because they are surprising, perhaps by opening new perspectives, and results that are found beautiful because they fit into the current 'aesthetic canon'. This paper addresses the last type of aesthetic considerations. I thank Michael Stölzner for pressing me to make the latter distinction explicit.

[3] This is not to suggest that standard examples of aesthetic features mentioned by physicists do not play a role in biology. For example, Gould mentions order a number of times and, as Sober (2000) points out, simplicity in the form of parsimony plays a considerable role in taxonomy.

[4] For the gravedigger example and some other ones, illustrating the same point, see also Weinberg (1993, p. 119).

Having observed the variation of aesthetic criteria, McAllister's main claims are as follows. First, scientists normally use aesthetic criteria in addition to empirical criteria for theory evaluation. Second, and most importantly, the aesthetic criteria of the time, the 'aesthetic canon', is based on 'aesthetic induction' regarding non-empirical features of paradigms of empirically successful theories which scientists have come to appreciate as beautiful. Third, aesthetic criteria can play a crucial, schismatic role in scientific revolutions. Since they may well be wrong, they may, in the hands of aesthetic conservatives, retard empirical progress and even truth approximation, but this does not occur in the hands of aesthetically flexible, 'revolutionary' scientists.

9.1.2 Outline of the Paper

In this article I present an analysis of the relation between the truth and the beauty of scientific theories that, in the end, elaborates and supports McAllister's claims. Like McAllister, I will concentrate on non-empirical aesthetic features, that is, features with aesthetic value but without empirical content, although empirical features may also be aesthetically valued. In Sect. 9.2 I will first argue, in the spirit of naturalized epistemology, that the phenomenon of aesthetic induction may be a variant of the so-called 'mere-exposure effect', and then decompose the notion into aesthetic induction proper and a related cognitive (meta-)induction. Together they lead to correlations between non-empirical features which are found beautiful on the one hand and empirically successful theories on the other. Such correlations will be called 'beauty-success correlations'. The corresponding received or 'canonical' aesthetic features are non-empirical features that have acquired (positive) aesthetic value and (empirical success related) inductive support. Moreover, this makes it plausible to explicate the notion of an 'aesthetic feature' as an aesthetically (positively) valued non-empirical (objective) feature.

In the rest of the paper I will argue that the co-production of the two types of induction is functional for empirical progress and even for truth approximation as far as the cognitive meta-induction is reliable. For this purpose I will present, in Sect. 9.3, the basic definition of greater truthlikeness, which, roughly, refers to a theory allowing more 'nomic possibilities' and fewer 'nomic *im*possibilities' than another. Subsequently I will argue that this definition can be rephrased as, again put roughly, a theory having fewer undesirable and more desirable features than another. In Sect. 9.4 the difference between empirical and non-empirical features will be explicated, followed by an elaboration of how claims to truth approximation can be judged in terms of empirical criteria, especially in terms of observed (and therefore nomic) possibilities, representing instantial successes, on the one hand and established desirable observational features, representing explanatory successes, on the other. In Sect. 9.5 the title question will come under discussion when I examine the relative importance of empirical and aesthetic considerations. I will do this by comparing the importance of relevant differences between two theories in the light of the hypothesis that one is closer to the truth than the other: a difference in explanatory success, a difference in instantial success, and a difference in 'aesthetic success', that is, (not) having a received aesthetic feature. One of the outcomes will be that an aesthetic success can be just as good a signpost to the truth as an extra case of explanatory success, albeit in a more modest degree. The relevant difference is that the justified desirability of such an explanatory success can be more reliably established than that of an aesthetic feature, which is why the latter should be approached with more care. In Sect. 9.6 this comparative analysis will enable us to point out the heuristic-methodological use of aesthetic features in seven different problem situations, amongst which is a typical 'revolutionary' one. Some suggestions for further research will be given in Sect. 9.7, followed by several conclusions in Sect. 9.8.

9.1.3 Some Limitations and Specifications

As I said, the formal analysis is based on the basic theory of truthlikeness and truth approximation by empirical progress. Some limitations and specifications should be mentioned beforehand.

First, in view of the fact that the basic theory, and for that matter the refined version too, has, up until now, mainly been restricted to the natural sciences, this study will be too.

Second, 'the truth' will be understood as the strongest, that is to say the most informative, true theory of what is physically (chemically, biologically) possible, within the scope of a domain and a vocabulary that have been chosen beforehand. Thus 'the truth' is always conceived as the product of language and reality; therefore there can be many truths. Because these truths are connected in many ways (one truth can, for instance, be reducible to another), this is not an extremely relativistic position, but one that has a realistic tendency. Neither is it an extreme metaphysical, essentialist position, for the language in question is a human construction, not an ideal language that is supposed to be somehow inextricably connected to the natural world. The name 'constructive realism' covers these two aspects. Beyond this, it is important to note that the definition of 'the truth', though it depends on the idea of a 'true theory', is not circular. I presuppose a definition of a true theory, namely as true for all (bio-)physical possibilities. Here 'true for a physical possibility' is defined in accordance with Tarski's truth definition.

Third, our truth approximation claims regarding aesthetic features are, at least in this paper, restricted to a certain formal type of aesthetic features. More precisely, the 'underlying' objective non-empirical features of aesthetic features, and objective (non-empirical and empirical) features of theories in general, will be restricted to a certain formal type. A feature of a theory is called 'distributed' when it corresponds to an objective property of all (formal representations of) the conceptual possibilities admitted by the theory. Note first that aesthetic features of theories are not supposed to be associated with (the set of) its real world instances, but with the corresponding (set of) conceptual possibilities. However, it may well be that the aesthetic appreciation concerns a non-formal type of representation of certain conceptual possibilities. The famous Feynman diagrams in quantum electrodynamics provide an example. But also in such a case, it is assumed that there is in addition a formal, i.e., logico-mathematical, representation of the conceptual possibilities, such that the aesthetic feature is co-extensional with an objective property of the relevant formal conceptual possibilities. The corresponding distributed feature is called the objective feature underlying the aesthetic feature. Aesthetic features of which the objective nature cannot be explicated in the suggested distributed way fall outside the scope of my truth approximation claims, and demand further investigation. However, it should be stressed that some standard aesthetic features are of the distributed type. Regarding simplicity, for example, it is important to note that the members of the set of conceptual possibilities satisfying a simple formula all share the property to 'fit' in this simple formula. Regarding symmetry, representing a kind of order, we may note that a theory is frequently called symmetric because all its possibilities show a definite symmetry. For example, all admitted orbits may have a certain symmetrical shape. Regarding inevitability and its opposite contingency (see below), it is also plausible to assume that at least certain types of both properties can be localized within conceptual possibilities.

Fourth, and finally, in this article I will not elaborate on the historical examples of truth approximation and aesthetic considerations. For the former I refer to Kuipers (2000, Chap. 10 and 11), for the latter I refer to McAllister's book and articles. But now and then I will refer to features that are widely considered beautiful, like simplicity and order in the form of symmetry.

In summary, the answer to the question of the title will be: yes, beauty can be a road to the truth, namely as far as the truth is beautiful in the specific sense that the truth has distributed features that we have come to experience as beautiful due to (a variant of) the mere-exposure effect. It will become clear that this is a nontrivial answer. Though it may be a disenchanting one, this conclusion has heuristic-methodological use for truth approximation, provided that the aesthetic criteria, in comparison to the empirical criteria, are handled with great caution. The answer can very well be considered as an explication of what McAllister can meaningfully have in mind when he speaks of the relation between beauty, empirical success, and truth in terms of aesthetic induction.

9.2 Aesthetic Induction and Exposure Effects

McAllister (1996, also 1989) introduces the notion of 'aesthetic induction'. It refers to the phenomenon that scientists tend aesthetically more and more to appreciate recurring non-empirical, objective features of successful theories. In this section I want to disentangle this phenomenon. In order to do so, I will first relate it to experimental psychological studies of emotive *e*ffects of repeated exposure to certain stimuli, that is, studies of exposure-*a*ffect relations in general and aesthetic appreciation in particular.

9.2.1 *Exposure Effects and Aesthetic Appreciation*

McAllister (1998) gives an architectural illustration of the phenomenon in question. To color photographs of the Washington Monument and the Eiffel Tower he adds the following telling caption:

> Architectural aesthetics, as embodied in the Washington Monument (left) and the Eiffel Tower (right), can change with the discovery of a material's structural utility – much like the aesthetic appeal of a scientific theory, which seems to grow with every empirical success. The Washington Monument (1884), a white marble obelisk, pays homage to the aesthetic canons of ancient Egypt. In contrast, the Eiffel Tower, which was built a mere five years afterward, is an iconoclastic cast-iron structure that displayed aesthetic properties unprecedented in its day. Cast-iron architecture eventually attained wide use, and a broad aesthetic appeal, as its utility was discovered.

That we grow to find something beautiful after acceptance of it is widely acknowledged where the arts are concerned. Marjoleine de Vos (1999) even goes as far as saying about canonized poetry:

> … but the not-beautiful, if accepted, in point of fact always changes into the beautiful. Time makes everything beautiful. And habituation helps.

In popular music the plug-effect is well known. By repeating the same song or video-clip time and again many people tend to notice by themselves that they like it more and more. This is evidently exploited for commercial purposes by popular broadcasting channels. Similarly, it is well known that the number of people appreciating the atonal music of Schönberg and their degree of appreciation initially was very low, but that both the number and the degree increased considerably with time. See (Mull 1957) for early experimental evidence on this.

All these phenomena seem to be related to the so-called 'mere-(repeated-)exposure effect' which has been studied in experimental psychology (Zajonc 1968, 2001; Temme 1983; Bornstein 1989; Ye 2000). Various experiments, for example with music, paintings, drawings, photos, Chinese characters, and advertisements, illustrate the fact that an increasing number of presentations of the same item tends to increase the aesthetic or, at least, affective appreciation of that item. However, frequently one observes not only first a phase of monotone increasing aesthetic appreciation with the number of confrontations, but also a second phase of decreasing appreciation, together producing the so-called 'inverse U' shape. The most plausible explanation is of course that in the end people get bored. After first having the excitement of recognizing more and more in combination with seeing or hearing more and more, one may become so used to a piece of art that both types of excitement fade away. The Bolero of Ravel might be an illustration. This explanation is known as the two-factor model of stimulus habituation and satiation. Although many comparative studies with varying experimental conditions prompting or retarding the switch have been done (see e.g. Bornstein 1989; Ye and Van Raaij 1997; Ye 2000), two conditions that are particularly interesting for our purposes, viz., successive variation of the same stimulus and introducing some kind of reinforcement, have not been studied, as far as I know. In music one might think of retardation of the switch point by introducing some variation, e.g., in a musical theme, or by presenting different performances of the same piece of music. This would be interesting because theory revision may frequently be considered as variation on a theme. Since 'mere exposure' is by definition unreinforced, experiments with various kinds of reinforcement would deviate from the paradigmatic type of research in this area. Reinforcement would of course be interesting because (increasing) empirical success evidently is a type of reinforcement in theory revision, in the same way as McAllister suggests in the caption quoted above that the utility of cast-iron architecture reinforces its ascribed aesthetic value. However this may be, it is not only established that the mere-exposure effect occurs under certain conditions and within certain limits, but also seems plausible that variation and reinforcement will retard satiation. The latter hypothetical variant of the mere-exposure effect will be called the (postulated) qualified-exposure effect.

Let me illustrate the mere-exposure effect by an informal, but nevertheless risky experiment, just initiated by my own experience. The television series mentioned above comprised 25 interviews with various celebrities in the sciences, the humanities, and the arts. Each instalment of the program started with the same intro of 1.5 min, namely, a tango danced by an elegant and charming older couple accompanied by light effects. Halfway through the series, I presented the main ideas of the present paper in Amsterdam for an audience of 30 people. After showing the intro by video, but before giving my expectations, I ascertained that about half the audience had never or almost never seen the program before. Only 40% of these said that they found the video particularly beautiful. In contrast, of the other half, that is, of those who had seen the program at least a number of times, 80% found the video particularly beautiful. In agreement with my own experience, most of the latter confirmed that their appreciation for the intro had gradually increased. It should be noted that there was no variation in the intro and, more importantly, something like intellectual reinforcement was not evidently the cause of the increasing appreciation of the series. For, like most commentators in the media, one had become more and more critical about the obsessive, non-stimulating way in which Kayzer interviewed his impressive guests.

The mere-exposure effect and the suggested qualified-exposure effect call for general explanations. There must be some kind of (related) psychophysical mechanisms producing them. Using Tinbergen's four well-known questions about biological behavioral patterns, there is reason to investigate the causal structure of the mechanisms, the direct or proximate functions of the effects, the ontological development of the mechanisms, and the phylogenetic, hence evolutionary, development, explaining the distal functions of the effects, that is, the indirect functions that more directly serve reproduction and survival. Regarding the mere-exposure effect itself, the two factors postulated by the two-factor theory, recall, habituation and satiation, certainly are the plausible psychological point of departure for searching for a general causal mechanism behind them. In the last decade it has been argued that there is a link with implicit learning and memory (Bornstein 1994; Seamon et al. 1995). Regarding the evolutionary background, Zajonc (2001) argues quite convincingly that it is plausible to think that the effect is related to safety experiences: if the same stimulus occurs again and again without problematic consequence it becomes a signal of safety, and hence preference will emerge. I will deal with only some of these and related questions, as far as aesthetic appreciation in science is concerned.

9.2.2 Aesthetic Induction in Science

It is evident that McAllister's notion of 'aesthetic induction' can be seen as a reinforcement variant of the mere-exposure effect. More specifically, McAllister claims that aesthetic induction is triggered by empirical success, i.e., in psychological terms, empirical success functions as a kind of reinforcement. If the number of empirically successful theories with a certain non-empirical feature increases the aesthetic appreciation of that feature increases. Similarly, if increasingly many empirically successful revisions of a theory have a constant non-empirical feature, that feature becomes aesthetically more and more appreciated. This phenomenon naturally leads to McAllister's idea of an 'aesthetic canon' of received aesthetic features in a certain phase of a discipline that may be replaced by a different one after a scientific revolution. And, I like to add, the canon may be different for coexisting research programs within one discipline and for different disciplines, depending on the specific non-empirical features of successful theories in the respective programs and fields.

From now on we should sharply distinguish between aesthetic features and merely non-empirical, but objectively, or at least intersubjectively, determinable features of theories. The latter will briefly be called 'non-empirical' features. Aesthetic features – more precisely, the canonical ones – are here conceived as non-empirical features which (certain) scientists (have come to) find beautiful, that is, to which they ascribe aesthetic value. Typical examples of frequently mentioned aesthetic features are symmetry, simplicity, visualizability, and inevitability.[5] From the present point of view it is perfectly possible that other scientists also call their opposites, viz., asymmetry, complexity/diversity, abstractness, contingency, aesthetic, especially in other periods or other disciplines.

The distinction between non-empirical features and their possible ascribed aesthetic value makes it possible to disentangle the kinds of induction that may be involved when a non-empirical feature accompanies empirical success. In fact, one can distinguish at least two kinds of induction, one of an emotive or affective and one of a cognitive nature.[6] The underlying idea of the suggested qualified-exposure effect in a scientific context is that there is a psychophysical mechanism that first

[5] McAllister (1996) deals in particular with symmetry, simplicity and visualizability, and their opposites, see below. Weinberg (1993) deals in Chap. 6, entitled "Beautiful theories", not only with inevitability or rigidity, as mentioned before, but also with simplicity and symmetry.

[6] Our terminology of cognitive, affective and (later) behavioral induction is inspired by Ye's (2000) distinction of cognitive, affective and behavioral priming.

arouses some kind of aesthetic appreciation for a non-empirical feature of a new, empirically successful theory, which, if positive, subsequently increases due to repetition of this feature in other successful theories. If the early response is negative, the idea is of course that this first diminishes and then switches to positive appreciation, which subsequently increases. But, for simplicity, I assume from now on that the start is positive. Note that the arousal as such is analogous to the generation of charge or current in electromagnetic induction, but note also that the latter does not have the increasing character which the former is supposed to have. Be this as it may, it is plausible to speak here of *proper aesthetic*, or more generally, emotive or *affective induction,*[7] by definition of an increasing nature. At the same time, the first co-occurrence of a non-empirical feature and success may arouse the cognitive expectation that it will also accompany the next successful theory and this expectation may be strengthened by subsequent co-occurrences of the feature and empirical success. This will be called *cognitive meta-induction*, where 'meta' refers to the fact that we are dealing with features of theories, rather than *object-induction* dealing with features of objects in the natural world. Whereas affective induction has only one (psychophysical) side, cognitive (meta-)induction has two sides: a psychophysical side and a logical side. The psychophysical side of cognitive induction in general is usually assumed to have some, albeit weak, justification at the logical side: inductive expectations and generalizations have some kind of support, called inductive support, and even on the meta-level of theories, there may be some truth in them of the following type: in order to be empirically successful, theories (in a certain area and perhaps even within a certain research program) may well require a certain non-empirical feature.

The result of both types of induction is that correlations gradually grow between success and beauty, mediated by features, called beauty-feature-success or, simply, beauty-success correlations, in the following sense: recurring non-empirical features of successful theories come to be found beautiful. According to this diagnosis, canonical aesthetic features are non-empirical features that have acquired both (positive) aesthetic value and (empirical success-related) inductive support. The analysis suggests the following related important questions with respect to the correlation-producing mechanism. (1) Is the correlation mechanism functional in evolutionary perspective? The answer is certainly 'yes', if the answer to the second question is so. (2) Is the correlation mechanism functional for empirical progress? The answer to this question is certainly positive as far as (the logical side of) cognitive meta-induction is reliable. The reason is that, in addition to cognitive meta-induction and perhaps other cognitive reasons, the aesthetic appreciation resulting from proper aesthetic induction further strengthens the search for theories having the relevant non-empirical features, which might be called *behavioral induction*. Of course, it is far from evident that, or to what extent, cognitive meta-induction is reliable. This is a general question, to which I need for present purposes only a very cautious positive answer. In the present, naturalistic approach, this positive answer may be based on the standard view of naturalized epistemology: in view of the widespread induc-

[7] Thagard (2005) has criticized this interpretation of (part of) the mere-exposure effect.

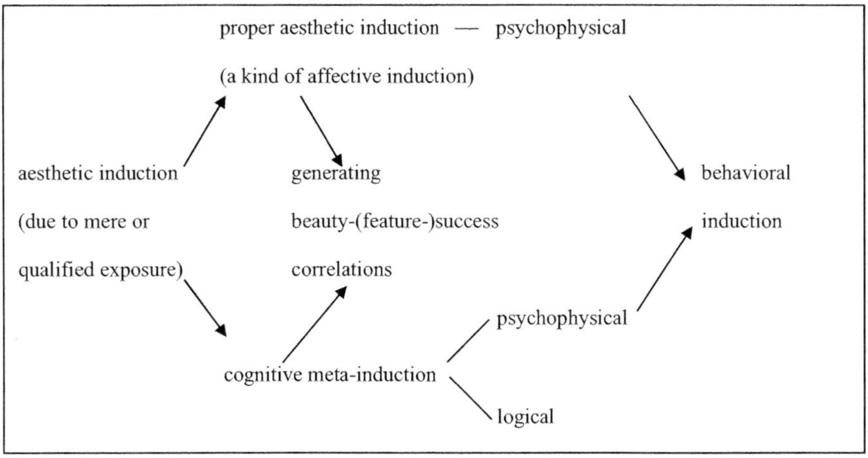

Scheme 9.1 A decomposition of 'aesthetic induction' in science

tive habits of humans and higher animals, these habits apparently have survival value. That is, they must bring us on average onto the right track, or at least more frequently than other types of systematic expectation formation, including random formation. For if they did not, other learning-from-experience strategies would have become dominant.

Scheme 9.1 summarizes the resulting disentanglement of McAllister's notion of aesthetic induction in science.

As far as some standard examples of aesthetic features in physics are concerned, viz. order and simplicity, it is possible to give, in addition, a priori reasons to expect them to be features of the truth, and hence, according to the formal analysis which is to follow, features of empirically successful theories. The first reason concerns the order in (the truth about) reality. Physics presupposes, to a large extent, the so-called principle of the uniformity of nature: not everything is physically possible and what is, does not depend on place and time. Assuming a certain vocabulary we try to grasp by means of our theories what is physically possible and, in view of the principle of uniformity, we assume this to have a certain order. If everything were physically possible, then there would be no, or at least less, order. Although the principle cannot be proven, it would be hard to explain the results of physics, unless the principle was true: scientific truths, i.e., laws of nature, are possible only when there is a certain order or system in reality.

Alongside order, simplicity is a non-empirical feature frequently mentioned by physicists as an aesthetic feature. This is at least partly a consequence of the fact that science is the work of humans. A vocabulary and domain cannot only have been chosen accidentally such that the resulting purported truth is simple, but it also frequently happens that after insight is acquired in the resulting truths of a number of combinations of related vocabularies and domains, that combination is deliberately chosen that leads to the most simple truth. In other words, because there is always some freedom in the choice of domain and vocabulary, we can, to some extent, make sure that the resulting truths are indeed simple.[8] Obviously the reason for expecting simplicity is made possible, or is at least supported, by the reason for expecting order.

Hence, for at least two non-empirical features we have a priori reasons for expecting them. The first, dealing with order, is of a principled nature, whereas the second, dealing with simplicity, presupposing the first, is of a pragmatic kind. It would certainly be possible to elaborate these a priori reasons for expecting these features and further to specify the relation between them. Moreover, for other non-empirical features, other a priori reasons to expect them, in physics or in other natural sciences, can probably be given. However, this is not important for my present purposes. As far as such a priori reasons are valid, they make sure that the processes of cognitive meta-induction and proper aesthetic induction will start and continue to work, but such reasons are of course not necessary for this purpose. It may well be a contingent fact, also supported by the principle of the uniformity of nature, that empirically successful theories, and the corresponding truths, have certain non-empirical features. Both types of induction will operate in these cases just as well.

In so far as cognitive meta-induction is reliable, a positive answer can be given to the resulting explication of the title question of this paper: (3) Is the beauty-success correlation mechanism functional for truth approximation? The conditionally positive answer to this main question requires first an exposition of the basic theory of truth approximation.

9.3 Truthlikeness

The starting point of the idea of truthlikeness[9] is a vocabulary and a domain. A conceptual possibility is a situation or state of affairs that is describable in the vocabulary, and therefore conceivable. Let U be the set of all conceptual possibilities describable in terms of the vocabulary. A theory will be associated with a subset of U. A basic assumption is that the representation of the chosen domain in terms of the vocabulary results in a subset **T** of U containing the nomic, e.g. physical, possibilities, and hence its complement c**T** represents the set of nomic impossibilities. We can equate this usually unknown subset with 'the truth' **T** for reasons that will become clear shortly. However, we will also assume that we can somehow characterize **T** in terms of the vocabulary, indicated by (non-bold) T. The aim of theory formation is the actual characterization of **T**.

---

[8]A telling example of this is the standard model of elementary particles, combining the electroweak theory with quantum chromodynamics. It is not uniquely determined from symmetry considerations alone, but it is further restricted in form by the requirement that certain infinities cancel in the calculation of physical quantities (the 'renormalizability condition' of non-abelian gauge theories applied by – 1999 Nobel Prize winners – Gerard 't Hooft and Martinus Veltman (see e.g. Weinberg 1993, pp. 95–96, 117).

[9]For the most recent and complete version of the basic (and refined) theory on truthlikeness and truth approximation, see the foregoing chapters.

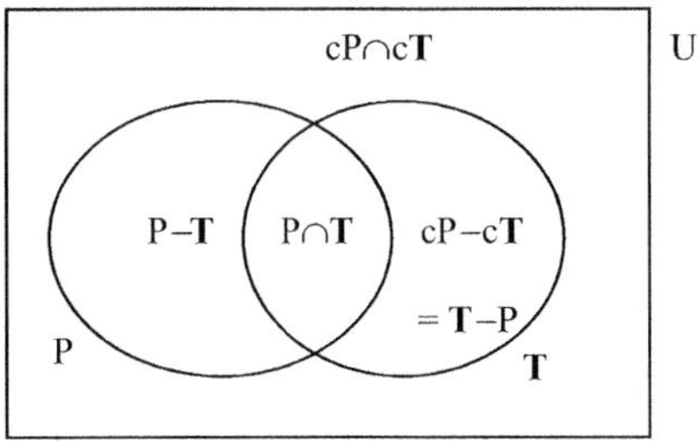

U: set of conceptual possibilities

T: set of nomic possibilities

P: set of admitted possibilities

P∩**T**: nomic possibilities admitted by P

P–**T**: nomic impossibilities admitted by P

cP∩c**T**: nomic impossibilities excluded by P

cP–c**T**=**T**–P: nomic possibilities excluded by P

Fig. 9.1 Four categories of possibilities

In this chapter we will restrict our attention to 'exclusion theories', for there does not seem to be a plausible way to transform the analysis to 'inclusion theories' in a meaningful way. Hence, we return to the notation of Chap. 2: a theory P consists of a subset P of U, with the exclusion claim "**T** ⊆ P", or, equivalently "cP ⊆c**T**". If this claim is true, P does not exclude nomic possibilities, only nomic impossibilities. If **T** ⊆ P* ⊆ P, P* excludes more nomic impossibilities than P and so the claim "**T** ⊆ P*", that goes with it, is stronger than "**T** ⊆ P", but nevertheless true. In this sense theory **T** itself is the strongest true theory, and may be called *the truth*. It seems useful to call the elements of P (its) non-excluded or admitted possibilities and those of cP the excluded possibilities (of P). Now it is important to note that the elements of P∩**T** are the nomic possibilities admitted by P, and P-**T** consists of the nomic impossibilities admitted by P. However, for our purposes, the remaining two are more important: cP∩c**T** is the set of nomic impossibilities excluded by P and cP-c**T** = **T**-P is the set of nomic possibilities excluded by P. In Fig. 9.1 all four resulting categories are depicted.

More specifically, we can call the elements of cP∩c**T** the rightly excluded possibilities of P and those of cP−c**T** the wrongly excluded possibilities of P. This brings us directly to the basic definition of (equal or greater) truthlikeness:

Definitions

P* is *at least as close to **T** as* P (or: P* resembles **T** as much as P) iff

(REP)	all rightly excluded possibilities of P are excluded by P*
(WEP)	all wrongly excluded possibilities of P* are also excluded by P

P* is *closer to **T** than* P (or: P* resembles **T** more than P) iff

(REP) & (REP⁺)	P* rightly excludes extra possibilities relative to P
and/or	
(WEP) & (WEP⁺)	P wrongly excludes extra possibilities relative to P*

Figure 9.2 indicates which sets must be empty (clause (REP) and (WEP): horizontal and vertical shading, respectively) and which sets have to be non-empty (clause (REP⁺) and (WEP⁺): area $+_{\mathrm{REP}}$ and/or area $+_{\mathrm{WEP}}$ non-empty, respectively) in the case that P* is closer to the truth than P.[10]

[10] For completeness I give a set-theoretic translation of the clauses:

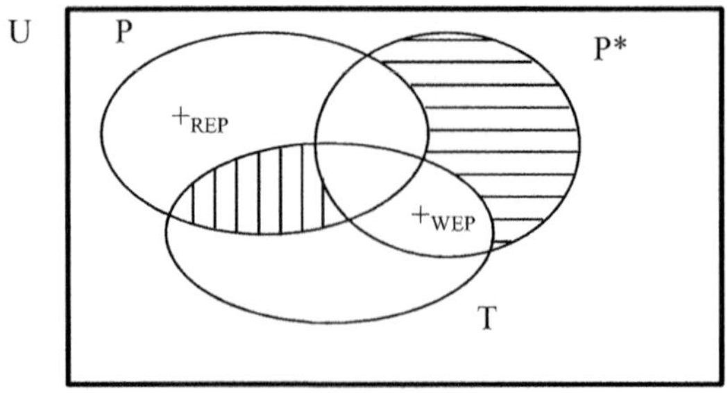

(REP): ≡-area empty

(WEP): |||-area empty

(REP$^+$): $+_{REP}$ -area non-empty

(WEP$^+$): $+_{WEP}$ -area non-empty

Fig. 9.2 $\underline{P^*}$ is closer to the truth T than $\underline{P}$

It is of great importance to our main question that the definitions can be reformulated in terms of desirable and undesirable features of a theory. The starting point consists of properties of possibilities. As announced, a feature of a theory will be understood as a 'distributed' feature, that is, a property of all the possibilities that the theory admits. Recall the example of symmetry. A theory is frequently called symmetric because all its possibilities show a definite symmetry.[11] According to this definition, a feature can be represented as a set of possibilities, namely as the set of all possibilities that have the relevant property. Note that this means that we could say that a feature excludes (exactly) all possibilities that do not have that property. A feature is a feature of a theory when the 'feature-set' contains the set of all possibilities that the theory admits.

By way of preview we can easily see already now that non-empirical features of theories may have something to do with truth approximation by theories, for both, that is, theories and their features are reconstructed as sets of possibilities. At first sight it may in particular have been surprising that aesthetic properties of theories can be cast in terms of sets of possibilities, since they seem to relate to form or structure. However, as has been indicated, as soon as it is a common characteristic of all the possibilities of a theory, which is frequently the case, it can be represented as a feature in our distributed sense.

By now it is obvious how we can formulate explicit definitions of desired, undesired, and remaining features in terms of the exclusion of nomic possibilities and impossibilities: *desired features* are features that admit all nomic possibilities or, equivalently, that exclude only nomic impossibilities; *undesired features* are features that admit all nomic impossibilities or, equivalently, that exclude only nomic

---

(REP)	$cP \cap c\mathbf{T} \subseteq cP^* \cap c\mathbf{T}$	(REP$^+$)	$(cP^* \cap c\mathbf{T}) - cP \neq \emptyset$
(WEP)	$cP^* - c\mathbf{T} \subseteq cP - c\mathbf{T}$	(WEP$^+$)	$(cP - c\mathbf{T}) - cP^* \neq \emptyset$

[11] However, it is easy to check that, for example, 'being true' of a theory $\underline{P}$ in the sense that **T** is a subset of P, is not a distributed feature of P, because it is not a property of P alone, let alone of all its members, but a relation between P and **T**.

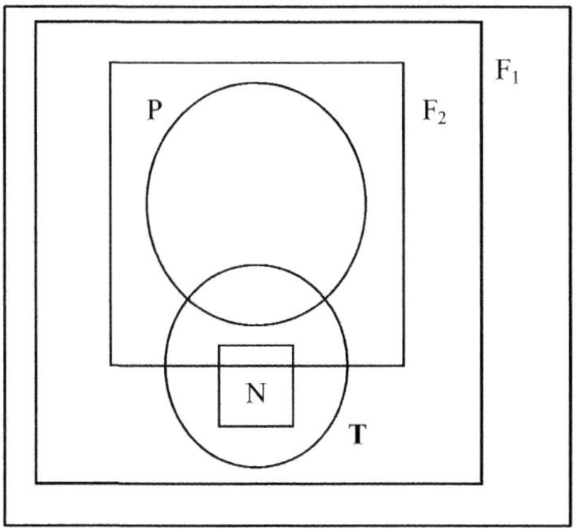

F_1: a desired feature of $\underline{P}$

F_2: a neutral feature of $\underline{P}$

$F_3 = cN$: an undesired feature of $\underline{P}$

Fig. 9.3 Three types of features

possibilities. All remaining features, as far as they can be represented as a subset of U, exclude nomic possibilities and impossibilities; that is, they do not admit either all nomic possibilities or all nomic impossibilities. These are features about which we can be neutral, for which reason I call them *neutral features*. However, they will play no role in the following analysis.[12] The three types of features are depicted in Fig. 9.3.

Note that a desired feature F of $\underline{P}$ is a true feature of $\underline{P}$ in the sense that not only P but also **T** is a subset of F, that is, the claim that may be associated with F, $\mathbf{T} \subseteq F$, is true. However, not only all undesired features of $\underline{P}$ are false in this sense but also all neutral features. The undesired features are false in a strong sense: they not only exclude some nomic possibilities, but only such possibilities.

The following theorems can now easily be proven (see footnote 15):

[12]As indicated in Chap. 2, Popper has given a definition of 'closer to the truth' in terms of more true and fewer false consequences, that was acknowledged later (also by Popper himself) to be unsound. In terms of features Popper's mistake can be rephrased as an exceedingly broad understanding of undesired features: not only the features that are defined undesired in the above, but also the neutral features fall under Popper's definition. For further analysis, see Kuipers (1997, 2000, Section 8.1) and also (Zwart 1998, Chap. 2), who has creatively reused part of Popper's intuitions (Chap. 6).

Equivalence Theses[13,14]

$\underline{P^*}$ is at least as close to **T** as $\underline{P}$ iff

(DF)	all desired features of $\underline{P}$ are also features of $\underline{P^*}$ (equivalent to (REP))
(UF)	all undesired features of $\underline{P^*}$ are also features of $\underline{P}$ (equivalent to (WEP))

$\underline{P^*}$ is closer to **T** than is $\underline{P}$ iff

(DF) & (DF$^+$)	$\underline{P^*}$ has extra desired features (equivalent to (REP$^+$))
and/or	
(UF) & (UF$^+$)	$\underline{P}$ has extra undesired features (equivalent to (WEP$^+$))

In Fig. 9.4, 'at least as close to the truth' is depicted in terms of features. The rectangle FEAT represents the 'universe' of all possibly relevant, distributed, features; hence all being subsets of U. Q($\underline{P}$) and Q($\underline{P^*}$) represent the set of features of $\underline{P}$ and $\underline{P^*}$, Q(**T**) represents the set of desired features (features of **T**) and Q(c**T**) represents the set of undesired features (the features of c**T**). Note that, Q(**T**) and Q(c**T**) have exactly one element in common, namely the tautology, which can be represented by U.[15]

Notice that (REP) and (DF) are not only equivalent but are also analogous in logical form. The same holds for the pairs (WEP) and (UF), (REP$^+$) and (DF$^+$) and (WEP$^+$) and (UF$^+$).[16]

---

[13] Let us give, by way of example, a proof of the claim that (REP) entails (DF). Assume (REP). If F is a desired feature of P, this means: P∪**T** ⊆ F. Now REP, cP∩c**T** ⊆ cP*∩c**T**, is equivalent to c(P∪**T**) ⊆ c(P*∪**T**) and hence to P*∪**T**) ⊆ P∪**T**). Since P∪**T** ⊆ F it now follows that P*∪**T** ⊆ F, that is, F is also a (desired) feature of P*. All proofs are of this elementary nature.

[14] This definition can formally also be read as: all true consequence of $\underline{P}$ are (true) consequences of $\underline{P^*}$ and all 'strongly-false' consequences of $\underline{P^*}$ are (strongly-false) consequences of $\underline{P}$. Here a true consequence C of $\underline{P}$ amounts to P∪**T** being a subset of C and a strongly-false consequence C′ of $\underline{P}$ to P∪c**T** being a subset of C′. For further analysis and comparison with Popper's original, failing definition see (Kuipers 2000, Chap. 8.1).

[15] The set-theoretic interpretation of the universe of features and of the common element and the set-theoretic characterization of the new clauses can simply be given in terms of 'powersets' and 'co-powersets'. The powerset P(P) of P is defined as the set of all subsets of P. The rectangle FEAT representing the 'universe' of all possibly relevant, distributed, features, can now be interpreted as the powerset P(U) of U. Like a kind of mirror notion to that of powerset, the co-powerset Q(P) of P is the set of all subsets of U that include P, also called the supersets of P (within U). Q(P) then represents the features of P, Q(**T**) the desired features and Q(c**T**) the undesired features. As noted, Q(**T**) and Q(c**T**) have exactly one set as common element, namely U, that corresponds with the tautology, and that is of course included in the set of features of every theory. This results in the following formal translations of the four feature clauses:

(DF)	$Q(P) \cap Q(\mathbf{T}) \subseteq Q(P^*) \cap Q(\mathbf{T})$	(DF$^+$)	$(Q(P^*) \cap Q(\mathbf{T})) - Q(P) \neq \emptyset$
(UF)	$Q(P^*) \cap Q(c\mathbf{T}) \subseteq Q(P) \cap Q(c\mathbf{T})$	(UF$^+$)	$(Q(P) \cap Q(c\mathbf{T})) - Q(P^*) \neq \emptyset$

Proving the equivalence theses in terms of sets now becomes an exercise in 'set calculation'.

[16] In the original paper, which assumed maximal theories (M = P), we started from the inclusion claim for P, with the consequence that the possibility clauses and the feature clauses, though having a similar logical form, were crosswise equivalent. This had the nice consequence that the two principal definitions can also be given in a mixed or 'dual' form, in terms of rightly *included* possibilities and desired features.

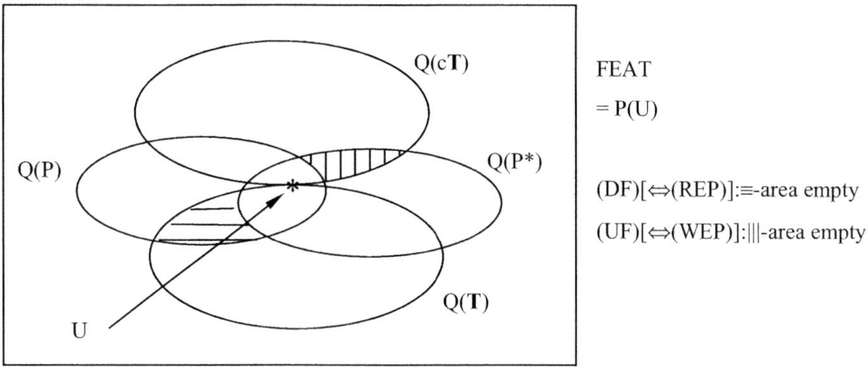

Fig. 9.4 P* is at least as close to T as P, in terms of features

The above provides us with all ingredients of the basic theory of truthlikeness that are needed to answer our main question.

9.4 Truth Approximation with the Aid of Empirical Criteria

In the preceding section I have discussed objective features of theories in general. Of course there are different kinds of features. An obvious classification is the division into empirical and non-empirical features. The non-empirical features are dividable into logical and non-logical or conceptual ones. The best-known and most important logical feature is consistency. Notice that every admitted possibility of a theory points to the consistency of the theory, so that consistency, like many other features, is distributed, that is, transposed from admitted possibilities to the level of theories. As I have argued in Sect. 9.2, non-empirical, more specifically, conceptual features such as symmetry, 'circularity' or satisfying a conservation principle, may or may not be part of the current aesthetic canon. We will see that this is of no great importance, provided and as long as we expect the truth to have this particular feature. Of course, empirical features can, at the same time, also have acquired aesthetic value. A certain observational implication can have both empirical importance and aesthetic value, for instance when it is of a symmetrical nature or when it can be derived, and hence explained or predicted in a very nice way. But in cases such as this we will take the line that the empirical feature is primary and the aesthetic value secondary. Here I will focus on non-empirical aesthetic features, such that their importance and role can be examined in the purest form. But first we will study empirical criteria for theory evaluation.

There are two main categories of empirical criteria of a theory. On the one hand there is the question whether or not the theory implies a certain established observational law or regularity, which, if so, can be explained or predicted by the theory. The implication of an observational law can thus be conceived as an established desired observational feature of the theory. Observational laws are of course established by 'object induction' on properties recurring in repeated experiments. Instead of speaking of implication or explanation and/or prediction of the theory, in what follows I will simply speak of explanation of such laws. Besides the 'explanation criterion' there is the 'instantial criterion', viz., the exclusion or admission of an observed possibility, that is, the result of a particular experiment being or not being a counterexample of the theory. So an observed possibility can be regarded as an established nomic observational possibility, preferably not to be excluded.

Assuming that empirical criteria are primary, relative to their aesthetic value, they are the only relevant criteria as long as only observational and no theoretical terms are involved, for in that case there are no non-empirical (distributed) features that can have aesthetic value. In other words, non-empirical features only exist if a (relative) distinction between observational and theoretical terms can be made. I suppose that, in the present context, this distinction holds. Of course such a distinction between theoretical and observational terms leads to the distinction between an observational level of conceptual possibilities Uo and a theoretical (cum observational) level of conceptual possibilities U_t. By means of this distinction a precise definition of empirical versus non-empirical features can be formulated: features of the first kind exclude possibilities on the observational level, features of the second kind do not. Formally, e.g. for the second kind, a subset F of U_t represents a non-empirical feature iff for all x in Uo there is at least one y in U_t such that y has x as its 'projection' in Uo. In other words, the projection of F equals Uo. This definition may suggest that non-empirical features of theories, in particular aesthetic ones, cannot be indicative of the empirical merits and prospects of a theory. However, by way of meta-induction they can become to be conceived as indicative in this respect. In this sense, aesthetic criteria may be seen as indirect empirical criteria, though formally quite different from the two categories of empirical criteria introduced above. From now on, we will speak only of empirical criteria (and features) in the direct sense explained above.

Truth approximation by means of empirical criteria can now be defined and founded on the basis of the following, easy to prove,

Combined Projection & Success Theorem[17]
If $\underline{P}^*$ is closer to **T** than $\underline{P}$ then $\underline{P}^*$ is at least as successful as $\underline{P}$, in the sense that:

(DF-Success:) Explanatory clause
All established observational laws explained by $\underline{P}$ are also explained by $\underline{P}^*$ (or: all established desired observational features of $\underline{P}$ are also features of $\underline{P}^*$)

(WEP-Success:) Instantial clause
All observed counterexamples of $\underline{P}^*$ are also counterexamples of $\underline{P}$, "unless $\underline{P}$ is lucky" (or: all established wrongly excluded possibilities of $\underline{P}^*$ are excluded by $\underline{P}$, "unless $\underline{P}$ is lucky")

---

[17] The projection theorem (see e.g. Sect. 2.3) states that if $\underline{P}^*$ is at least as close to the truth **T** as $\underline{P}$ then the observational theory corresponding to $\underline{P}^*$, $\underline{P}_o^*$, is at least as close to the observational truth T_o as the observational theory corresponding to $\underline{P}$, $\underline{P}_o$, provided $\underline{P}$ has no lucky misses. The success theorem (see e.g. Sect. 2.2), applied to the observational theories, states now unconditionally that if $\underline{P}_o^*$ is at least as close to the observational truth T_o as $\underline{P}_o$ then $\underline{P}_o^*$ is at least as successful as $\underline{P}_o$, amounting to the two clauses in terms of the observational theories.

The subclause "unless $\underline{P}$ is lucky" will be clarified later. The underlying assumption for the proof of this theorem is the correctness of the empirical data, that is to say, the observed possibilities and the observational laws that are (through an inductive leap) based on them are correct.

By means of this theorem the following argument can be defended. Assume that theory $\underline{P}^*$ at time t is more successful than $\underline{P}$ in the sense suggested above: not only are the two clauses (DF- and WEP-Success) fulfilled, but also $\underline{P}^*$ explains at least one extra observational law and/or $\underline{P}$ has at least one extra observed counterexample (in other words: $\underline{P}$ has an extra established wrongly excluded possibility). This evokes the *empirical progress hypothesis* that $\underline{P}^*$ will be lastingly more successful than $\underline{P}$. This hypothesis is a neat empirical hypothesis of a comparative nature that can be tested by deriving and testing new test implications. As soon as this hypothesis has, in the eyes of some scientists, been sufficiently tested, they can tentatively draw the conclusion that $\underline{P}^*$ will remain more successful than $\underline{P}$, i.e., that empirical progress has been achieved. It can be proven[18] that this is equivalent to concluding that the observational theory that follows from $\underline{P}^*$ is closer to the observational truth $\mathbf{T}_o$ (the strongest true theory that can be formulated with the observational vocabulary, thus as a subset of Uo) than $\underline{P}$. But this conclusion is in its turn a good argument for the *truth approximation hypothesis* (TAH) on the theoretical level: $\underline{P}^*$ is closer to the (theoretical) truth $\mathbf{T} = \mathbf{T}_t$ than $\underline{P}$. In other words, empirical progress is functional for truth approximation. For this, three specific reasons can be given,[19] albeit under a proviso: (1) TAH explains, according to the theorem, why $\underline{P}^*$ is at least as successful as $\underline{P}$, (2) the reversed hypothesis, $\underline{P}$ is closer to the truth than $\underline{P}^*$ (RTAH), is excluded by the theorem, and (3) the denial of TAH (which is much weaker than RTAH) calls for a specific explanation of the difference in success.

9.5 The Importance of Empirical and Aesthetic Considerations

Following the above reconstruction and justification of empirical considerations for the choice between theories, the role of non-empirical considerations can now be analyzed. Although in principle all possible non-empirical considerations are involved, the analysis to be presented seems applicable especially to non-empirical aesthetic features because their formal role can be fully treated. Recall that non-empirical aesthetic features are features without empirical content that have acquired (meta-)inductive support and (hence, due to aesthetic induction) aesthetic value. It is easy to see how aesthetic value of empirical features could come into play.

Before indicating the role of aesthetic features, we must try to determine their objective importance, including their relative importance in comparison with empirical considerations. Whereas for aesthetic criteria as yet only the desired features play a part (later on I will briefly return to undesired aesthetic features),

[18] See the Forward Theorem in Kuipers (2000, p. 164) or in Chap. 11 of this book.

[19] See the end of Sect. 2.2. for a more elaborate presentation of these reasons.

empirical criteria are concerned with established desired features, the implication of an observational law, that is, explanatory success, on the one hand, and established nomic possibilities, of which it is of course undesired to be excluded, on the other hand.

Our point of departure is the assumption that a distinction has been drawn between a theoretical and an observational level and the expectation, based on inductive grounds, that all successful theories, and hence the (relevant) truth will have a certain aesthetic feature. This feature is therefore 'held to be desired'. I also presuppose that some observed possibilities and observational laws have already been established.

Firstly I will define three kinds of similarities and differences between theories P and P*. A difference is to be understood as a difference in favor of P*, unless the contrary is stated explicitly. Later we will see that the three differences, seen as advantages of P*, come with very specific qualifications. An observed regularity that is explained by both theories P and P* or by neither will be called an E(xplanatory)-similarity; by an E-difference I will mean an observed regularity that is explained by P* but not by P. Analogously, an I(nstantial)-similarity is an observed possibility that is admitted by both or neither theory, and an I-difference is an observed possibility that is admitted by P* but excluded by P. Lastly, an A(esthetic)-similarity is a supposedly desired aesthetic feature that both P and P* or neither have, and an A-difference is a desired aesthetic feature that P* has, and P does not.

Especially where the differences are concerned, it is important to distinguish between possible hardness and reliability as signposts to the truth. I will argue as follows. I-differences are harder than E-differences in the sense that the correctness of I-differences is easier to determine than that of E-differences. In the same sense E-differences are harder than A-differences. But, if correct, E- and A-differences are reliable, be it modest, signposts to the truth, whereas I-differences are not.[20]

The correct determination of an I-difference requires only a correct description of an observed possibility that suffices to prove that P does exclude this possibility but P* does not. The correct determination of an E-difference is less simple. It presupposes proofs to the effect that the alleged law is not explainable by means of P but can be explained by P*, and secondly that the alleged law is true (for all nomic possibilities). The assumption that the alleged law is true must first of all be based on a feature that a number of observed possibilities have in common, assuming that no mistakes in the observation and measurement have been made, and secondly on 'object-(level-)induction' or, more precisely, on 'observational induction' or 'inductive generalization' to the (time- and place-independent) general validity of that feature for all nomic possibilities. If the law concerned gets falsified after all, the E-difference is nullified.

The correct determination of an A-difference, lastly, is even more difficult. It presupposes not only proofs that P* has an aesthetic feature that P does not have, but also that this feature is a feature of the truth and therefore a desired one. Generally the latter is less easy to determine than in the case of an observational feature,

[20] The hardness of differences will be defined in relation to the kind of inductive generalization involved, whereas their reliability as signposts to the truth will be interpreted in relation to the transition from the observational level to the theoretical one.

1. *By experiment,* leading to established nomic possibilities not to be excluded, and opening the way for an instantial difference and hence for instantial empirical progress.

2. *By object-induction* on these experiments, leading to established observational laws to be explained, and opening the way for an explanatory difference and hence for explanatory empirical progress.

3. *By meta-induction* on empirically successful (consecutive and/or related) theories (all belonging perhaps to one research program), leading to non-empirical, notably aesthetic, features to be satisfied, and opening the way for a non-empirical, notably an aesthetic, difference and hence for non-empirical, notably aesthetic, progress.

Scheme 9.2 Methods of establishing nomic possibilities and desired features

because the desirability of an aesthetic feature cannot be based on 'object-induction' but at the most on (cognitive) 'meta-induction', as discussed in Sect. 9.2. It is evident that, however tempting or even plausible, this is always very risky. In summary, an E-difference is less hard than an I-difference but harder than an A-difference. Note that the differences in hardness are not so much a matter of the required proofs, but rather related to the empirical justification of being a nomic possibility or of being an observational or an aesthetic feature.

Scheme 9.2 summarizes the three ways in which nomic possibilities and desired features can be established and hence the corresponding similarities and, more importantly, differences and types of progress.

Assuming we have obtained the correct similarities and differences, what then is their relation to claims of truth approximation on a theoretical level, i.e., the truth approximation hypothesis TAH ("$\underline{P}^*$ is closer to **T** than $\underline{P}$") and the reversed hypothesis RTAH ("$\underline{P}$ is closer to **T** than $\underline{P}^*$")? For the sake of convenience I start with E-similarities and E-differences. It is evident that an E-similarity confirms the DF-clause and therefore TAH, but also the 'reversed DF-clause' and therefore RTAH. An E-difference on the other hand verifies the DF$^+$-clause, and thus confirms TAH, but falsifies the *reversed* DF-clause and therefore RTAH. For these reasons I call an E-difference an indicator of presumably being closer to the truth or, simply, a signpost to the truth. The signpost metaphor is a bit misleading in that a signpost on a crossroads, if correctly situated, indicates the one out of four directions that is the right one. In this manner of speaking, an E-difference is more similar to a signpost that tells you: do not take this direction, in this case $\underline{P}$! In a word, it is a modest indicator, due to its Popperian, negative flavor. However, in as far as it is correctly determined, an E-difference is a reliable, albeit modest, signpost to the truth, that is, a good argument in favor of TAH.

Turning to aesthetic features it is important to note first that the explanatory clause of the combined projection and success theorem can be generalized to all established desired features: if $\underline{P}^*$ is closer to **T** than $\underline{P}$ then all established desired (observational and non-observational) features of $\underline{P}$ are also features of $\underline{P}^*$. Hence, since a presumably desired aesthetic feature is just as much a feature of a theory as a desired observational one, the formal roles of E- and A-similarities are fully analogous. The same goes for E- and A-differences. Correct A-differences can therefore be considered, just like correct E-differences, as indicators of presumably being closer to the truth or, simply, as signposts to the truth. To recapitulate, an A-difference is (much) less hard than an E-difference. However, as far as it is correctly determined, an A-difference, too, is a reliable, modest, signpost to the truth, that is, a good argument in favor of TAH, in principle just as reliable as an E-difference.

Let us now turn to I-similarities and -differences. Here the situation is more complex. Because of the 'one-many' character of the relation between the observational and theoretical levels of conceptual possibilities, a theory can have an observational feature on the observational level only if it has the corresponding one on the theoretical level. The non-exclusion of an observational possibility on the theoretical level, however, cannot be based only on the non-exclusion of a suitable nomic theoretical possibility, but can be based also on a suitable theoretical impossibility. If the observed nomic possibility can be based on some admitted theoretical nomic possibility, it may be called a real success of the theory. However, if the observed nomic possibility can only be based on admitted theoretical impossibilities it is some sort of *lucky miss* of that theory. For I-similarities there are all kinds of possibilities for this to occur, but it is not worth the effort of spelling them all out. I-differences, on the other hand, are very interesting. An I-difference can be based on a lucky miss of $\underline{P}^*$, in which case the WEP$^+$($\equiv$ UF$^+$)-clause, on the theoretical-cum-observational level, will not be verified and so TAH will not be confirmed (and therefore the reversed WEP-clause and RTAH are not falsified). Of course, if it is a real success of $\underline{P}^*$, the WEP$^+$-clause is verified, TAH is confirmed, and the reversed WEP-clause and RTAH are falsified. These possibilities are depicted in Fig. 9.5.

In Fig. 9.5, the observed nomic possibility **a**, being a counterexample of the observational theory $\pi\underline{P}$ (the 'projection' of $\underline{P}$ on the observational level) but not of $\pi\underline{P}^*$, is a real success of $\underline{P}^*$ if there is a theoretical version in area 4 and it is a lucky miss if there is not, in which case there must be versions in 1 and 2. If (DF) [($\Leftrightarrow$ (REP)] holds, area 2 is empty, so **a** must be a real success. It is clear that whether an extra success of $\underline{P}^*$ is real or only apparent cannot be ascertained on the basis of the observed nomic possibility. We can say, though, that if TAH (especially (DF/REP)) is true, the example must be real. As said, in that case the WEP$^+$-clause is verified and TAH confirmed. Although this is not a completely circular confirmation, it is a '(DF)-laden' and therefore 'TAH-laden' confirmation. So an I-difference is not reliable as a (modest) signpost, even when it is correctly determined. In summary, we can say that an I-difference is, though harder than an E-difference and a lot harder than an A-difference, even when correctly determined, an unreliable (modest) signpost, for the difference can be based on a lucky miss.[21]

In total, E- and A-differences are, if correctly determined, equally good arguments in favor of TAH, but E-differences are generally much better supportable than A-differences. I-differences, on the other hand, though even better supportable than E-differences, do not constitute, if correctly determined, good arguments in favor of TAH because their confirmation is TAH-laden. I-differences are therefore difficult

[21] This nature of I-differences makes it possible for realists who want to defend TAH in a concrete case, to relativize reversed I-differences: after all, an instantial extra success of $\underline{P}$, that is, a counterexample of $\underline{P}^*$ admitted by $\underline{P}$, could be due to a lucky miss on the part of $\underline{P}$.

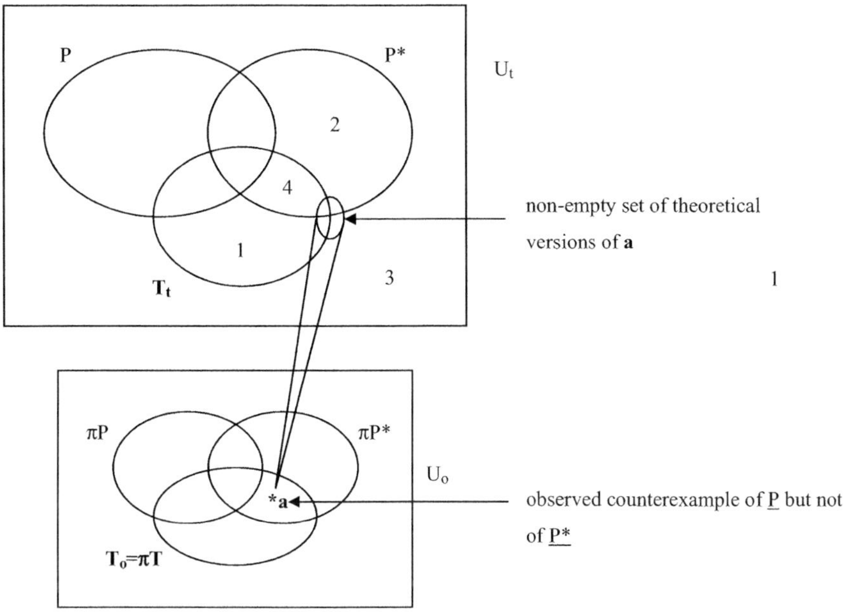

Fig. 9.5 I-difference: observed counterexample of P (but not of P*) as a real success or as a lucky miss. Further explanation in the text

to weigh against E- and A-differences. Or, to return to the signpost metaphor: A-differences are more difficult to support than E-differences, but, if correctly determined, they are equally reliable signposts. I-differences, in contrast, are relatively easy to support, but, even if correctly determined, they are not reliable signposts. The consequence of all this is that, in terms of hardness and reliability, E-differences are more important than A-differences, because of the greater hardness, that is, less risky induction, and equal (positive) reliability, but the relative importance of I-differences in relation to both E- and A-differences cannot be characterized unambiguously, because they combine greater hardness, no induction being involved, with structural unreliability.

Scheme 9.3 summarizes the hardness and reliability of the resulting three types of differences, in the order of decreasing hardness, depending on the needed assumptions, leading to first increasing and then non-decreasing reliability.

9.6 The Role of Aesthetic Features

In order to describe the role of aesthetic features I will outline a number of problem cases in which the question about this role comes up. Except for the last, 'revolutionary' example, the starting-point will be that theory P* has all desired aesthetic features that P has plus one extra, which means that there is an A-difference in favor of P*. Let us denote the feature in question by B(eautiful). Since B will be counted

I(instantial)-difference in favour of P*
- assumptions: correct proofs and one description
- possible lucky miss, confirms TAH only in a TAH-laden way: unreliable, modest signpost
E(xplanatory)-difference in favour of P*
- assumptions: correct proofs, descriptions and one object-induction
- confirms TAH: reliable, modest signpost
A(esthetic)-difference in favour of P*
- assumptions: correct proofs and one meta-induction
- confirms TAH: reliable, modest signpost

Scheme 9.3 A comparison of empirical and aesthetic considerations

among the desired features, the assumption is that the corresponding meta-induction is correct, that is, that **T** has feature B, denoted by B(**T**).

In the first problem case I will assume that P and P* are in a logical sense empirically, or observationally, equivalent, meaning that they have exactly the same observational consequences. The result is that they will always have equal empirical success. The claim that we can now attach to the A-difference is that it gives a specific reason for TAH. In terms of confirmation, the A-difference confirms TAH. The argument is as follows. Because of the presupposed observational equivalence of P and P*, TAH amounts, according to the equivalence theorem, to the claim that P* is at least as good as P when it comes to desired and undesired non-empirical features (first part), and is better than P as far as some desired and/or undesired non-empirical features are concerned (second part). In other words, TAH implies (explains and predicts) both parts of the claim deductively, where the second part concerns an unspecific implication. Because of the presupposition B(**T**), the A-difference verifies the unspecific implication in the form of a desired feature, hence (DF^+), and thus (non-deductively and theoretically[22]) confirms TAH, or, informally put, provides a specific reason for TAH. Still assuming B(**T**), we can add that the A-difference makes the reversed TAH (RTAH, roughly "P is closer to **T** than P*") impossible and

[22] The confirmation of TAH is called 'non-deductive' because the specific A-difference does not follow deductively from TAH, though it does make TAH more plausible. See (Kuipers 2000, Part I) for a coherent analysis of deductive, non-deductive, and inductive confirmation. The confirmation is called 'theoretical' because B is a non-empirical feature.

that a specific explanation of the A-difference is needed for the denial of TAH (which, as has been mentioned earlier, is much weaker than RTAH).

Related to this problem case is the explication of so-called inference to the best explanation (IBE). This rule is explicated in Chap. 11 for empirically non-equivalent theories. The consequences of taking aesthetic features into account for the explication of this rule are elaborated in Chap. 16 in terms of aesthetic progress among empirically equivalent theories.

In the next problem cases I shall speak only informally of (extra) specific reasons for TAH, keeping in mind that a translation in terms of confirmation is always possible, whereas the additional considerations in terms of the reversed TAH and the denial of TAH will be indicated only where specifically relevant.

For the second problem I will assume $\underline{P}$ and $\underline{P}^*$ to be, though not in the logical sense, de facto observationally equivalent, because they are currently (observationally) just as successful, meaning that they share at the moment the same established desired empirical features and observed counterexamples. According to the success theorem, TAH entails that $\underline{P}^*$ must explain at least all observational laws that $\underline{P}$ explains, whereas $\underline{P}$ could have fewer counterexamples only by a lucky accident. Also following from TAH, according to the equivalence theorem, $\underline{P}^*$ must be at least as good as $\underline{P}$ when it comes to non-empirical (desired and undesired) features. Another consequence of TAH is that at some time $\underline{P}^*$ must be better regarding desired features, which could be empirical or non-empirical features. All three cases have to do with deductive implications, the third being of a non-specific nature. The result is that the equivalence of explanatory success can be explained by TAH, that the equivalence of instantial success can also, though with some reservation ($\underline{P}$ has no lucky misses), be explained by TAH and lastly that B can be seen, as in the first problem case, as a specific reason for TAH. The only difference is that now not only non-empirical desired features, like B, can serve as specific reasons for TAH, but also empirical features, had they been determined, which they have not. As we have seen before, the possible E-differences yet to be determined have a greater bearing than the A-difference. Hence, unlike in the first problem case, here the A-difference is a relatively weak specific reason, because stronger, empirical reasons are still feasible.

In the third problem case I take $\underline{P}^*$ to have more explanatory success than $\underline{P}$ and the two theories to be instantially just as successful. The argumentation is to a high degree analogous to the previous case. The only difference is that B is now only an extra (again relatively weak) reason for TAH, on top of the extra explanatory success.

In the fourth situation I presuppose that $\underline{P}^*$ is explanatorily just as successful as $\underline{P}$ but instantially more successful than $\underline{P}$: there is at least one I-difference to $\underline{P}^*$'s advantage, i.e., a counterexample of $\underline{P}$, admitted by $\underline{P}^*$. This situation is more complex than the last. Note first how TAH, as with the desired features, predicts only non-specifically that $\underline{P}^*$ admits extra nomic possibilities that are excluded by $\underline{P}$, but that this does not automatically imply extra instantial success because extra admitted nomic possibilities do not necessarily lead to extra *observable* nomic possibilities. Even so, the extra instantial success of $\underline{P}^*$ is surely compatible with TAH, though this is less revealing than explanatory success. More specifically formulated, only when the absence of lucky misses of $\underline{P}^*$ is assumed, they exclude the reversed TAH. From now on, I will call an I-difference an instantial (empirical) reason and, for later use, an E-difference will be called an explanatory (empirical) reason. In these terms we can now say that the A-difference provides, besides the instantial reasons, an extra (theoretical) reason for TAH. While an E-difference constitutes a stronger reason for TAH than an A-difference, this is not so obvious for an I-difference, as we have seen. If an I-difference were in favor not of P* but of P, it would be debatable which of the two differences, if either, would be more significant. But in the given situation, both differences point in the same direction.

From the last two situations we can easily compose the next, fifth situation: P* is explanatorily as well as instantially more successful than P. In this case, the A-difference supplies a reason for TAH additional to empirical reasons of explanatory and instantial kind.

In the sixth problem case, P* is explanatorily more successful than P but instantially not at least as successful as P. That is, against E-differences in favor of P* and no E-differences in favor of P, there are I-differences in favor of P, possibly alongside I-differences in favor of P*. As we have seen before, the I-differences in favor of P do not exclude TAH because the I-differences could be the result of lucky misses on the part of P. It is not observable whether, on the theoretical level, it is a matter of counterexamples of P, whereas it is observable that (at the observational level and therefore) on the theoretical level it is a matter of counterexamples of P*. This situation can arise because all 'P-versions' of such observed possibilities admitted by P, that is, all theoretical versions of them within P may lie outside **T** and may therefore be theoretical impossibilities. As was mentioned earlier in Note 21, this possibility leads to a general relativization of additional counterexamples (of P*), especially in the case that P* is explanatorily more successful than P. An A-difference that is considered to be desired can make this relativization more plausible. When all theoretical P-versions of all extra observable possibilities admitted by P are 'outside B', they must, on the theoretical level, be counterexamples of P if **T** has B, for then 'outside B' would imply 'outside **T**'. The result is that in this situation B can supply an extra specific reason for TAH, additional to that of the third situation. It would be interesting to check whether in the history of science we can find evidence for this theoretical possibility of using aesthetic arguments, namely as support for the relativization of extra counterexamples of a theory that is explanatorily more successful.

Let us lastly consider a situation, the seventh, which according to McAllister is more or less typical of scientific revolutions. Let a new theory P* be explanatorily and possibly also instantially more successful than P, but without P* having all the desired aesthetic features of P. Supporters of P*, the revolutionary scientists, will of course be the first to cast doubt on the conviction that the extra aesthetic features of P should be retained. Those who cling to P, the conservatives, will remain committed to those features because they cannot imagine that the truth does not possess them. But given the weak nature of the meta-induction they may well be wrong in this, but they may also be right. However this may be, at most one of the two parties can be right. Since they assign different weights to empirical versus non-empirical criteria, they set themselves to different tasks. The challenge to the conservatives is to repair their favorite theory P, preserving its aesthetic features, in order to level the empirical scores; and they may succeed in this. The revolutionaries, who let themselves basically be guided by empirical success, have only to accommodate themselves to the adjustments of the aesthetic canon that appear to be necessary.[23] In this way, the prevailing aesthetic criteria, the aesthetic canon, may well play a schismatic role, in the long run in favor of the revolutionaries if and only if the conservatives do not succeed in their task.

McAllister goes as far as to say that he thinks that we can speak of a scientific revolution only if the theory change in question implies a rupture with the aesthetic canon. The result is, for instance, that he, in line with Kuhn, sees the transition to quantum mechanics as a revolutionary step. However, contrary to Kuhn, he also sees the transition from Copernicus's circular orbits to Kepler's elliptic orbits as a revolutionary step. Moreover, he does not see Einstein's relativity theory as a revolutionary step. Not only is there no change in the aesthetic canon, but the transition is even initiated by Einstein's attempt to apply this canon consistently. In the end this is of course a matter of definition: when does one speak of scientific revolution? In his review of McAllister's book, De Regt (1998) elaborates on this point. The only thing that is of consequence to my account is the relation between beauty and truth in the case of revolutionary transitions that include a change of the aesthetic canon. It seems that in this case at least some aesthetic induction underlying the canon and, more specifically, the corresponding cognitive meta-induction was not correct, for on further investigation it turns out that the (relevant) truth does not satisfy the induced aesthetic expectations.

To be sure, 'the truth' changes in all revolutionary transitions as soon as the (theoretical) vocabulary changes. The analysis in Sects. 9.3 and 9.4 essentially implies that vocabulary change does not prevent the relevance of truth approximation considerations. For example, the most extreme case is that the old theoretical vocabulary is no longer supposed to have referring terms at all, in which case its theoretical truth coincides as a matter of fact with its observational truth. The new theoretical vocabulary introduces in fact a new hypothesis about a theoretical truth that goes beyond, but is relevant for, the common observational truth, corresponding to the common observational vocabulary as far as that is not disputed (See Kuipers 2000, Chap. 9). The foregoing discussion about 'beauty and revolution' shows in addition that aesthetic considerations can play a role in revolutionary disputes which is relevant from the truth approximation perspective.

9.7 Suggestions for Further Research

Various matters await further research. I shall first touch on some possible experiments. Then I will suggest some formal elaborations. Finally, I will suggest some possible connections with other perspectives and problems that seem worth investigating.

---

[23] The temporal opposite of the seventh situation is also interesting. Assume that a new theory $\underline{P^*}$ is very beautiful, but has extra empirical problems compared to $\underline{P}$. Then one can defend the position that the problems do not actually arise from $\underline{P^*}$, but from additional presuppositions that need to be made. Dirac's thought experiment concerning negative results for general relativity theory (McAllister 1996, pp. 93–94) is an example of this.

In Sect. 9.2 I have argued that the aesthetic induction may be a variant of the mere-exposure effect, viz., the qualified-exposure effect. The naturalized approach suggests several experiments with normal and toy pieces of art and with scientific examples to establish the conditions and limitations of the effect. In particular, the presumably strengthening and lengthening role of variation on a theme, comparable to theory revisions, and, similarly, the role of a counterpart to empirical success, as a kind of extra reward, seem worth investigating. For the first, one might think of Bach's successive two-part and three-part Inventionen, for the second of some kind of utility, as in McAllister's case of cast-iron architecture. Moreover, further evidence for the varying character of the aesthetic canon when different phases or different research programs of the same discipline and different disciplines are compared would strengthen the basic ideas around aesthetic induction as such and its diagnosis as a variant of the mere-exposure effect. Finally, my refined claim about aesthetic induction can be falsified: determine a non-empirical, (not necessarily) distributed feature which happens to accompany all increasingly successful theories in a certain area from a certain stage on and which is not generally considered beautiful, and increasingly so, by the relevant scientists.

The formal analysis that followed after Sect. 9.2 was based on conceiving the beauty of theories and of the truth in terms of their non-empirical features, and they were exclusively interpreted in a distributed way, namely as shared properties of all possibilities admitted by them. Although I have argued that this leaves room for many more features than one might think at first sight, it is worth to look for examples of features that cannot be reconstructed in this way. One type concerns so-called constraints. According to the structuralist view of theories, in terms of which the formal analysis also in this chapter is presented, theories are not only seen as sets of conceptual possibilities, but many additional restrictions are involved, notably so-called constraints. For instance, an identity constraint says that a certain function, such as the mass function, must in every application assign the same (rest) mass to the same object. Constraints can also have acquired inductive support and aesthetic value and can be represented as a certain kind of set of sets of conceptual possibilities. It seems perfectly possible to extend the basic truth approximation theory for this kind of restrictions in general and for their aesthetic value in particular. The reason is that constraints have the so-called 'subset property': if a set of possibilities satisfies a constraint, all its subsets do so, including the singleton sets. Hence, constraints are distributed features of sorts.

Of course, there may well be aesthetic features that can neither be represented as a set of conceptual possibilities nor as a set of such sets. They may be of a more holistic kind. For example, a theory may be called symmetric not only because of its symmetric possibilities, but also because it is closed under a certain operation: given a model, applying the operation leads again to a model of the theory. Other examples of holistic, at least non-distributed, features of theories are diversity (of admitted possibilities) and convexity. In general, all formal features that postulate membership claims in response to given members cannot be distributed. For such non-distributed features an alternative formal analysis will have to be found to complete the naturalistic analysis to a full-fledged naturalistic-cum-formal analysis of such features.

One plausible option is immediately suggested by the strong, but nonetheless partial, analogy between truth-oriented research (of a nomological nature) and 'design research', as elaborated in (Kuipers et al. 1992; Kuipers 2001, Chap. 10). Design research concentrates on making or improving certain products or processes. The equivalence theorem not only makes possible a better formulation of this analogy, but also strengthens the analogy because the logic of design research is defined, as a matter of course, in terms of features of the target product and of a realized prototype. As we have seen in Sect. 9.3, if only distributed features are taken into account, truth approximation can be redescribed as an attempt to improve a theory by enlarging its set of desired features and diminishing its set of undesired features, where the (un-) desired features are based on the (unknown) strongest true theory. This suggests defining truthlikeness in this way without the restriction to distributed features, which would of course imply the restricted redescribed one, and hence the original 'possibility definition'. If such a stronger definition is formally possible it might leave room for the above suggested holistic features representing, for example, (a kind of) symmetry, diversity and convexity. This is a matter of further research.

The analysis was further restricted to (distributed) desired aesthetic features, but of course, at first sight, there are also undesired aesthetic features: features that we find ugly and strive to avoid. The question is whether those can be accounted for by the (UF)-clause in accordance with the informal use. This seems plausible as far as negative aesthetic appreciation for a non-empirical feature can be generated in a similar systematic way as its positive counterpart by aesthetic induction. But this is not obvious, as far as such a feature is not the formal counterpart, in one way or another, of a positively appreciated feature, and if this is the case separate treatment seems redundant.

In the truth approximation theory the reference of theoretical terms and the idea of a better approach of 'the referential truth' (Kuipers 2000, Chap. 9) can be expressed as well. Here, too, aesthetic considerations can be involved, for instance in the form of ontological economy considerations (Occam's razor) as a well-known form of simplicity. Within this framework it is relevant to note that there appears to be a connection between the equivalence theorem of Sect. 9.3 and (the strong version of) Leibniz's theory of identity (two objects are identical if (and only if) they have the same features) applied to sets.

There is also a 'refined' version of the theory, see Chap. 6, according to which not all admitted undesired possibilities are equally unfavorable, with the consequence that improvement of theories is possible by replacing nomic impossibilities by less unfavorable ones. Since, according to the equivalence theorem, nomic impossibilities have to do with desired features, this also means that a refinement of aesthetic features and their role seems possible.

The current treatment of aesthetic features is strongly determined by the qualitative nature of the truth approximation theory used. It is difficult to imagine how the quantitative theory of Chap. 5 or that of Niiniluoto (1987) could be employed for aesthetic features, but the present chapter certainly offers a challenge to do so. However this may be, the application of the quantitative theory is much more limited, namely to those areas where a meaningful function of the distance between the relevant conceptual possibilities can be defined. But when this can be done, a quantitative definition results in a linear ordering of theories whereas a qualitative definition leads only to a partial ordering.

According to Thagard (1988, 1992) simplicity can compensate for empirical shortcomings as a result of which on closer inspection it becomes possible that a theory is chosen that is more simple but empirically less successful. In the light of the above analysis, this does not seem justifiable for explanatory success, but because lucky misses are possible, it may be reasonable for instantial success. Nevertheless, it seems difficult to formalize this within a qualitative approach, but it should be possible within a quantitative approach.

So far a number of suggestions for further formal elaboration of this article. But other questions crop up too. Laudan (1977) distinguishes between empirical and conceptual problems. It seems reasonable to take empirical problems of a theory as observed counterexamples and unexplained laws. The suggestion of this article is that conceptual problems can be taken as a lack of certain desired non-empirical features or perhaps as the presence of certain undesired non-empirical features. Especially for the former interpretation it is obvious that the role of conceptual problems will then be formally analogous to the role of aesthetic problems, that is to say the lack of certain (desired) aesthetic features. Analogously the formal role of a conceptual problem that is solved by a new theory can be compared with having a certain desired aesthetic feature.

Besides the partial analogy mentioned above between truth-oriented research and design research, another basic form of scientific research, viz., explication of concepts, turns out also to be partially analogous to truth approximation research. Obvious examples and non-examples of an intuitive concept function as (fixed) desired, and undesired possibilities, respectively, whereas conditions of adequacy can be seen as (fixed) desired features of the intended concept (see Kuipers 2007). Thus a 'general logic of research', with different specifications, appears possible (Kuipers 2001, Chap. 9). In view of the present paper, all these types of research leave room for a modest role of aesthetic considerations.

Lastly the connection between 'simplicity', 'expected predictive success', and 'truth approximation' must be examined for so-called curve fitting. In the literature on simplicity for curve fitting (Sober 1998) the Akaike's theorem plays a central role. It specifies an unbiased estimate of the predictive accuracy of a family of (e.g. linear or parabolic) curves, also called its closeness to the truth, based on past performance. The simplicity or, rather, the complexity of a family is measured in terms of the number of parameters that is characteristic of that family. As far as complexity is concerned, Akaike's theorem states that the estimated predictive accuracy decreases with increasing complexity. Though the breadth of this theorem is heavily criticized (Kieseppä 1997) and the type of simplicity is not a straightforward non-empirical distributed feature of families of curves, it seems logical to wonder what its relation is to truth approximation in the sense of this chapter.

9.8 Conclusion

Even without further investigations we can draw the following conclusions from the presented naturalistic-cum-formal analysis. First of all, as could be expected, scientists' intuitions are correct. Aesthetic considerations can usefully be put into service, even though they are less hard than empirical considerations of an explanatory nature. They can nevertheless function as modest signposts to the truth. Secondly, McAllister could very well be right concerning the origin of aesthetic considerations in general: they may arise from aesthetic induction. This was argued to be partly a matter of cognitive meta-induction and partly a variant of the mere-exposure effect, called affective induction. As for their role in scientific revolutions in particular, aesthetic considerations can be far from the truth and therefore obstruct scientific progress among aesthetically conservative scientists.

The core of the answer to the title question "Beauty, a road to the truth?" hence is: yes, beauty can be a road to the truth as far as the truth is beautiful in the specific sense that it has distributed features that we have come to experience as beautiful. This is a nontrivial answer because it is not immediately obvious that a common feature of a theory and the truth can be considered as a (modest) signpost to the truth. The answer is also a demystifying and perhaps disenchanting one because the analysis does not specifically relate to the (acquired) aesthetic value of the feature, but to its formal aspect, namely which conceptual possibilities it includes and excludes, and to its meta-inductive support, that is, the legitimate inductive reasons there are to assume that the truth has this formal feature. As a consequence, every non-empirical feature can indicate the road to the truth if there are inductive grounds for assuming that the truth probably has this feature. Contrary to Dirac and Einstein (McAllister 1996, Chap. 6) one does not need to assume that the connection between truth and beauty is intrinsic. As for every form of cognitive induction, it is sufficient that the meta-induction involved in aesthetic induction is on average justified at least as often as it is beside the mark.[24]

The outline of the seven problem cases shows that the answer is also useful. Depending on the situation, an aesthetic feature that is assumed to be desired can play a specific role. These features must be handled prudently, though. First, one must keep a close watch over the necessary relativization of the feature in terms of the meta-induction. Second, an aesthetic feature can, for the same reason, hardly compete with explanatory success that points in a different direction. So one should realize that one might retard some revolutionary change by sticking to them. But, as we have seen in the sixth problem case, an aesthetic feature can very well raise an objection against instantial success that points in a different direction.

---

[24] Contrary to the relativization that Derksen (1999) attempts to perform on McAllister's inductive argument for aesthetic features in a narrower sense, that argument appears to have nothing to do with the specific aesthetic value of aesthetic features, but exclusively with the formal and meta-inductive aspects of such features and of other non-empirical features.

Put differently, though there is a clear difference in weight between desired empirical and non-empirical features, the case is different when it comes to the relative importance of differences in instantial success. If the relevant data and the accompanying inductive generalization are correct, the truth has the empirical feature in question. Though only correct data are needed for a difference in instantial success, such a difference is not a reliable signpost to the truth when a distinction between theoretical and observational terms is operative. As for aesthetic and, more generally, non-empirical features, presupposing them remains a meta-inductive gamble waiting for decisive empirical arguments. That is, subsequent empirical success may make them superfluous or they may downplay their weight. Such new empirical arguments may or may not be the result of a well-motivated change in the difference between theoretical and observational terms. The standard form of such a change is the transformation of theoretical terms into observational ones, by accepting certain theories as true. In this way the former theoretical terms become observable or at least measurable.[25]

References

Bornstein, R. (1989). Exposure and affect: Overview and meta-analysis of research, 1968–1987. *Psychological Bulletin, 106*(2), 265–289.

Bornstein, R. (1994). Are subliminal mere exposure effects a form of implicit learning? *Behavioral and Brain Sciences, 17*, 398–399.

de Regt, H. (1998). Explaining the splendour of science. *Studies in History and Philosophy of Science, 29*(1), 155–165.

de Vos, M. (1999). Diep in het geheim gestoken. Over duistere poëzie. In Mooi, special issue, De Gids 3–4: 166–172.

Derksen, T. (1999). Schoonheid als argument. Over James W. McAllister, beauty and revolution in science. *Algemeen Nederlands Tijdschrift voor Wijsbegeerte, 91*(3), 168–173.

Kayzer, W. (2000). *Het boek over de schoonheid en de troost*. Amsterdam: Contact.

Kieseppä, I. (1997). Akaike information criterion, curve-fitting, and the philosophical problem of simplicity. *The British Journal for the Philosophy of Science, 48*(1), 21–48.

Kuipers, T. (1991). Dat vind ik nou mooi. In R. Segers (Ed.), *Visies op cultuur en literatuur. Opstellen naar aanleiding van het werk van J.J.A. Mooij* (pp. 69–75). Amsterdam: Rodopi.

Kuipers, T. (1997). The dual foundation of qualitative truth approximation. *Erkenntnis, 47*(2), 145–179.

Kuipers, T. (2000). *From instrumentalism to constructive realism. On some relations between confirmation, empirical progress, and truth approximation* (Synthese Library 287). Dordrecht: Kluwer Academic Publishers.

Kuipers, T. (2001). *Structures in science. Heuristic patterns based on cognitive structures. An advanced textbook in neo-classical philosophy of science* (Synthese Library 301). Dordrecht: Kluwer Academic Publishers.

Kuipers, T., Vos, R., & Sie, H. (1992). Design research programs and the logic of their development. *Erkenntnis, 37*(1), 37–63.

Laudan, L. (1977). *Progress and its problems*. Berkeley: University of California Press.

McAllister, J. (1996). *Beauty and revolution in science*. Ithaca: Cornell University Press.

[25] For this long-term dynamics in science, see Kuipers (2000, Sections 9.2 and 13.3).

McAllister, J. (1998). Is beauty a sign of truth in scientific theories? *American Scientist, 86*, 174–183.

McAllister, J. (1999). Waarheid en schoonheid in de wetenschap. *Algemeen Nederlands Tijdschrift voor Wijsbegeerte, 93*(1), 153–167.

Mull, H. (1957). The effect of repetition upon the enjoyment of modern music. *The Journal of Psychology, 43*, 155–162.

Niiniluoto, I. (1987). *Truthlikeness* (Synthese Library 185). Dordrecht: Reidel.

Seamon, J., et al. (1995). The mere exposure effect is based on implicit memory. *Journal of Experimental Psychology: Learning, Memory, and Cognition, 21*(3), 711–721.

Sober, E. (1998). Simplicity (in scientific theories). In E. Craig (Ed.), *Routledge encyclopedia of philosophy* (Vol. 8, pp. 780–783). London: Routledge.

Sober, E. (2000). *Philosophy of biology*. Oxford: Oxford University Press.

Temme, J. (1983). Over smaak valt te twisten. Sociaal-psychologische beïnvloedingsprocessen van esthetische waardering (Accounting for tastes. Social psychological influence processes on aesthetic appreciation). With a summary in English. Dissertation. University of Utrecht.

Thagard, P. (1988). *Computational philosophy of science*. Cambridge: MIT Press.

Thagard, P. (1992). *Conceptual revolutions*. Princeton: Princeton University Press.

Thagard, P. (2005). Why is beauty a road to the truth? In R. Festa, A. Aliseda, & J. Peijnenburg (Eds.), *Cognitive structures in scientific inquiry. Essays in debate with Theo Kuipers* (Vol. 2, pp. 365–370). Amsterdam: Rodopi. (Reply by T. Kuipers, 371–374).

Weinberg, S. (1993). *Dreams of a final theory*. London: Vintage.

Ye, G. (2000). *Modeling the unconscious components of marketing communication: familiarity, decision criteria, primary affect, and mere-exposure effect*. Dissertation. University of Tilburg.

Ye, G., & van Raaij, F. (1997). What inhibits the mere-exposure effect: Recollection or familiarity? *Journal of Economic Psychology, 18*, 629–648.

Zajonc, R. (1968). Attitudinal effects of mere exposure. *J Pers Soc Psychol Monogr Suppl, 9*, 1–27.

Zajonc, R. (2001). Mere exposure: A gateway to the subliminal. *Current Directions in Psychological Science, 10*(6), 224–228.

Zwart, S. (1998). *Approach to the truth. Verisimilitude and truthlikeness*. PhD-thesis, Amsterdam: ILLC, 1998. Revised version: Zwart, Sjoerd. 2001. *Refined verisimilitude* (Synthese Library 307). Dordrecht: Kluwer Academic Publishers.

Part IV
Reasoning Around Nomic Truth Approximation

Chapter 10
Abduction Aiming at Empirical Progress or Even Truth Approximation: A Challenge for Computational Modelling

Abstract This chapter primarily deals with the conceptual prospects for generalizing the aim of abduction from the standard one of explaining surprising or anomalous observations to that of empirical progress or even truth approximation. It turns out that the main abduction task then becomes the instrumentalist task of theory revision aiming at an empirically more successful theory, relative to the available data, but not necessarily compatible with them. The rest, that is, genuine empirical progress as well as observational, referential and theoretical truth approximation, is a matter of evaluation and selection, and possibly new generation tasks for further improvement. The chapter concludes with a survey of possible points of departure, in AI and logic, for computational treatment of the instrumentalist task guided by the 'comparative evaluation matrix'.

Keywords Abduction · Abduction task · Instrumentalist task · Empirical progress · Truth approximation · Truthlikeness · Computational modelling · Comparative evaluation matrix

10.1 Introduction

In the literature there is more or less agreement about the general nature of abduction. Following Aliseda (1997, 2006), it amounts to *the search for an acceptable explanatory hypothesis for a surprising or anomalous (individual or general) observational fact*. To get a sharper idea of it, we need to make two distinctions. First, there is the distinction between the *process* of abduction and the *product*. Second, the process can be subdivided into three phases: the *generation* of one or more hypotheses, their *evaluation*, and the *selection* of one of them as the best one. The second subdivision may be merely analytical, for the three aspects of the process may in fact be integrated. Combining the two distinctions, we get the following

Revised version of: "Abduction aiming at empirical progress or even at truth approximation, leading to challenge for computational modelling", *Scientific Discovery and Creativity*, eds. J. Meheus, T. Nickles, special issue of *Foundations of Science*. Vol. **4**.3, 1999, 307–23. *Acknowledgements*: I would like to thank Atocha Aliseda, Johan van Benthem, Alexander van den Bosch, Hidde de Jong and the editors for their helpful and interesting comments.

T. A. F. Kuipers, *Nomic Truth Approximation Revisited*, Synthese Library 399,
https://doi.org/10.1007/978-3-319-98388-2_10

division of the resulting total product in kinds of subproducts: the generated hypotheses, the evaluation reports of them, and the selected hypothesis. There has been a lot of dispute about whether Peirce meant to include the evaluation and selection phases in his notion of abduction, and his preference in this respect seems to have changed. Neglecting this exegetical point further, we propose to speak only of (the process of) abduction when the generation, evaluation, and selection of hypotheses are all included, and to speak of abduction *in the strict sense* when only the generation is meant.

Our investigations of types of scientific processes or products are governed by the working hypothesis that the analysis of a type of product may suggest an analysis of the process, and vice versa. For example, in her discussion of abduction aiming at the explanation of a surprising fact, Aliseda starts from the following general characterization of the (total) product: a hypothesis H, such that B, H ⇒ E, where B represents the background knowledge, H the generated hypothesis, E the surprising fact to be explained, and '⇒' a formal relation which has to be specified further, e.g. as deductive consequence. In line with this product searched for Aliseda's hypothesis generation algorithm systematically seeks, for given B and E, hypotheses that close the deductive(ly specified) gap between B and E. In the following we will use the same type of product/process interaction.

From the general characterization in the beginning of this section, it is clear that there are at least two more specific abduction tasks, where we limit ourselves to the standard, 'deductive' versions.

Task I. In the case of a surprising observation, i.e., the situation that E is not entailed by B, the task is to expand B with some H such that B&H entails E, but H alone does not. This is called *novelty guided abduction*.

Task II. In the case of an anomalous observation, i.e., the situation that E contradicts B, the task is to revise B to B* such that B* entails E. This is called *anomaly guided abduction*. There are at least two different types of revision in this case. One is similar to the so-called belief revision program: first contract B to some weaker B', then expand B' with some H such that B'&H&E are consistent, neither B' nor H entails E, and B'&H (=B*) entails E. The other type of revision, presumably dominating in scientific practice, consists of 'concretization' of B to B*, where B can be seen as an 'idealization' of B*. This second type of revision falls as yet outside the reach of (most of) the algorithms considered. Hence, we focus on the first, contraction-followed-by-expansion, type of revision.

To call not only the first but also the second, anomaly guided, abduction task 'standard', is only defensible as far as approaches are concerned that connect abduction with belief revision or with applications in philosophy of science. As far as I know, logical AI approaches and abductive logic programming, for example, only deal with the first, novelty guided, abduction task.

The main point of this paper is to generalize the standard abduction tasks to theory revision aiming at empirical progress (Task III) and truth approximation (Task IV). Here, 'theory revision' may or may not be conceived by definition as 'theory revision within a certain research program'. Below we will further explicate these tasks in such a way that the following claims can be argued for. The first claim is that Task I and II are special cases of Task III. The second claim is that Task IV amounts to Task III and some (additional) evaluation tasks. Hence, Task III or, more precisely, the subtask here called *instrumentalist abduction* can be viewed as the main abduction task. The paper concludes with a survey of possible points of departure, in AI and logic, for computational treatment of the instrumentalist task guided by the 'comparative evaluation matrix'.

10.2 Analysis of the Empirical Progress Task

Task III, that is, theory revision aiming at empirical progress, requires first of all an explication of the idea that a theory X* is an empirical improvement over theory X. Our point of departure for this explication is the *evaluation report* of a theory in terms of its successes, failures and lacunae, according to the data at a certain time. More specifically, *successes* of a theory are conceived as explanatory successes, that is, observations that can be explained by the theory, which include observations that have been successfully predicted. Here, 'explanation' is used in the liberal sense of deductive or some other type of entailment of the relevant statement by the theory, without assuming that the theory is true, or at least not known to be false. On the contrary, from the present point of view, the truth or falsity of a theory is not so important. What matters is the success domination (see below) when compared with other theories. This is directly related to the truth approximation point of view: a false theory may well be very close to the truth, and if it is closer to the truth than another one, it will be at least as successful. The *failures* of a theory amount to so-called anomalous observations or counterexamples, the descriptions of which contradict the theory. Finally, *lacunae* pertain to surprising observations, that is, observations that cannot be explained by the theory, let alone predicted, but are at the same time not in conflict with the theory.

Below we will define the idea that theory X* is more successful than X at time t in terms of their respective evaluation reports. Presupposing such a definition, Task III can be divided into two subtasks:

Task III: Search for a revision X* of X such that

1. X* is more successful than X, relative to the available data, and
2. X* remains more successful than X, relative to all future data

Whereas Task III.1 is really a subtask of Task III, Task III.2, *the* (further) *evaluation task*, essentially is a strengthening of the former.

Task III.1 may be called *the instrumentalist task* since 'more successful' is, of course, defined in observation terms (relative to the background knowledge) and X* may well be 'born refuted', that is, X* may have failures. If so, they should be (at least as serious) failures of X as well.

Clause 2 amounts to a further condition on X* relative to X. Suppose we have constructed a theory X* that is, relative to the available data, more successful than X. Then the question remains whether it will remain at least as successful as X relative to all future data. This amounts to what might be called the *empirical progress hypothesis* (EP-hypothesis). It is important to realize that the EP-hypothesis is a decent empirical, that is, falsifiable, hypothesis, which is serious even if $\underline{X}^*$ is known to be false. The EP-hypothesis can be tested by suitable experiments and will be accepted, for the time being, when 'sufficiently many and varied experiments' have failed to falsify it. In that case, empirical progress has been made, or at least it seems so.

10.3 Analysis of the Truth Approximation Task

Recall that Task IV dealt with theory revision aiming at truth approximation:

Task IV: Search for a revision $\underline{X}^*$ of $\underline{X}$ such that $\underline{X}^*$ is closer to the truth than $\underline{X}$

The condition will be called the *truth approximation hypothesis* (TA-hypothesis). Note again that the TA-hypothesis is serious even if $\underline{X}^*$ is (known to be) false. To be sure, the condition presupposes a definition of 'closer to the truth' and hence, first of all, a definition of 'the truth'. The latter is defined relative to the (union of) vocabularies of $\underline{X}$ and $\underline{X}^*$ and the relevant domain as the strongest, hence most informative, true hypothesis that can be formulated within the vocabulary about the domain. Hence, 'the truth' is conceived in a constructive, non-essentialistic way. In the natural sciences the domain pertains to all physical or, more generally, nomic possibilities of a certain kind, in which case the truth will not be complete in the logical sense (at least as long as we do not include modal operators in the vocabulary). Assuming this nomic type of 'the truth', it is possible to define 'closer to the truth' (or 'more truthlikeness') in a naïve or basic and a refined way such that everybody will at least agree that the defining condition is a sufficient one. Unfortunately, it would lead too far afield to elaborate these definitions here in a formal sense, not least because one can give three versions of increasing strength. For this purpose, the reader is referred to (Kuipers 2000) and Part I of this book for details. Here it will be sufficient to give verbal formulations of the three (basic) versions, in the context of a further analysis of Task IV.

Task IV (differentiated)[1]: Search for a revision X* of X such that X* is

1. closer to the observational truth than $\underline{X}$ and
2. closer to the referential truth than $\underline{X}$ and
3. closer to the theoretical truth than $\underline{X}$

In Task IV.1 the truth is restricted to the observational vocabulary. In the next section we will argue in some detail that the first clause amounts to the conjunction of

[1] It is plausible to add a fourth subtask, essentially alongside 2: search for a revision $\underline{Y}$ of $\underline{X}$ such that $\underline{Y}$ is closer to 'the structural truth' than $\underline{X}$. However, so far we have not succeeded in making sense of this notion of truth, let alone of 'closer to structural truth'.

clauses 1 and 2 of Task III. For this reason it will be called *the (constructive) empiricist task*, where we freely, but plausibly, presuppose a nomic version of (constructive) empiricism.

Clause 2 of Task IV, *the referential realist task*, adds the requirement that $\underline{X^*}$ should be referentially better than $\underline{X}$ in the sense that the referential claims of X* regarding the joint theoretical vocabulary are at least as adequate as those of $\underline{X}$ and superior in at least one respect. Here 'referential claims' are interpreted as 'contributing to the restriction of models of the theory or not', and they are true when this is so according to 'the referential truth', that is, the truth regarding the reference of the theoretical terms, and false otherwise.

Finally, clause 3, to be called *the constructive realist task*, adds that $\underline{X^*}$ should be closer to the theoretical truth. Of course, the theoretical truth is defined in such a way that it implies the observational as well as the referential truth in the sense that it reduces to these truths when we restrict the attention to the observational vocabulary and to referential claims of the theoretical vocabulary, respectively.

The general idea of the (basic, qualitative) truth approximation theory has two versions, an inclusion and an exclusion one and they can also be combined. According to the inclusion version, a theory claims that its models only concern nomic possibilities. In this 'representational' view, $\underline{M^*}$ is at least as close to the truth as $\underline{M}$ when $\underline{M^*}$ shares all possibilities that are rightly included by $\underline{M}$ and $\underline{M}$ shares all possibilities that are wrongly included by $\underline{M^*}$. Here the truth is the strongest true theory, that is, the theory that includes precisely all nomic possibilities. This idea leaves room for observational and theoretical specifications in the case of a stratified vocabulary.

According to the exclusive approach, a theory claims that its models include all nomic possibilities and hence that its non-models only concern nomic *im*possibilities. In this 'Popperian' view, assuming theories postulates that exclude, $\underline{P^*}$ is at least as close to the truth as $\underline{P}$ when $\underline{P^*}$ excludes all possibilities that are rightly excluded by $\underline{P}$ and $\underline{P}$ excludes all possibilities that are wrongly excluded by $\underline{P^*}$. Here the truth is again the strongest true theory, that is, the theory that excludes precisely all nomic impossibilities. This idea again leaves room for observational and theoretical specifications in the case of a stratified vocabulary.

Formally, they lead to the same symmetric difference condition: the symmetric difference between (the corresponding sets) M* (P*) and the truth is a subset of the symmetric difference between M (P) and the truth. As suggested, the two versions can also be combined in so-called two-sided theories. In part I, such a theory is of the form <M, P>, where M indicates the set of models of the inclusion side and P the set of models of the exclusion side, suggesting the models of the Postulates that exclude certain possibilities.

10.4 Relations Between Empirical Progress and (Kinds of) Truth Approximation

In (Kuipers 2000) we have extensively argued that there are strong relations between the various kinds of truth and successfulness and hence between the corresponding abduction tasks. To begin with, Task IV.3, searching a revision closer to the theoretical truth, is likely to be fulfilled when Task IV.2, searching a revision closer to the referential truth, is fulfilled, but it is not guaranteed. Moreover, Task IV.2 may well be fulfilled when Task IV.1, searching a revision closer to the observational truth, is, but it is far from likely. At most one might say that it is more likely than not. Hence, when Task IV.1 is fulfilled, additional testing of the referential truth approximation hypothesis is required. When and only when evidence from different directions converges (a happy situation called triangulation of evidence) scientists tend to accept referential claims.

We noted already that Task IV.1 essentially amounts to Task III.2. In one direction it is almost obvious in the following sense: 'closer to the truth', when construed in a safe way, not only implies 'at least as successful' but even the possibility of some extra success. But it is also plausible in the other direction. For 'remaining more successful' could be violated when $\underline{X}^*$ is 'not closer to the observational truth' than $\underline{X}$, since the latter would leave room for an extra success of $\underline{X}$ relative to $\underline{X}^*$.[2]

So let us turn to Task III.2, assuming that Task III.1 has delivered a suitable $\underline{X}^*$. The second clause requires further comparative testing of $\underline{X}^*$ against $\underline{X}$. This asks for the generation and testing of distinctive test implications. Relative to Task III.1, Task III.2 amounts to the evaluation phase of a hypothesis generated by the former. Hence, there remains as the abduction task in the strict sense, Task III.1, the instrumentalist task.

In sum, abduction aiming at empirical progress amounts to the generation Task III.1 and the further evaluation task required by Task III.2. Moreover, abduction aiming at truth approximation requires in *addition* only some further evaluation tasks, in particular regarding reference claims. To be sure, selection is not yet included in this picture, but it is fairly obvious how that could be done.

Suppose that the generation task has produced two alternative revisions of $\underline{X}$, viz. $\underline{X}^*$ and $\underline{X}^{**}$, and suppose that both survive the additional evaluation tasks, i.e., both seem to remain more successful then $\underline{X}$. Then the selection question is of course whether one is a suitable revision of the other. To begin with, e.g. is $\underline{X}^{**}$ more successful than $\underline{X}^*$, relative to the available data? If so, the other evaluation questions have to be raised. If not, and if $\underline{X}^*$ also fails to be more successful than $\underline{X}^{**}$, that is, if $\underline{X}^*$ and $\underline{X}^{**}$ have 'divided success', there arises a new generation task, viz. to construct a hypothesis that is more successful than $\underline{X}^*$ and $\underline{X}^{**}$, and hence, than $\underline{X}$. Of course, this selection story can be generalized to finitely many theories generated as improvements of $\underline{X}$. The conclusion remains that the primary abduction task

[2] Note the return to the general symbolization in this section, where X may indicate an inclusion, an exclusion or a two-sided theory.

aiming at empirical progress or even at truth approximation is the instrumentalist task III.1. The rest is evaluation and selection, and possibly new generation tasks for further improvement.

10.5 The Remaining Instrumentalist Task

Recall that Task III.1 amounts to:

> search for a revision X* of X (possibly within one research program) such that X* is more successful than X at time t

For the crucial definition of 'more successful' we first have to specify the kind of theories we are dealing with. Let us first deal with exclusion theories, where the tests may lead to counterexamples and explanatory successes. Then there are still two related alternatives, an asymmetric and a symmetric one. Both definitions may be called 'dominance' definitions, providing at least sufficient conditions, and hence leaving room for more liberal definitions. Moreover, in both cases there are basic and refined versions, but we restrict the attention to the basic versions. The asymmetric one reads:

P* is (at t) *more successful than* P iff (at t)

- P shares P*'s *individual* counterexamples
- P* shares P's *general* explanatory successes
- at least once, not vice versa

This definition seems in accordance with scientific practice and is in close harmony with a similar, intuitively appealing, asymmetric definition of 'closer to the truth'. However, there are also two symmetric definitions, an individual and a general one, that are at least as much in accordance with scientific practice, but less straightforwardly related to the intuitively appealing definition of 'closer to the truth'. The symmetric definitions are in terms of individual or general explanatory successes, lacunae and failures (individual counterexamples and falsifying general hypotheses,[3] respectively) and are suggested by the comparative evaluation matrix for individual or general data, presented in Table 10.1.

Table 10.1 The (comparative) evaluation matrix

		P		
		Failures	Lacunae	Successes
P*	Failures	0	–	–
	Lacunae	+	0	–
	Successes	+	+	0

[3] For a precise specification of such hypotheses, see Kuipers (2000), Sect. 5.2.

Besides neutral results (0), there are three types of results which are favorable (+) for P* relative to P, and three types of results which are unfavorable (−) for P* relative to P. Now P* is more successful than P when there are, besides neutral results, some favorable results for P* and no unfavorable results for P*.

It is not difficult to check that both symmetric definitions and the asymmetric one are essentially equivalent, that is, equivalent assuming some boundary conditions. Moreover, in particular in the symmetric set-up, it is clear that the different types of favorable results could be given different weights, which may be useful for non-dominance definitions. Another point of refinement is that failures, lacunae and successes may be given different weights even if they belong to the same comparative category. E.g. as Hanne Andersen (2009) notes, anomalies may be graded as 'mere' anomalies, important anomalies or even severe ones.

Finally, it is clear from the symmetric definition that Task I and II are special cases of Task III.1. In the case of Task I, dealing with a surprising event, we have an individual or general fact which is a lacuna for P, but a(n) (explanatory) success of P*. In the case of Task II, dealing with an anomalous event, we have an individual or general fact which is a failure for P and a success of P*. Hence, the two standard abduction tasks are included in the instrumentalist abduction task. In other words, the latter is a generalization of the standard tasks. However, in the two standard cases the revised theory has to be compatible with all the available data, whereas this need no longer be the case in the non-standard cases. This is one of the fundamental differences, if not the most fundamental one, between standard and non-standard cases.

In the case of inclusion theories a symmetric and individual definition of 'more successful', neglecting neutral results, seems the most plausible one for Task 3.1[4]:

M* is (at t) more successful than M iff (at t)

- M shares M*'s realized (individual) examples
- M* shares M's induced (individual) counterexamples
- at least once, not vice versa

The resulting invitation for abduction aiming at empirical progress and even truth approximation reads: design instrumentalist abduction along symmetric or asymmetric lines. At first sight, it may seem to be a generalization of the standard tasks with fewer constraints, since no consistency between the new theory and the data is required; hence the threat of becoming intractable seems real. However, since the required piecemeal improvements need to avoid drawbacks, every success and every lacuna (not being the target) provide an extra constraint.

[4] See Sect. 4.4.2.

10.6 AI-Prospects for Computational Approaches to Instrumentalist Abduction

There are several computational ideas and computer programs dealing with (standard) abduction and theory revision.[5] Here we will give an (incomplete)[6] survey of them. It is only meant as a starter for a research project that first tries to complete the present survey as much as possible and next selects one or two possibilities for computational elaboration in the direction of the above described instrumentalist abduction task. At first sight, all indicated programs seem worthy of further investigation for this purpose, in particular when guided by the (symmetric) comparative evaluation matrix, for in most programs the data to be explained and the anomalous data are of the same, individual or general, nature.

We start by reviewing some of the leading AI-ideas/programs for abduction and theory revision, viz. Thagard's program PI and those presented in (Shrager and Langley 1990). In Sect. 10.7 we will deal with some logic oriented approaches. In the final section (Sect. 10.8) we will say a few words about evaluation and selection. Paul Thagard's program PI (Processes of Induction) (Thagard 1988) is based on a distinction between six forms of induction, four of which are called abduction: individual, rule, existential, and analogical abduction. All of them are basically cases of standard novelty guided abduction. PI is grafted upon models of heuristic search developed in cognitive psychology. Hence, abductive explanatory problems are approached by PI, in a quasi-connectionist way, by activation of the concepts occurring in the problem, followed by the activation of concepts coupled to the first ones by rules ('rule firing'). PI does not deal with anomaly guided abduction, and, more generally, in its present form PI has serious limitations in the sphere of evaluation and selection, since it is restricted to still unfalsified theories (see below). However, it seems certainly worthwhile to try to revise PI in order to overcome these limitations (for some suggestions, see Kuipers 1993).

Let us now turn to AI-programs for abduction and theory revision presented in (Shrager and Langley 1990). The 'Abduction Engine' (AbE) of O'Rorke, Morris and Schulenburg pretends to make revolutionary revision of a theory possible, starting from an observational anomaly between the theory, the knowledge base, and new data. AbE uses Forbus's Qualitative Process language (Forbus 1984) for describing qualitative changes due to processes acting on quantities. After the recognition of an anomaly, the theory, conceived as a hierarchically ordered set of principles, is weakened (contraction) to a basic theory that is compatible with the new data. Next, this basic theory is strengthened (expansion), roughly along similar lines as in Thagard's PI, to a theory that is able to explain the anomalous data. The

[5] For a general survey of computational philosophy of science, see Chap. 11 of Kuipers (2000). It includes also an impression of Paul Thagard's later quasi-connectionist approach to theory selection with the program ECHO (Thagard 1992). For a constructive critical evaluation, see also Vreeswijk (2005).

[6] It should be realized that this paper was written around 1998. So Sections 10.6 and 10.7 are certainly very incomplete. However, these sections may still have some stimulating value.

program is developed on the basis of the transition of the phlogiston theory to the oxygen theory of combustion. Hence, it is no surprise that the program is able to reproduce that episode. It is easy to recognize in this step-wise procedure an overall case of anomaly guided abduction, and a subcase of novelty guided abduction. Hence, it deals with standard cases of instrumentalist abduction.

In the program COAST of Rajamoney, also using Forbus's Qualitative Process language, theory revision is approached step-wise: (1) registration of an anomaly between new data and a theory in the knowledge base, (2) design of revision proposals that transform the anomalous data into explanatory successes, where revision operators act on components, domains and effects, (3) suggestion of (new) experiments, (4) first selection on the basis of the anomaly to be explained, the outcome of the experiments, and the degree in which previous successes are retained, (5) further selection on the basis of simplicity and predictive power (strength). It is easy to recognize in these steps standard abductive steps and consecutive evaluation and selection phases. COAST, and the following two programs, were guided in their development by examples from biochemistry.

The program KEKADA of Kulkarni and Simon is globally similar to COAST, apart from differences indicated below. It registers surprising phenomena by comparing new experimental results, obtained for 'normal' purposes, with expectations on the basis of the knowledge base, and proceeds by dealing with such phenomena. That is, it tries to revise the relevant theory and proposes experiments to test such revisions. For this purpose, five strategies are built into the program, each of which consists of an hypothesis generator, an experiment proposer and an evaluator. For example, the core idea of the first strategy is to try to strengthen the surprising phenomenon by independently manipulating the variables of the system.

Karp's program HYPGENE is conceptually rather different from the previous ones. It considers hypothesis formation and revision explicitly as a matter of design, with constraints (a profile of desired properties) and operators that revise the provisional prototype (the profile of factual properties) in order to let it satisfy the constraints better and better. Here for the first time, non-standard cases are naturally included for the revised prototype may well have failures. As was pointed out in (Kuipers et al. 1992), the symmetric definitions of 'more successful' (and of 'closer to the truth') are completely analogous to the claim that one prototype is an improvement of another, relative to a fixed intended profile. In other words, the (ultimately) intended profile, in case of instrumentalist abduction, is a theory for which all data are successes. Van den Bosch (1999, 2001, and related, 2005) shows a similar strategy can be based on Anderson's ACT-R.

Finally, Darden prepares in the volume of Shrager and Langley the conceptual ground for a program still to be written by considering the solution of an anomaly of a theory as a task for diagnostic reasoning, which has been developed for expert systems, guided by the tracing of a defect in a technical system. She shows with Mendelian examples (see also Darden 1991) that the hierarchical decomposition of all presuppositions of a theory may provide the points of departure for solving the anomaly, and that its solution, as in the famous analysis of some mathematical examples by Lakatos (1976), may or may not lead to fundamental theory revision, that is, a revision staying within a research program or breaking through the barriers of the relevant program.

10.7 Logical Prospects for Computational Approaches to Instrumentalist Abduction

One challenge for logic is the question to what extent the AI-approaches sketched in the previous section can be reproduced in a logically decent way. However, we will not pursue this further here, and will concentrate on already existing purely or primarily logical approaches. As far as we know, there are only a few logical approaches to standard abduction and theory revision, viz. so-called belief revision developed by Peter Gärdenfors and others (Gärdenfors 1988), non-monotonic logic approaches by Konolige (1990, 1996), adaptive logic approaches (Meheus 2005)[7] and the semantic tableau approach introduced by Mayer and Pirri (1993) (see also Pirri 1995), which was redirected in a more efficient way by Atocha Aliseda (1997, 2005, 2006). The precise relation between belief revision and non-monotonic logic approaches on the one hand and instrumentalist abduction aiming at empirical progress and truth approximation on the other has still to be investigated. For example, *prima facie*, there are some important limitations of belief revision, as seen from the instrumentalist abduction task. To begin with, belief revision is rather 'actual world' oriented, instead of oriented toward 'physically, or nomically, possible worlds' or, simply, 'real world' oriented. Moreover, it aims at a consistent product of the available data and the resulting belief set. The latter feature it shares, however, with almost all approaches, including the one by Aliseda.[8]

Let us turn to a brief indication of Aliseda's semantic tableau method of novelty and anomaly guided abduction. As remarked already earlier, her conceptual approach guided our presentation in several respects. Here we only want to indicate briefly the basic idea behind her algorithms, which are as yet restricted to propositional languages. Recall that a semantic tableau is a way to find out whether a set of premises logically entails a conclusion by systematically trying to construct a counterexample, that is, a propositional model or structure in which the premises are true and the conclusion is false. If all branches generated by the (formula decomposing) tableau rules from the set of premises and the negated conclusion become 'closed', that is, contain a formula and its negation, the argument is valid. Open branches describe counterexamples to the argument. In the case of novelty guided abduction, there are of course open branches in the tableau starting from the theory and the negation of the phenomenon description. Systematically closing these branches

[7] Meheus (2005), presents a more ambitious approach than Aliseda (below): "search the maximally successful hypotheses from the set of background assumptions ant the set of observational statements".

[8] See Chaps. 14 and 15 for basic and refined elaborations of the idea of belief revision aiming at truth approximation.

produces equally many consistent proposals for abductive explanation of the novel phenomenon. In the case of anomaly guided abduction, the theory apparently entails the negation of the anomaly, that is, the corresponding tableau starting from the theory and the phenomenon description is closed. Now the task is to systematically open the branches of this tableau by deleting parts of the theory (contraction), followed by adding new formulas (expansion) that lead to the closing of all branches when combined with the negation of the phenomenon description.

The above description neglects all nuances and versions of Aliseda's approach, but it may nevertheless already be plausible to conjecture that the approach can be generalized to instrumentalist abduction purposes. The main thing is to operationalize in tableau terms the possibility that one theory is 'more incompatible' with the data than another. Then the task is to generate theories that are not more incompatible with the available data than the original one, explain at least as much as the latter, and improve the latter in at least one of these respects. Some other adaptations will be required to make it suitable for scientific purposes. First, there should be left room for general facts as data and, second, the original and resulting theories should be of a general nature. When restricted to propositional languages, simple implicative formulas will do the job in both cases, provided there is left room for three kinds of truth: actually true, 'physically or nomically true' and logically true. One plausible way to do this is using a modal propositional language (Zwart 1998, 2001), but there may well be other ways.

Let me conclude this section with some specific remarks. The first deals with theory structure. In several approaches there is a natural tendency to conceive theories in a hierarchical way (notably O'Rorke c.c. and Darden). Starting from intuitive hierarchies as used by scientists, this is easy to reproduce in syntactic as well as semantic or structuralist axiomatizations of theories. However, when such intuitive hierarchies are not available, the method of 'natural axiomatization' developed by Gemes (1994) may well be helpful in producing hierarchies.

The second remark pertains to the restriction to the 'basic' instrumentalist abduction task, viz. all examples of theory revision presented so far deal with the three possible relations between a

theory and one individual or general fact: entailment, incompatibility, and no logical relation. The 'refined' instrumentalist abduction task asks for refinement of the second and third alternative: one theory may approach or explain a (possibly incompatible) fact better than another. It is likely that programs aiming at quantitative theory generation and revision are more appropriate for these purposes. As a matter of fact, it is plausible to conjecture that the BACON-programs, developed by Simon and others, can be reconstructed as refined revision programs of quantitative (quasi-true, that is, at most approximately true) laws. Although it will be worthwhile to pursue this line further, and also to develop something like refined theory revision in the sense of 'concretization' of quantitative theories, the basic instrumentalist abduction task is at least as important since it deals with qualitative theories. Before refining propositional instrumentalist abduction, it will be necessary to fully develop a basic version. Moreover, the generalization of the suggested basic propositional instrumentalist abduction in tableau terms, to first order logic may be considered as at least as important as the indicated refinement.

10.8 Concluding Remarks

Sections 10.5, 10.6, and 10.7 may be seen as a description of a research project aiming at the computational realization of instrumentalist abduction. We close by indicating some general features of such programs. First, although the terminology of theory revision suggests that there should be a contingent initial theory, this is not so. It is possible to start from the theory representing *ignorance*, that is, a tautology, in which case the task is to improve upon ignorance. Second, one may or may not aim at *incremental* abductive programs. A program is incremental, starting from a certain theory and a certain set of data, when the revised version dealing with these data, is the point of departure when the set of data further increases. The non-incremental alternative is that the second revision process starts again from scratch, that is, from the original theory. Third, when the generation process generates alternative theories, of which one is more successful than all the other ones, the selection is clear. Depending on the further aims of the program, the selection may then be followed by further evaluation subprograms dealing with empirical progress or truth approximation. When the generated theories cannot be ordered in the suggested way due to 'divided success' between any pair of theories, the generation task has not yet been completed. Finally, when the result contains a couple of theories that are at least as successful as all other theories, one may or may not want to have further selection. However, as long as the data remain the same, one will have to use non-empirical features, such as *simplicity* and *strength*. The strength criterion is straightforwardly functional for empirical progress and hence for truth approximation, in the sense that new data may well lead to the conclusion that a stronger theory is more successful than a weaker one. However, for the simplicity criterion such straightforward functionality does not seem to exist as long as it does not imply the strength criterion. On the contrary, as McAllister (1996) argues and illustrates simplicity or, more generally, beauty considerations may well retard empirical progress.[9]

References

Aliseda, A. (1997). *Seeking explanations: Abduction in logic, philosophy of science and artificial intelligence*. Dissertation Stanford. Amsterdam: ILLC Dissertations Series (1997-04).

Aliseda, A. (2005). Lacunae, empirical progress, and semantic tableaux. In A. Aliseda, R. Festa, & J. Peijnenburg (Eds.), *Confirmation, empirical progress and truth approximation. Essays in debate with Theo Kuipers* (Vol. 1, pp. 169–189). Amsterdam: Rodopi (Reply by T. Kuipers, 190–192).

Aliseda, A. (2006). *Abductive reasoning. Logical investigations into discovery and explanation*. Synthese Library 330. Dordrecht: Springer (Extended and revised version of Aliseda, 1997).

[9] See Chap. 9 for the way in which aesthetic considerations may be functional for empirical progress and truth approximation.

Andersen, H. (2009). Unexpected discoveries, graded structures, and the difference between acceptance and neglect. In J. Meheus & T. Nickles (Eds.), *Models of discovery and creativity* (pp. 1–27). Dordrecht: Kluwer Academic Publishers.
Darden, L. (1991). *Theory change in science: Strategies from Mendelian genetics*. Oxford: Oxford University Press.
Forbus, K. (1984). Qualitative process theory. *Artificial Intelligence, 24*, 85–186.
Gärdenfors, P. (1988). *Knowledge in flux: Modeling the dynamics of epistemic states*. Cambridge: MIT Press.
Gemes, K. (1994). A new theory of content. *Journal of Philosophical Logic, 23*, 596–620.
Konolige, K. (1990). *A general theory of abduction*. In Automated abduction, Working notes. Spring symposium series of the AAA (pp. 62–66). Stanford: Stanford University.
Konolige, K. (1996). Abductive theories in artificial intelligence. In G. Brewka (Ed.), *Principles of knowledge representations* (pp. 129–152). Stanford: CSLI Publications, Stanford University.
Kuipers, T. (1993). On the architecture of computational theory selection. In R. Casati & G. White (Eds.), *Philosophy and the cognitive sciences* (pp. 271–278). Kirchberg: Austrian Ludwig Wittgenstein Society.
Kuipers, T. (2000). *From instrumentalism to constructive realism. On some relations between confirmation, empirical progress, and truth approximation*. Synthese Library 287. Dordrecht: Kluwer Academic Publishers.
Kuipers, T., Vos, R., & Sie, H. (1992). Design research programs and the logic of their development. *Erkenntnis, 37*(1), 37–63.
Lakatos, I. (1976). *Proofs and refutations: the logic of mathematical discovery*. Cambridge: Cambridge University Press.
Mayer, C., & Pirri, F. (1993). First order abduction via tableau and sequent calculi. *Bulletin of the IGPL, 1*, 99–117.
McAllister, J. (1996). *Beauty and revolution in science*. Ithaca: Cornell University Press.
Meheus, J. (2005). Empirical progress and ampliative abductive logics. In A. Aliseda, R. Festa, & J. Peijnenburg (Eds.), *Confirmation, empirical progress and truth approximation. Essays in debate with Theo Kuipers* (Vol. 1, pp. 193–217). Amsterdam: Rodopi (Reply by T. Kuipers, 218–220).
Pirri, F. (1995). *Abduction*. PhD Dissertation. Paris: Université de Paris VI, Pierre et Marie Curie.
Shrager, J., & Langley, P. (Eds.). (1990). *Computational models of scientific discovery and theory formation*. San Mateo: Kaufmann.
Thagard, P. (1988). *Computational philosophy of science*. Cambridge: MIT Press.
Thagard, P. (1992). *Conceptual revolutions*. Princeton: Princeton University Press.
van den Bosch, A. (1999). Inference to the best manipulation – A case study of qualitative reasoning. *Foundations of Science, 4*(4), 483–495.
van den Bosch, A. (2001). *Rationality in discovery*. Amsterdam: ILLC Publications.
van den Bosch, A.. (2005). Structures in neuropharmacology. In A. Aliseda, R. Festa, & J. Peijnenburg (Eds.), *Cognitive structures in scientific inquiry. Essays in debate with Theo Kuipers* (Vol. 2, pp. 343–359). Amsterdam: Rodopi (Reply by T. Kuipers, 360–364).
Vreeswijk, G. (2005). Direct connectionistic methods for scientific theory formation. In A. Aliseda, R. Festa, & J. Peijnenburg (Eds.), *Cognitive structures in scientific inquiry. Essays in debate with Theo Kuipers* (Vol. 2, pp. 375–403). Amsterdam: Rodopi (Reply by T. Kuipers, 404–406).
Zwart, S. (1998). *Approach to the truth. Verisimilitude and truthlikeness*. PhD-thesis. Amsterdam: ILLC.
Zwart, S. (2001). *Refined verisimilitude*. Synthese Library 307. Dordrecht: Kluwer Academic Publishers.

Chapter 11
Inference to the Best Theory, Rather Than Inference to the Best Explanation. Kinds of Abduction and Induction

Abstract An interesting consequence of the theory of nomic truth approximation, as developed in my *From Instrumentalism to Constructive Realism* (Kuipers T, From instrumentalism to constructive realism. On some relations between confirmation, empirical progress, and truth approximation. Synthese Library 287. Kluwer Academic Publishers, Dordrecht, 2000), and enriched in this book, concerns so-called 'inference to the best explanation' (IBE). It can be argued that this popular rule among scientific realists can better be replaced by, various kinds of, 'inference to the best theory' (IBT). This chapter provides a survey of observational, theoretical, and referential kinds of IBT and discusses when and in what sense they may be seen as cases of 'abduction', that is, arguments for which an 'abductive' formulation and, hence, justification can be given in the minimal sense of Peirce.

Whereas it is difficult to imagine an abductive justification for IBE, especially due to its asymmetric character, it appears that abductive justifications can be given for IBT *as the closest to the relevant truth*, where 'the (relevant) truth' is defined as the strongest true claim that can be made with the relevant vocabulary about the relevant domain.

The relation of IBT with the instrumentalist or evaluative use of the hypothetico-deductive (HD-)method is also indicated. In particular, it is argued that the theoretical kind of IBT may well be seen as the proper realist extension of the instrumentalist

---

Revised version of: "Inference to the best theory, rather than inference to the best explanation. Kinds of abduction and induction". In: F. Stadler (ed.), *Induction and Deduction in the Sciences*, Proceedings of the ESF-workshop *Induction and Deduction in the Sciences*, Vienna, July, 2002, Dordrecht, Kluwer Academic Publishers, pp. 25–51, followed by a commentary of Adam Grobler, pp. 53–56. *Acknowledgements*: I would like to thank in particular Adam Grobler, Ladislav Kvasz, Joke Meheus, Ilkka Niiniluoto, Wolgang Spohn and Igor Douven for their stimulating comments on the draft version.

The idea of 'Inference to the best theory' is also discussed by Ilkka Niiniluoto (2005) and by Igor Douven (2005) in a 'volume in debate'. Niiniluoto's paper mainly deals with the unstratified kind and Douven's paper with the theoretical kind. In both Douven's paper and Chap. 9 (based on Kuipers 2002) arguments are given for including non-empirical, e.g. aesthetic, criteria in the comparative assessment of theories, with consequences for the precise formulation of IBT. These consequences are elaborated in Chap. 16 in terms of aesthetic progress among empirically equivalent theories. See further my replies to Douven and Niiniluoto in the volume in debate referred to and my reply to Grobler and Wisniewski (2005) in the companion volume.

T. A. F. Kuipers, *Nomic Truth Approximation Revisited*, Synthese Library 399,
https://doi.org/10.1007/978-3-319-98388-2_11

rule of success, prescribing to choose the most successful theory, whether already falsified or not.

The chapter also introduces weak and strong forms of observational, theoretical, and referential induction. The weak forms merely conclude that a theory is true in the relevant sense and the strong forms can be seen as extreme special cases of the corresponding kinds of IBT, viz. 'inference to the best theory *as the relevant truth*'. Finally, it is argued that the most benevolent and defensible reading of IBE combines IBT with the corresponding weak kind of induction.

Keywords Kinds of abduction · Inference to the best explanation · IBE · Inference to the best theory · IBT · Hypothetico-deductive method · HD-method · Abductive justification · Kinds of induction

11.1 Introduction

An interesting consequence of the theory of nomic truth approximation, as developed in my *From Instrumentalism to Constructive Realism* (Kuipers 2000), henceforth ICR, and enriched in Part II of this book, concerns so-called 'inference to the best explanation' (IBE). It will be argued that this popular rule among scientific realists can better be replaced by, various kinds of, 'inference to the best theory' (IBT).

11.1.1 Background

I will start this chapter with a survey of the most relevant conclusions of ICR. It will provide the background of the present chapter, which is itself partly drawn from ICR, to be indicated, and partly supplementary, in particular the theme of 'abductive justification'. The survey of ICR that follows is a selective adaptation of ICR-section 13.1.[1] The selected conclusions will be elaborated to some extent in the rest of the chapter.

*The *first* conclusion, from Part I and II of ICR, is that a useful distinction can be made between two kinds of applying the Hypothetico-Deductive (HD-)method, viz., testing of hypotheses, in order to try to establish their truth-value, leading to falsification or confirmation, and evaluation of theories, in the first place, in order to establish their observational merits and failures.

**Third*, the evaluation of theories by the HD-method, studied in Part II, concerns, in the first place, the separate evaluation in terms of general successes and counter examples, but immediately suggests a comparative sequel, leading to the possible establishment of empirical progress. The resulting evaluation methodology can be seen as an explication of the instrumentalist methodology, according to which counterexamples play a crucial role, but not the dramatic one that is associated with the

[1] For an adapted, but otherwise complete survey of these conclusions, see Section 1 of the introductory chapter.

falsificationist methodology. This explains and even justifies non-falsificationist, i.e., dogmatic, behavior of scientists. However, in contrast to pseudoscientists, they nevertheless aim at the improvement of their theories, which is frequently possible within a set of dogmas.

**Fourth*, empirical progress cannot be seen only as an aim in itself; it can also be seen as a means to truth approximation, as argued from Chap. 7 onwards. Here, 'the truth' is conceived as the nomic truth, provided by the so-called Nomic Postulate (Nomic Principle, in the present book), i.e., the strongest true theory about what is nomically possible for a given domain of reality assuming a certain vocabulary. Using the structuralist (or set-theoretic) approach to theories, the Nomic Postulate assures the existence of a subset of nomic possibilities within the set of conceptual possibilities generated by the vocabulary. Within this framework, a *prima facie* plausible definition of the idea that one theory is more truthlike than another can be given. The latter can be seen as a hypothesis, the truth approximation hypothesis, which can be tested due to its prediction of empirical progress, which makes empirical progress functional for truth approximation.

**Sixth*, the (shifting) distinction between observational and theoretical terms not only naturally leads, together with the Nomic Postulate, in Chap. 9, to the distinction between theoretical and observational truthlikeness, but also to definitions of reference, referential claims of theories, and referential truthlikeness. Whereas observational truthlikeness implies empirical progress, theoretical truthlikeness only implies observational truthlikeness as far as true consequences are concerned. This leads to a further relativization of the role of counterexamples. Whereas the loss of a true observational consequence is a genuine drawback of a new theory, an extra counterexample of a new theory may well be an accidental (observational) merit of the old one. Moreover, since there are no deductive relations between referential truthlikeness and the other comparative notions, other kinds of (theoretical and experimental) arguments may also have to be taken into account for the assessment of referential claims.

Many other conclusions followed from the above ones, but I will not present them here, but confine myself to one, *tenth*, overall epistemological conclusion, which I will first introduce. ICR opens and closes with general epistemological chapters.

Chapter 1 introduces the main epistemological positions, instrumentalism,[2] constructive empiricism, referential realism, constructive (theory) realism and essentialist (theory) realism.[3] Chapter 13 provides a further articulation of constructive realism. The reason is that at the end of Part III the following conclusion was already drawn provisionally and was further strengthened in Part IV:

**Tenth*, the instrumentalist methodology provides good reasons for the transition of epistemological positions from instrumentalism, via constructive empiricism and referential realism, to constructive realism. Here, the intermediate step from

[2] Instrumentalism as an epistemological position should be distinguished from the instrumentalist or evaluation methodology. The latter is claimed to serve also other epistemological positions.

[3] For 'structural realism', see Chap. 16, Sections 4.2 and 5.

constructive empiricism to referential realism turned out to be the hardest one, whereas the step from constructive to essentialist realism had to be rejected.

In the rest of ICR-Chap. 13 I have drawn the main lines of the resulting favorite epistemological position of *constructive realism*. It is a conceptually relative, hence non-essentialist, nomic truth approximation version of theory realism, accounting for objective truths. They can be approached by an intersubjective method, viz., the evaluation methodology for theories, in which the role of (truth-)testing of hypotheses primarily concerns testing test implications of theories as well as testing comparative success and truth approximation hypotheses regarding theories.

11.1.2 Terminology

Before giving an outline of the chapter, I will first delineate some crucial terms and types of conclusions. The terms 'induction' and 'inductive argument' will be used in the broad sense (Thagard 1988; Salmon et al. 1999) of an argument that is ampliative.[4] Following Salmon (p. 11), it is an argument of which the conclusion goes beyond the content of its premises. Hence, it is not necessarily truth preserving, it is not erosion-proof, that is, new premises may (completely) undermine an inductive argument,[5] even a strong one, and inductive arguments come in different (informal or even formal) degrees of strength.

The terms 'abduction' and 'abductive argument' will be used to refer to the special case that an inductive argument can be presented as a straightforward case of abduction in the minimal sense of Peirce captured by the famous quote [Peirce, CP, 5.189]:

> The surprising fact C is observed.
> But if A were true, C would be a matter of course.
> Hence, there is reason to suspect that A is true.

This quote represents a minimal kind of abduction, even if we assume, as Peirce seems to have intended, that the second premise is to be interpreted as stating that A would explain C in a deductive sense.[6] Hence, abduction is a provisional, explanatory kind of induction, that is, it is an induction of a deductive-explanatory nature of which the conclusion has to be further tested. Frequently a couple of additional conditions are suggested, also by Peirce. To begin with, the hypothesis A should be (further) empirically testable, which I will assume throughout. Other possible extra conditions are: (1) the generation of hypothesis A is also part of abduction, (2) A has to be 'economic' in some absolute and/or comparative sense, and (3) the conclusion

[4] See Flach and Kakas (2000) for an interesting survey and discussion of the various uses of the terms 'induction' and 'abduction'.

[5] Hence, an inductive argument is a non-monotonic argument with a default character: its conclusion is supposed to hold as long as there is no counterevidence.

[6] However, weaker kinds, that is, non-deductive ones, are certainly possible, see e.g. Aliseda (1997, p. 27, 2006).

can be made stronger, e.g. "Hence, we may assume that A is true". Accordingly, the suggested modest sense does neither imply a discovery claim, nor an economic claim regarding A, and suggests the need of further testing A.[7]

Inductive arguments and, more generally, rules of inductive inference, that is, general patterns of inductive arguments, will be said to have an 'abductive justification' if a (re-)formulation of the argument as an abductive argument is possible. Otherwise I will speak of merely an 'inductive justification'.

Regarding types of conclusions of inductive arguments, it is important to note that in the context of theories and truth approximation there are at least four interesting conclusions. It may be concluded that (there is reason to suspect that) a theory is.

1. true
2. the closest to the truth of the available alternatives
3. true and the closest to the truth of the available alternatives
4. the truth

Here, 'the truth' is conceived as the strongest true theory about a given domain in a given vocabulary. More specifically, 'the truth' (and 'true') may concern some actual possibility (world), 'the actual truth', or all 'nomic possibilities' in the domain, 'the nomic truth'. The latter target will here be assumed throughout, except when otherwise stated.

Of course, conclusion 4 entails conclusion 3, which entails, by definition, conclusions 1 and 2. Moreover, conclusions 1 and 2 are logically independent. Note also that conclusion 4 can be given, like conclusion 2, but now somewhat artificially, a comparative, even superlative, formulation: the closest to the truth of all possible theories. Conclusion 1, in contrast, is merely of a classificatory nature, and conclusion 3 is of a mixed nature. Instead of referring to numbered conclusions, I will also speak of conclusions of 'type-1, -2, -3, -4'.

11.1.3 Outline of the Chapter

I will first introduce, in Sect. 11.2, the unstratified rule of 'inference to the best theory' (IBT), that is, IBT with the type-2 conclusion 'as the closest to the truth'. It turns out to have a straightforward abductive justification, in contrast to (the rule of) 'inference to the best explanation' (IBE), that is, IBE with the type-1 conclusion 'as

[7] In the course of the chapter, I will nevertheless frequently assume that (not only C, but also) A has already severely been tested. However, in this chapter I will not deal with the generation of hypotheses, nor with economic factors. See Chap. 10 (based on Kuipers (1999)), for an analysis of the main abductive generation and evaluation tasks that result from ICR. See Kuipers (2001, Section 11.1.4) for a sketch. See Chap. 9 (based on Kuipers (2002)) for a way to take non-empirical factors, in particular aesthetic ones, into account. The consequences for the explication of 'inference to the best explanation' are elaborated in Chap. 16 in terms of aesthetic progress among empirically equivalent theories.

true', with which it will be compared. In Sect. 11.3, observational and theoretical kinds of both rules will be defined, compared, and discussed in the light of the epistemological positions, with essentially similar results. Section 11.4 deals with the prospects for referential truth approximation and the IBT-like rule of 'inference to the referential claim of the best theory' (IrcBT), that is, IrcBT 'as the closest to the referential truth'. It will be compared with 'inference to the likeliest cause' (ILC), that is, ILC 'as true'. In Sect. 11.5, I will introduce and discuss weak and strong forms of three kinds of induction with type-1 and type-4 conclusions, respectively: observational, theoretical, and referential induction, and compare them with corresponding versions of IBE/ILC in their type-1 reading as well as in their type-3 (= type-1 + type-2) reading. I will close with some general conclusions and surveys (Sect. 11.6) and a discussion (Sect. 11.7) of points raised by Adam Grobler and other participants of the relevant conference.

As far as possible, I will formulate the main points and arguments independent of the specific formal analysis of nomic truth approximation in ICR. Ideally, they apply, perhaps in some modified way, to any comparative theory of *nomic truth* approximation, and perhaps even to any comparative theory of *actual truth* approximation as well.

11.2 Unstratified Inference to the Best Theory

11.2.1 Preliminaries

In ICR-section 7.5.3., the first plea for reshaping 'Inference to the best explanation' (IBE) as 'Inference to the best theory' (IBT) is given. It is still unstratified in the sense that there is not yet made a distinction between observational and theoretical terms. It presupposes explications of the notions of 'being at least as (more) successful' and 'being as close (closer) to the truth'. The first can be explicated in at least three, strongly related, ways in terms of successes and problems of separate theories resulting from straightforward applications of the HD-method. In the present context I restrict the attention to exclusion theories (see Chap. 2) and prefer more specifically:

Theory $\underline{P}^*$ is (at time t) *at least as (more) successful as (than)* theory $\underline{P}$ iff (at t)

- all individual problems of $\underline{P}^*$ are individual problems (i.e., counterexamples) of $\underline{P}$
- all general successes of $\underline{P}$ are general (explanatory) successes of $\underline{P}^*$

(– P has at least one extra individual problem or P* has at least one extra general success)

From this perspective, the crucial methodological steps are the following. Suppose that P* is, at t, more successful than P. This suggests the *Comparative Success Hypothesis*[8]:

CSH: P* (is and) remains more successful than P

CSH has of course to be further tested, now by the comparative application of the HD-method. At last we may come to the conclusion that it has been sufficiently tested, which provides the condition for applying the *Rule of Success*:

RS: If CSH has been sufficiently confirmed to be accepted as true, i.e., if P* has so far proven to be more successful than P, then eliminate P, in favor of P*, for the time being.

By 'rRS' I will indicate RS restricted to not yet falsified theories. The phrase "if P* has so far proven to be more successful than P" may be made more precise in Popperian style by adding: "despite severe attempts to falsify CSH".[9]

Fitting to the explication of 'more successful' is the following definition of 'closer to the truth':

Theory[10] P* is *at least as close (closer) to the nomic truth **T** as (than)* theory P iff

- all wrongly excluded possibilities by P* are (wrongly) excluded by P
- all true consequences of P are (true) consequences of P*[11]

(– P has at least one extra incorrect model or P* has at least one extra true consequence)

As already alluded to, 'the (nomic) truth T' is here interpreted as the unknown but fixed set of *nomic* possibilities, which is posited by the *'Nomic Principle'*, given a fixed domain and a fixed vocabulary. The vocabulary is designed to characterize the domain and gives rise to the set of *conceptual possibilities*. In this stage I assume that the vocabulary is observational, that is, all non-logical terms can be applied in an intersubjective way. Of course, observational terms may well be, and usually are, laden with background theories.

Like CSH, the *truth approximation (TA-)hypothesis*, P* is at least as close (similar) to the truth **T** as P, is a decent comparative hypothesis. Assuming that the data are correct, it can even explain that P* is at least as successful as P, for the following theorem is easy to prove:

---

[8] In this chapter we will keep the distinction in ICR between the comparative success hypothesis and the rule of success because it makes the further analysis more transparent.

[9] Igor Douven has suggested an interesting weaker version of RS, by not prescribing to eliminate *P* in favor of *P** when CSH has been sufficiently confirmed, but by prescribing something like: "spend more recourses on *P** than on *P* to the extent that CSH has been confirmed." This would legitimate some continued research attention to *P*.

[10] In this chapter all theories are exclusion theories: e.g. the claim of theory *P* is that all nomic possibilities are members of *P*, and hence that all non-members of *P* are nomic impossibilities.

[11] This is equivalent to "all correctly excluded possibilities by P are (correctly) excluded by P*.

Success Theorem
If theory $\underline{P}^*$ is at least as close to the truth $\mathbf{T}$ as $\underline{P}$ and if the data are correct then $\underline{P}^*$ (always) remains at least as successful as $\underline{P}$

By this theorem, HD-testing of the TA-hypothesis immediately leads to (further) testing of CSH.

Finally, we have a theorem in the other direction:

Forward Theorem
If CSH (which speaks of *remaining* more successful) is true, this implies the TA-hypothesis.

Given an observational vocabulary, this theorem arises from the fact that, if $\underline{P}^*$ is *not* closer to the truth than $\underline{P}$, (further) testing of CSH will sooner or later lead to an extra counterexample of $\underline{P}^*$ or to an extra success of $\underline{P}$. The Forward Theorem leads to the claim that RS, and hence the HD-method, are functional for approaching the observational truth.

By now we have all the ingredients for ICR-section 7.5.3., which I will just repeat here, only with some marginal adaptations.

11.2.2 Inference to the Best Theory, Rather Than the Best Explanation

As suggested, an interesting consequence of the truth approximation analysis concerns so-called 'inference to the best explanation'. I will argue that it has to be replaced by 'inference to the best theory'. Let me first define the two crucial notions: 'the best theory' is the best, i.e., the most successful, theory among the available theories, assuming that there is one, whereas 'the best explanation' is the best theory which has not yet been falsified. Adapted to the present context, the standard idea of 'inference to the best explanation'[12] amounts to the following:

Inference to the Best Explanation (IBE)
If an explanation has so far proven to be the best one among the non-falsified available theories, then (choose it, i.e., apply rRS, and) conclude, for the time being, that it is true (i.e., that it is true as hypothesis).

In our setting, 'true as hypothesis' is plausibly interpreted as 'true with respect to all nomic possibilities', which amounts in the present context to the claim that the truth, that is, the set of nomic possibilities, is a subset of the set of possibilities not excluded by that explanation.[13] IBE is problematic for at least three reasons. First, it does not say anything about falsified theories. It is restricted to the case that the best theory has not yet been falsified; hence counterexamples are not taken into account. Second, it has an asymmetric character in the sense that it couples a non-comparative

---

[12] Harman (1965) coined the term. See (Lipton 1991) for an extensive exposition and defense of IBE.

[13] Or, in the terms of Chap. 2 and 4, true if the explanation is conceived as an exclusion theory.

(type-1) conclusion, being true, to a comparative premise, being the best unfalsified theory. Third, and finally, it is difficult to see how it could be justified in terms of general considerations concerning the true/false-distinction.

Consider the plausible alternative, suggested by RS and the Forward Theorem:

Inference to the Best Theory (IBT)
If a theory has so far proven to be the best one among the available theories, then (choose it, i.e., apply RS, and) conclude, for the time being, that it is the closest to the nomic truth **T** of the available theories

IBT does not have the three shortcomings of IBE. It applies to unfalsified as well as to falsified theories. It couples a comparative (type-2) conclusion, being the closest to the truth, to a comparative premise, being the best theory. Last but not least, it is directly justifiable in terms of truth approximation, viz., by the Forward Theorem. To be precise, the comparative hypothesis that the best theory remains the best, a hypothesis that apparently seems to be true, implies that the best theory is the closest to the truth. Hence, IBT can be seen as a, for good reasons, severely corrected version of IBE.

11.2.3 IBT as a Kind of Abduction

We might call the previous justification of IBT an 'inductive justification', because it is essentially non-deductively arguing in favor of the truth of the (proper) conclusion. Now I will turn to an (explanatory or) 'abductive justification', as defined in the first section. Recall that Peirce's basic scheme for abduction reads as follows:

> The surprising fact C is observed.
> But if A were true, C would be a matter of course.
> Hence, there is reason to suspect that A is true.

Let us see whether IBE and IBT can be viewed as cases of abduction in this sense. In the case of IBE we have to interpret C as the (surprising) observation that a certain theory has so far proven to be better than any of its available alternatives *and has not yet been falsified*.[14] Moreover, A has to be interpreted as this theory. However, it is immediately clear that the second premise of the abductive argument does not hold: although 'being true' entails 'not yet being falsified', it does not at all entail being, relative to its available alternatives, the best explanation of the available data, not even that it is at least as good an explanation as its alternatives.

Turning to IBT, we have to interpret C 'merely' as the observation that a certain theory has so far proven to be the best one among the available theories and A has to be interpreted as the hypothesis that this theory is closer to the truth than any of its alternatives. In this case, the second premise ("But if A were true, C would be a

[14] In contrast to Peirce's favorite examples, the surprising observation is here not an individual fact, but a kind of general fact, in the present IBE-directed interpretation *partly* of a comparative nature, in the IBT-directed interpretation, see below, *in full*.

matter of course.") holds almost perfectly, mainly due to the Success Theorem. More specifically (ICR, p. 162):

> "the theorem enables the attachment of three conclusions to the fact that $\underline{P^*}$ has so far proven to be more successful than $\underline{P}$; conclusions which are independent of what exactly 'the nomic truth' is:
>
> - first, it is still possible that $\underline{P^*}$ is closer to the truth than $\underline{P}$, a possibility which, when conceived as hypothesis, the TA-hypothesis, according to the Success Theorem, would explain the greater success,
> - second, it is impossible that $\underline{P^*}$ is further from the truth than $\underline{P}$ (an hence $\underline{P}$ closer to the truth than $\underline{P^*}$), for otherwise, so teaches the Success Theorem, $\underline{P^*}$ could not be more successful,
> - third, it is also possible that $\underline{P^*}$ is neither closer nor further from the truth than $\underline{P}$, in which case, however, another specific explanation has to be given for the fact that $\underline{P^*}$ has so far proven to be more successful."

The first conclusion needs some qualification (which the second conclusion does not need). In the strict sense, the Success Theorem merely explains deductively that $\underline{P^*}$ is at least as successful as $\underline{P}$. However, although it does not entail that $\underline{P^*}$ will have an extra success, it does entail this very possibility. Put in a different way, the surprising observation is not so much that $\underline{P^*}$ is at a certain moment more successful than $\underline{P}$ (the assumed point of departure for testing CSH), but that it remains more successful and this is a matter of course, assuming the TA-hypothesis.[15] Moreover, as the third conclusion makes clear, without this hypothesis, this observation is far from a matter of course. On the contrary, it will be a hard job to explain it without this hypothesis.

In sum, unlike IBE, IBT is obviously a kind of abduction, even in the strong sense that the conclusion, that is, the TA-hypothesis, has been further tested, viz., via CSH. Of course, the inclination to draw the TA-conclusion may be seen as a good reason for still further testing of the TA-hypothesis, if possible also in other ways than via CSH.

Let me briefly further comment on the different types of conclusions of IBE and IBT. The typical and justifiable conclusion of IBT is a type-2 conclusion, that is, 'being closest to the truth'. We have also seen, suggested by IBE, that the type-1 conclusion, 'being true', cannot easily be seen as the conclusion of an abductive argument, assuming a comparative surprising observation. However, this does not exclude that such a conclusion can be 'sufficiently' supported by observations. But in that case, it is merely a kind of 'induction'. I will elaborate the various possibilities, after introducing the distinction between observational and theoretical terms. Then we will also see that the type-4 conclusion, 'being the truth', cannot only be seen as a kind of induction but even as an extreme special case of IBT.

[15] In the IBE-directed interpretation the surprising observation would then be that the relevant theory 'remains more successful (than its alternatives) *and unfalsified*'. Again, the assumption that the theory (i.e., A) is true would make the second part of this observation a matter of course, but not the first part.

11.3 Stratified Inference to the Best Theory

11.3.1 Stratification

In Chap. 9 of ICR, I have presented a refined version of the theory of truthlikeness and truth approximation by introducing a (relative) distinction between theoretical and observational terms. More specifically, assuming a (fixed) domain of interest D,[16] I first introduce an *observational (o-)level* and a theoretical *(t-)level* for D. Let Vt indicate the richest vocabulary of non-logico-mathematical terms that play a role in one of the theories involved, that is, all relevant observational and theoretical terms which are in consideration. Let V_0 indicate the subset of V_t of observation terms.

Let $U(V_0)$ and $U(V_t)$ indicate the set of conceptual possibilities on the o- and t-level, respectively. The *projection function* π is generally defined as the function from a set of conceptual possibilities U(V) of some vocabulary V onto that set U(V′) of a sub-vocabulary V′ for which holds that, for all x in U(V), π(x) is the result of dropping the extra components in x and those clauses that specify x and use those components. By πP I indicate the set of 'projections' of all members of P, and by $\pi^{-1}\pi P$ the set of 'reproductions' of these projections, of which P must be a subset, possibly a proper one, as is easy to check.

Applying the Nomic Principle on both levels, these sets generate, in combination with D, unique, time-independent subsets of nomic possibilities on each level, called *the observational (nomic) truth* $\mathbf{T}_o$ and *the theoretical (nomic) truth* $\mathbf{T}_t$, respectively. I will further assume that $\pi\mathbf{T}_t = \mathbf{T}_o$, the *Truth Projection (TP-)Principle* (from the t-level to the o-level). That $\pi\mathbf{T}_t$ is a subset of $\mathbf{T}_o$ is a semantic fact, for if a conceptual possibility on the t-level is nomically possible, it will remain nomically possible when one skips some components. For the motivation for the substantial side of the TP-Principle, viz., $\mathbf{T}_o$ is a subset of $\pi\mathbf{T}_t$, see (ICR, pp. 215–7).

It would be nice when, not only 'the truth' but also 'more truthlikeness', of e.g. P̲* relative to P̲, would be projected unconditionally. However, although it does hold to a large extent, we need a condition to really guarantee this projection. The condition amounts to the supposition that theory P̲ does not have *extra* 'potential lucky misses', that is, observational nomic possibilities not excluded by P̲, but excluded by P̲*, that have no theoretical nomic possibilities in P-P*. Formally, $(\pi P \cap \pi\mathbf{T}_t)-\pi P^*$ has to be empty. A general, P*-independent, hence stronger, way of excluding such potential lucky misses is assuming that P̲ has no potential lucky misses at all or, in the terms of ICR, P̲ is *relatively correct*, which formally amounts to the condition: $\pi(P\cap \mathbf{T}_t)= \pi P \cap \pi\mathbf{T}_t$. But here I prefer to use the weakest condition in the relevant theorems.

[16] Here we assume a fixed domain, as elsewhere in this book, except in Chap. 8, in which domain revision is studied.

Projection Theorem
if <u>P*</u> is at least as close to the theoretical truth $\mathbf{T_t}$ as <u>P</u>, and if <u>P</u> has, relative to <u>P*</u>, no extra potential lucky misses, then πP* is at least as close to the observational truth $\mathbf{T_o}$ as πP

Together with the Success Theorem, that is, a generalized version, formally taking theoretical terms into account (see the next subsection for a similar adaptation of (observational) IBT), we now get:

Success Corollary
if <u>P*</u> is at least as close to $\mathbf{T_t}$ as <u>P</u>, and <u>P</u> has no extra potential lucky misses, then <u>P*</u> is always at least as successful as <u>P</u>

The Forward Theorem remains of course as before, apart from formally taking theoretical terms into account. Now we have in addition:

Upward Theorem
if <u>πP*</u> is closer to $\mathbf{T_o}$ than <u>πP</u> then <u>P*#</u> is closer to **Tt** than <u>P#</u>, where

$P^{\#} =_{def} P-\{x\in P-P^*/\pi(x)\in \pi P^*\cap \mathbf{T_o}\}$ and $P^{*\#} =_{def} P\cap P^*$

Note that both <u>P#</u> and <u>P*#</u> are strengthenings of their corresponding departure theory, i.e., they exclude more than their 'originals'. P# omits all P's extra potential lucky misses, if any, and P*# restricts P* to its intersection with P. Of course, the latter may well be a drawback relative to <u>P*</u> itself, but the fundamental point is that we can construe in this way a genuine example of theoretical truth approximation starting from a case of observational truth approximation. This leaves room for the possibility that <u>P*</u> is also an example of theoretical truth approximation relative to <u>P*#</u>, but this we cannot be sure of.

The Upward Theorem leads, together with (a generalized version of) the Forward Theorem, to the claim that RS, and hence the HD-method, are functional for approaching the observational truth, which in its turn is functional for approaching the theoretical truth (see ICR, p. 214, for a detailed motivation).

11.3.2 *Stratified Inferences*

Now I will extend the previous analysis of IBT and IBE to the more standard stratification by an observational and a theoretical level. I start with a generalized version of the unstratified IBT. I call it now the observational IBT (IBT-o). Whereas the unstratified version assumes that the whole vocabulary is observational, this is no longer necessary, by jumping, whenever relevant, to the so-called 'corresponding observational theory', which refers of course to the projection of a theory on the observational level.

Inference to the Best Theory as the Closest to the Observational Truth
IBT-o: If a theory has so far proven to be the best one among the available theories, then (choose it, i.e., apply RS, and) conclude, for the time being, that the

corresponding observational theory is the closest to the (nomic) observational truth of the observational theories corresponding to the available theories.

Its inductive justification in terms of the (equally generalized version of the) Forward Theorem remains the same. Similarly for its abductive justification in terms of the Success Theorem. But now we have of course also a rule on the theoretical level:

Inference to the Best Theory as the Closest to the Theoretical Truth
IBT-t: If a theory has so far proven to be the best one among the available theories, then (choose it, i.e., apply RS, and) conclude, for the time being, that it is the closest to the (nomic) theoretical truth of the available theories.

For the inductive justification of this theoretical IBT we not only need to appeal to the Forward Theorem but also to the Upward Theorem, which leaves room for the possibility that the relevant truth approximation claim only applies to strengthened theories. However, I will not spell this out in further detail. For the abductive justification of IBT-t an appeal to both the Success Theorem and the Projection Theorem is required. The conditional character of the latter has an important relativizing consequence. For the explanation of 'being more successful' we need to assume not only 'being closer to the truth' but also that the worse theory has no extra 'potential lucky misses'.

Recall the severe corrections IBT contains in comparison to what is standardly called 'Inference to the best explanation (as true)' (IBE), where the best explanation is conceived as the best not yet falsified theory. Again one can distinguish an observational and a theoretical kind, where I restrict myself to the latter, for the former is again merely a generalized version of the unstratified one.

Inference to the Best Explanation as True (even on the Theoretical Level)
IBE-t: If an explanation has so far proven to be the best one among the available theories, then (choose it, i.e., apply rRS, and) conclude, for the time being, that it is true (i.e., it is entailed by the theoretical truth).

The corrections of IBT-t compared to IBE-t are the following: unlike IBE-t, IBT-t is not restricted to unfalsified theories, and IBT-t amounts to a comparative conclusion attached to a comparative premise, whereas IBE-t amounts to an absolute conclusion attached to a comparative premise. Finally, IBT-t is directly justifiable in terms of truth approximation, viz., inductively by the Forward Theorem and the Upward Theorem, and abductively by the Projection Theorem and the Success Theorem. Neither justification of IBE-t is easy to imagine.

11.3.3 Epistemological Positions and IBT

Let us now investigate to what extent IBT may be acceptable for what I see as the four main epistemological positions, viz., instrumentalism, constructive empiricism, referential realism and theory-realism (of which I indicate in ICR a defensible, constructive version and a problematic essentialist version). In view of the

Forward Theorem, IBT-o may well be acceptable for all positions as far as they are willing to take the notion of the nomic truth on the observational level seriously, which however does not seem to be the case for, for example, Van Fraassen (1980, 1989), who wants to avoid '(physical or nomic) possibility-talk'. IBT-t will be acceptable for the theory-realist, assuming that he is willing to speak of true or false claims when there may be fictions involved, see below.

What about the referential realist? As suggested before, he, or she, may be willing to subscribe to IBT-o, but it is unlikely that somebody like Cartwright (1983, 1989) will subscribe to IBT-t with some enthusiasm. On the basis of my analysis, she might conclude that IBT-t is defensible and hence innocent, but she will see it nevertheless as unimportant, for she will think that there are good reasons that it will almost never have interesting applications. Her view seems to suggest that the fundamental laws of interest are not only false, but, at least when separately conceived, that they are far from the truth. Moreover, for the referential realist, the crucial question concerns, of course, whether there is something like 'inference to the most likely cause' (Cartwright 1983, 1989), or in my ICR-terminology, and more generally, 'inference to the best referential claim'.

The background is the following one. In ICR I have introduced an intermediate level determined by the terms in V_t that refer, together denoted by V_r, where V_o-terms are supposed to refer anyway, but the reference of genuine theoretical terms is determined by whether they really matter in the transition from $\mathbf{T_o}$ to $\mathbf{T_t}$, that is, whether they are merely fictions or not. This enables the definition of the referential claim of a theory, of 'the referential truth' as the strongest true referential claim, and of the notions of referential truthlikeness and truth approximation. This gives rise to a corresponding definition of the idea of inference to the best referential claim.

11.4 Inference to the Best Referential Claim

11.4.1 Referential Truth Approximation

I will start with the most relevant passages from ICR (parts of Section 9.1.2, pp. 215–6, and Section 9.2.1, pp. 219–20, with only minor adaptations).

Between the observational and the theoretical level, I introduce the intermediate level of terms that really refer to something real concerning the domain, called the *referential (r-)level*. For term τ in V_t, let $V_{t'}$ indicate $V_t - \{\tau\}$ and let π project $U(V_t)$ onto $U(V_{t'})$. Then τ is said to refer according to $\mathbf{T_t}$, or simply *really refers* or $\boldsymbol{T_t}$-*refers*, if and only if $\mathbf{T_t}$, being on formal grounds a subset of $\pi^{-1}\pi\mathbf{T_t}$, is a *proper* subset of $\pi^{-1}\pi\mathbf{T_t}$. That is, $\mathbf{T_t}$ is a proper subset of the reproduction of the projection of $\mathbf{T_t}$. This clause, the $\boldsymbol{T_t}$-*reference clause*, is not yet to be interpreted as a complete definition of reference. It is at most a partial definition, providing a reference criterion. It amounts to the claim that τ makes a difference or plays a substantial role in the theoretical truth $\mathbf{T_t}$ in the sense that $\mathbf{T_t}$ is constrained by τ in the sense defined.

With this partial definition, I formally operationalize the idea that there is in reality some (type of) item that plays the role τ is ascribed to by $\mathbf{T_t}$.

The criterion still needs some qualification to function as a definition of reference. It seems adequate primarily for theoretical terms purporting to refer to attributes (properties, relations, and functions); *attribute terms* for short. However, for theoretical terms purporting to refer to classes of entities, used as domain-sets for attribute terms – *entity terms* for short – this definition does not seem to work. However, it is plausible to say that a (theoretical) entity term in V_t $\mathbf{T_t}$-refers if and only if there is at least one (theoretical) attribute term that $\mathbf{T_t}$-refers and that uses the entity term as (one of) its domain-set(s). The consequence is that an entity term does not $\mathbf{T_t}$-refer if there are no attribute terms using it as domain-set. This is plausible, for in this case it is difficult to see how that entity term could play in some other sense a substantial role in $\mathbf{T_t}$, which is precisely the reason for the detour via attribute terms. If this detour is not possible, the theoretical entities hang in the air as unconstrained entities, not distinguishable from genuine fictitious entities, at least not with the means provided by D and V_t.

In sum, my definition of reference amounts to a direct reference criterion for attribute terms and an indirect one for entity terms. Note that the very possibility to define reference of entity terms on the basis of the definition of reference of attribute terms is a good reason to extend the idea of 'entity realism' to 'referential realism'.

The given definition of reference is explicitly $\mathbf{T_t}$-relative. We might abstract from this by the definition that a term τ refers in an absolute sense if there is at least one domain and if τ belongs to at least one vocabulary such that τ refers relative to the resulting theoretical truth. The resulting theory of reference is rather non-standard. It would be interesting to compare it with the standard theories. Although it evidently is not a causal theory, it is also not exactly a causal-descriptive theory in the sense of David Lewis.

For the moment I stick to the definition given above and define the referential vocabulary $V_r(\mathbf{T_t})$ or simply V_r, called *the referential truth*, giving rise to the *referential (r-)level*, as the union of Vo and those members of V_t that $\mathbf{T_t}$-refer. Hence, by definition, V_0 is a subset of V_r and V_r is a subset of V_t. So I assume that it has already been established that the observation terms refer in some other context. They may or may not refer in the present context, i.e., they may or may not $\mathbf{T_t}$-refer.

For completeness, I introduce the notion of substantial truth. Let $U(V_r)$ indicate the set of conceptual possibilities on the r-level. By applying the Nomic Principle on this level, $U(V_r)$ generates, in combination with D, a unique, time-independent subset of nomic possibilities, called the *substantial truth* $\mathbf{T_r}$. However, here I will neglect the topic of substantial truth approximation and its relation to observational and theoretical truth approximation (see the second part of ICR-section 9.1.2) and turn immediately to referential truthlikeness and truth approximation (ICR-section 9.2.1).

It is crucial to note that, whereas $\mathbf{T_t}$ determines which terms really or $\mathbf{T_t}$-refer and which do not, that is, which are fictions, every theory implies in a similar way one or more specific $\mathbf{T_t}$-reference claims. For theory $\underline{P}$, being a subset of $U(V_t)$, with the exclusion claim "$\mathbf{T_t} \subseteq P$", implicitly claims that τ in $V_t - V_o$ $\mathbf{T_t}$-refers if τ plays a substantial role in P̲, that is, if P, being on formal grounds a subset of $\pi^{-1}\pi P$ (where π again projects $U(V_t)$ again onto $U(V_t\text{-}\{\tau\})$), is a proper subset of $\pi^{-1}\pi P$.

As in the case of the definition of $\mathbf{T_t}$-reference, I have again to qualify this definition because it only seems appropriate for theoretical *attribute* terms. For specific $\mathbf{T_t}$-reference claims for theoretical entity terms we may again use the detour via specific $\mathbf{T_t}$-reference claims for attribute terms: if τ in V_t-V_o is an entity term it $\mathbf{T_t}$-refers according to P̲ iff there is at least one attribute term, using that entity term as (one of) its domain-set(s), which is $\mathbf{T_t}$-referring according to P̲.

Let us indicate the subset of V_t consisting of V_o united with the set of theoretical terms that $\mathbf{T_t}$-refer according to P̲ by V_r(P̲). Hence, V_o is a subset of V_r(P̲), which is a subset of V_t. However, we do not know how V_r(P̲) is situated with respect to V_r. I call the conjunctive claim of P̲ that all τ in V_o refer and that all τ in V_r(P̲)-V_o $\mathbf{T_t}$-refer *the referential claim* of P̲, and it will be indicated by RC-P̲. In this way, the referential truth becomes the referential claim of $\mathbf{T_t}$, hence the strongest true referential claim about V_t.

For subsets P and P* of $U(V_t)$ I define, of course, the comparative claim "*theory P̲* is at least as close to the referential truth V_r as P̲*", or "*P̲* is referentially at least as close to the truth as P̲*", if RC-P̲* is at least as close to the referential truth as RC-P̲ in the following plausible sense:

for all τ in V_r if τ belongs to V_r(P̲) then it belongs to V_r(P̲*) (formally: $V_r(\underline{P}^*) \cap V_r \supseteq V_r(\underline{P}) \cap V_r)$)

for all τ in $V_t - V_r$ if τ belongs to V_r(P̲*) then it belongs to V_r(P̲) (formally: $V_r(\underline{P}^*) - V_r \subseteq V_r(\underline{P}) - V_r)$)[17]

It amounts to a weak case of more actual truthlikeness in propositional terms, as defined in Chap. 7 of ICR, ranging over all elementary propositions of the form: "τ refers" for all τ in V_o and "τ $\mathbf{T_t}$-refers" for all τ in V_t-V_o.

11.4.2 Inference to the Referential Claim of the Best Theory

In the first part of this section I terminologically deviate somewhat from the corresponding part of ICR-section 9.3.1.

In the light of the analysis above, it is now plausible to formulate the rule:

Inference to the Referential Claim of the Best Theory as the Closest to the Referential Truth (IrcBT)

If a theory has so far proven to be the best one among the available theories, then conclude, for the time being, that its referential claim is the closest to the referential truth

[17] Note the formal correspondence with the definitions of 'at least as close to the truth' in earlier chapters.

Hence, if P* is the best theory, in some V_t, then IrcBT suggests to conclude that Vr(P*) is closer to Vr than V_r(P), as defined above, for any available alternative theory P in V_t. My previous analysis suggests that IrcBT is plausible, though its conclusion is certainly, like the IBT-o/t conclusions, not compelling. However, it is without doubt more plausible than the straightforwardly generalized rule suggested by Cartwright's 'inference to the likeliest cause', where I can and will now leave room for already falsified theories:

Inference to the Referential Claim of the Best Theory (the Likeliest Cause) as True (ILC)

If a theory has so far proven to be the best one among the available theories, then conclude, for the time being, that its referential claim is true

There is only one exception to the superiority of IrcBT over ILC. They coincide when there is only one theoretical term at stake, which is $\mathbf{T}_t$-referring according to the best theory. Otherwise, ILC is (much) stronger than IrcBT.

Let us turn to the more detailed inductive and abductive justification of IrcBT, starting with the second. Hence, the question is, what does the conclusion of IrcBT, that the referential claim of the best theory is the closest to the referential truth, explain? The problem is that this conclusion itself does not explain very much. For this reason, referential realists have a hard job to provide abductive arguments for their position. The problem is that we have to take the conclusion of IrcBT together with the conclusion of IBT-t itself into account, a move that is only available to the theory realist.

One might think that the conclusion of IBT-t, viz. the best theory is the closest to the theoretical truth, does entail the conclusion of IrcBT. A specific reason to expect this is the fact that the theoretical claim of a theory, e.g. "$\mathbf{T}_t \subseteq$ P", entails the partial referential claim "V_r(P)$\subseteq V_r(\mathbf{T}_t)$". However, a proof for the suggested conjecture is not possible, for interesting reasons. Suppose that P* is closer to $\mathbf{T}_t$ than P. What we would like to prove amounts to the following: all τ in V_r(P*)-V_r(P) belong to V_r and all τ in V_r(P)-V_r(P*) do not belong to V_r. Let us suppose that τ in V_r(P*)-V_r(P) does not belong to V_r. Hence, P* wrongly claims that it does, whereas P rightly claims that it does not. However, P's claim may be based on a wrong aspect of its theoretical claim, whereas P* may not yet be so good that its theoretical claim implies the reference claim for the right reasons. It is important to note that the suggested proof already fails in the case of just one theoretical term. Similar arguments can be given for the possibility that τ in V_r(P)-V_r(P*) belongs to V_r. In both cases we might say that P has lucky reference claims relative to P*, that is, where P* has unlucky ones. In sum, the conclusion of IBT-t does not entail the conclusion of IrcBT.

This does not mean, of course, that the first conclusion does not provide good arguments for supposing that the relevant theory is closer to the referential truth. For, as can be learned from the suggested proof attempt, the suggested entailment is only violated when a theory bases its referential claim on wrong reasons in precisely the right direction, which would be rather exceptional because it is rather artificial. In this respect there is a similarity with the relation between truth approximation and novel facts (see ICR-section 7.5.1.): although truth approximation without novel facts is in principle possible, it is rather artificial and hence may be supposed to be exceptional.

Recall now that the Projection Theorem, projecting more theoretical truthlikeness on more observational truthlikeness, has some extra condition, notably the assumption that $\underline{P}$ has no extra lucky misses, relative to $\underline{P^*}$. Hence, the next question is whether the conclusion of IrcBT is of any help in filling this gap in the Projection Theorem. Since I do not have a proof I formulate the following conjecture:

Conjecture

If $\underline{P^*}$ is closer to $\mathbf{T}_t$ than $\underline{P}$ then the assumption that RC-$\underline{P^*}$ is closer to RC-$\mathbf{T}_t$ than RC-$\underline{P}$ narrows down (or even excludes) the possibility for potential lucky misses of $\underline{P}$ relative to $\underline{P^*}$

Although proof attempts were failing, I also did not succeed in proving that it does not narrow down the possibilities for lucky misses. On the contrary, it seems very much that, as a rule, the gap is substantially reduced, though not necessarily completely. The idea is, of course, that potential lucky misses of $\underline{P}$ relative to $\underline{P^*}$ may be due to lucky reference claims of $\underline{P}$ relative to $\underline{P^*}$, which are excluded by the closer to the referential truth assumption of $\underline{P^*}$ relative to $\underline{P}$. Hence, I provisionally conclude that the conclusion of IrcBT substantially strengthens the explanation of the superior success of the best theory on the basis of the Projection and the Success Theorem. That is, an abductive justification for IrcBT, in combination with that of IBT-t, seems possible. I will call such a justification an IBT(-t)-laden abductive justification of IrcBT.

As to the possible abductive justification of ILC, the situation is not as bad as that for IBE-t, for even if the best theory is already known to be false, its reference claim may well be true. However, an abductive justification remains impossible for it remains the case that ILC is essentially asymmetric, in such a way that its non-comparative conclusion is (much) more risky than the comparative one of IrcBT, for it is (much) stronger, except when only one term is in debate. If we assume the conclusion of IBE, that is, that the best theory, apparently not yet falsified, is true, then the conclusion of ILC does not have to follow, for the best theory may not be strong enough to entail the correct elementary reference claims. Only if there is just one term involved, and the best theory's reference claim is positive, which is plausible indeed, this has to be a correct claim if that theory is true. The final question is whether the surprising phenomenon, being the best theory, can be explained by assuming both the conclusion of IBE-t and that of ILC. That is, an IBE(-t)-laden abductive justification of ILC might be possible. However, even two comparative premises cannot entail a non-comparative one.

In sum, not only IBT-o and IBT-t, but also IrcBT is likely to have an abductive justification, albeit an IBT-laden one, whereas not only IBE-o and IBE-t lack one, but also ILC lacks one, even an IBE-laden one.

Let us now turn to the inductive justification of IrcBT and ILC. In ICR-section 9.2.2, I have given a number of theoretical and experimental arguments that may support specific and general, separate and comparative, reference claims. As a matter of fact, the emphasis is on the prospects for (specific and general) referential induction, see the next section, that is, whether a certain theoretical term of a theory refers or all its theoretical terms refer rather than not refer according to the old, less successful, theory. This is the relevant question when the new, presumably, more successful theory postulates extra theoretical entities and properties, relative to the old theory, for in that case the 'closer to the referential truth' claim for this new theory implies the claim that the extra referential claims of the new theory are true. In other cases, however, the old theory will also have 'positive' reference claims, which are rejected by the new theory. For such cases, the arguments in ICR-9.2.2 have to be adapted. In general, I see the explicit theoretical and experimental arguments in that section for 'positive' referential induction, and the implicit ones for 'negative' referential induction, merely as a beginning. The only claim is that my specific definition of reference in ICR is a plausible new point of departure.

However this may be, although one might think of other than abductive arguments in favor of IrcBT, it is difficult to imagine whatever argument in favor of ILC, simply due to its asymmetric character.

11.5 Three Types of Induction

It is not only plausible to distinguish three kinds of induction, viz. observational, theoretical and referential, but also weak (type-1) and strong (type-4) forms of it.[18] Although I will not systematically speak of inductive jumps, this terminology would stress more strongly that inductions are always fallible, and that the standards of induction are debatable. Moreover, it would withdraw the suggestion that inductions are merely extrapolations, as in the case of straightforward inductive generalizations. But even such generalizations are essentially fallible leaps. Finally, in ICR-sections 3.3 and 4.3, I have argued that, although the (threshold) problem of the acceptance of hypotheses, as true or as the truth, is still a difficult one, it is relatively clear how acceptance is guided by the general and comparative principles of confirmation. Here I will concentrate on aspects that can now be formulated more clearly.

11.5.1 Observational Induction

In the case of weak/strong *observational induction* it is concluded, for the time being, and for a certain domain and a certain observational vocabulary, on the basis of the available evidence, that a certain general hypothesis stated in that vocabulary is true for all nomic possibilities (weak, type-1), or is the strongest true hypothesis

[18] It might be argued that the three kinds of IBT discussed before are also kinds of induction, viz., comparative inductions, i.e., inductions with the comparative conclusion of being the closest to the relevant truth, in contrast to the separate or absolute kinds of induction discussed in this section, i.e., inductions with the 'absolute' conclusion of being true or of being the truth.

about all nomic possibilities (strong, type-4), respectively. It is evident that observational induction requires the Nomic Principle in order to bridge the gap between actual and nomic possibilities. To be sure, weak observational induction presupposes that the hypothesis has not been falsified, and strong observational induction presupposes in addition that the hypothesis is the best available theory. But these are just necessary conditions for observational induction, which requires, in addition, the much stronger and much more debatable assumption that enough attempts to falsify it, to improve it, respectively, have failed, in order to pass the relevant threshold of plausibility.

However, to call *weak* observational induction 'inference to the best explanation' (as true) would be misleading, since there is no comparison of hypotheses involved. This is not to say that observational induction cannot have abductive justification. On the contrary, this is perfectly possible. As a matter of fact, all 'deductively derivable' (more or less surprising) evidence, to begin with not being falsified, can be explained by its type-1 conclusion. Hence, cases of weak observational induction may well be called (cases of weak observational) abduction. Turning to *strong* observational induction, calling that 'inference to the best explanation' in the sense of 'inference to the best explanation as the (observational) truth' might be defensible as an extreme special case of IBT (-o), for strong observational induction amounts to 'inference to the best theory as the observational truth'. For this purpose, recall that, despite Note 20, a type-4 conclusion can be rephrased as a comparative, even superlative, conclusion: the closest to the truth of all possible theories. Moreover, 'being the truth' explains of course, even trivially, being not falsified and, by the Success Theorem, being the best theory of the available theories. Hence, cases of strong observational induction may well be called abduction.

11.5.2 Theoretical Induction

Let us now turn to weak and strong *theoretical induction*. In this case, it is concluded, for the time being, and for a certain domain and a certain theoretical (cum observational) vocabulary, on the basis of the available evidence, that a certain general hypothesis stated in that vocabulary is true for all nomic possibilities (weak, type-1), or is the strongest true hypothesis about all nomic possibilities (strong, type-4), respectively. Again, it is evident that theoretical induction requires the Nomic Principle. Moreover, weak theoretical induction presupposes again that the hypothesis has not been falsified, and strong theoretical induction presupposes, in addition, that the hypothesis is the best one of the available theories. As in the case of observational induction, these are now simply necessary conditions for theoretical induction, which require, in addition, the much stronger and much more debatable assumption that enough attempts to falsify it, to improve it, respectively, have failed, in order to pass the relevant threshold of plausibility. However, in ICR Chap. 2, I have argued that the guidelines provided by confirmation theory only fail in the context of theoretical induction in the case of observationally equivalent *and* (initial and hence posterior) equally plausible theories.

Again, cases of weak and strong theoretical induction may be called cases of abduction, for similar reasons as in the case of observational induction, except that for the strong case the Projection Theorem has to be used as well for explaining to be the best theory. However, again, calling weak theoretical induction 'inference to the best explanation' (as true) would be misleading, since there is no comparison involved. But calling strong theoretical induction 'inference to the best explanation' in the sense of 'inference to the best explanation as the (theoretical) truth' might again be defensible as an extreme special case of IBT-t, for strong theoretical induction amounts to 'inference to the best theory as the theoretical truth'.

11.5.3 Referential Induction

Let us finally turn to *referential induction*. Weak referential induction amounts to the conclusion, for the time being, that, under appropriate conditions, the conjunction of some specific referential claims of a theory is true. In the case of strong referential induction it is concluded that the entire referential claim of the theory is the strongest true referential claim, i.e., the referential truth. Recall that, although being closer to the theoretical truth did not guarantee being closer to the referential truth, the fact that being the strongest truth on the theoretical level, i.e., being the theoretical truth, guarantees that the referential claim is true. Moreover, when the claim of a hypothesis is true, this guarantees that all specific referential claims implied by that claim are true. Accordingly, weak and strong theoretical induction imply, and hence are sufficient conditions for, weak and strong referential induction, respectively. Although it may be possible to formulate other (pragmatically) sufficient conditions for referential induction of a theoretical or experimental nature (see ICR-section 9.2.2.), I confine myself to referential induction as implied by theoretical induction. However, one should realize that one has to supplement theoretical induction explicitly with referential induction as defined above to take such other types of referential induction into account.

Weak and strong referential induction based on the corresponding theoretical induction obtain of course abductive justification via, and hence are laden with, such theoretical induction. However, calling weak referential induction 'inference to the likeliest cause' (as true) would be misleading, since there is no comparison involved. But calling strong referential induction 'inference to the likeliest cause' in the sense of 'inference to the referential claim of the best theory as the (referential) truth' might again be defensible, now as an extreme special case of IrcBT, for strong referential induction amounts to 'inference to the referential claim of the best theory as the referential truth'.

11.5.4 Benevolent Explications of IBE and ILC

Finally, there is one (type of) way to explicate IBE and ILC that may be seen as the most benevolent and defensible. IBE might be interpreted as a combination of weak (observational/ theoretical) induction and IBT(-o/t), and hence with a type-3 conclusion: inference to the best theory as (observationally/ theoretically) true and as the closest to the (observational/ theoretical) truth of the available theories. Similarly, ILC may be interpreted as a combination of referential induction and IrcBT, again with a type-3 conclusion: inference to the referential claim of the best theory as true and as the closest to the referential truth of the referential claims of the available theories. In all three cases, the supposition of not being falsified and the symmetric comparative character are built in, and hence both kinds abductive justification, that is, that for weak induction and that for IBT and IrcBT, respectively. Since abductive justifications of conclusions are essentially based on their explanatory merits, we get in this way interpretations of IBE and ILC that do justice to their names, in particular IBE. However, this does not imply that these combined inference rules add anything to their component rules. The latter, together with strong inductions, form the basic type of rules. To be precise, the three kinds of IBT and those of weak and strong induction seem to constitute the basic inductive rules, with abductive justification, for scientific realists.

11.6 Concluding Remarks

The main conclusions of this chapter are:

(1) In the light of the nomic theory of truth approximation, 'Inference to the best explanation' (IBE) and 'Inference to the likeliest cause' (ILC) should be replaced by kinds of 'Inference to the best theory' (IBT), that is, either by 'Inference to the best theory as the closest to the (observational, theoretical, or referential) truth' or by their extreme versions 'Inference to the best theory as the (observational, theoretical, or referential) truth', where the latter may also be called 'strong (observational, theoretical, or referential) induction'. In Diagram 11.1 I have left out the referential kinds. The arrows have plausible interpretations.

(2)For the observational, theoretical, and referential (non-extreme and extreme) kinds not merely inductive justifications can be given but also abductive justifications based on the theory of truth approximation, albeit of decreasing strength.

(3) IBE (observational and theoretical) and ILC lack abductive justification for several reasons, of which their asymmetric character is the common one.

(4) In contrast, their non-comparative versions, here called 'weak (observational, theoretical, or referential) induction', have abductive justifications.

(5) However, (1) and (4) can be combined to explications of IBE/ILC, see Diagram 11.2. Again I have left out the referential kinds in the diagram and the arrows have plausible interpretations.

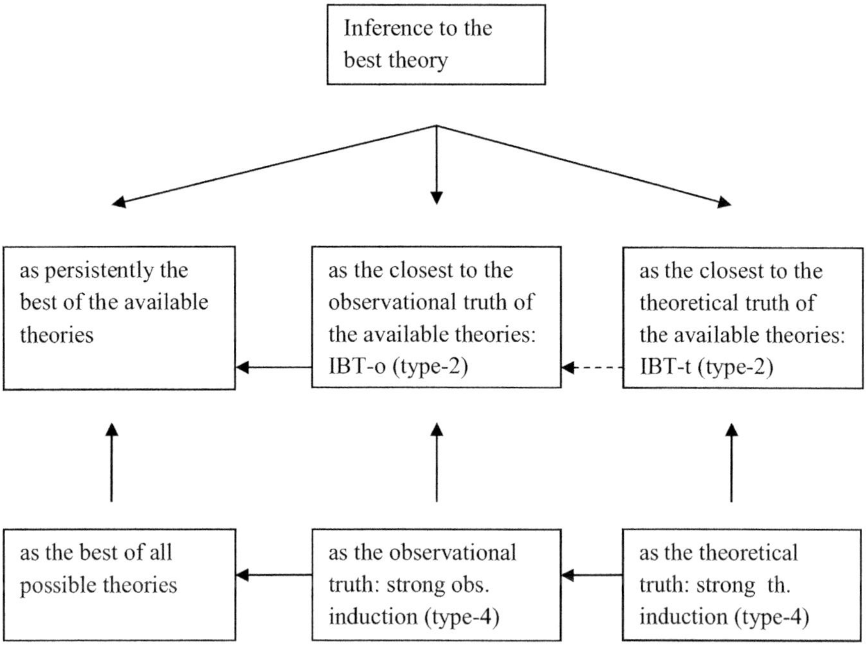

Diagram 11.1 Kinds of inference to the best theory

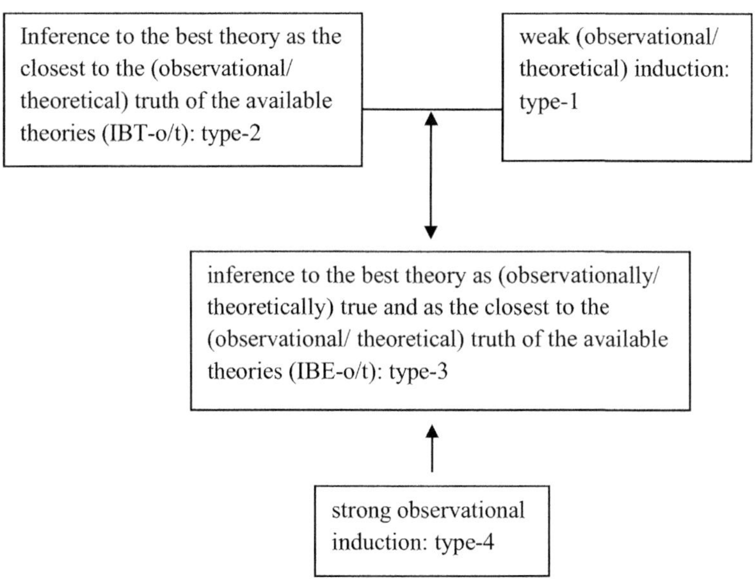

Diagram 11.2 Kinds of induction in relation to some kinds of inference to the best theory

Finally, I like to conclude this chapter with (an adapted version of the first paragraph of ICR-section 13.3) stressing the crucial role played by observational and theoretical induction in the construction and determination of terms and, hence, the long-term dynamics in science. Besides the formation of observation terms by straightforward explicit definition, observational induction may provide the necessary and sufficient conditions, e.g., in the form of existence and uniqueness requirements, for explicitly defining new observation terms. The quantitative notions of pressure and temperature are examples. Such new terms are unambiguously, hence intersubjectively, applicable, and they (may be supposed to) refer if they enable new observational inductions. Besides implying referential induction, theoretical induction may provide the necessary and sufficient conditions, e.g., in the form of existence and uniqueness requirements, for applying theoretical terms, that is, for identifying theoretical entities and for measuring theoretical attributes. E.g., the detection of electrons, and the measurement of their mass and charge, are based on such inductions. In this way, theoretical terms (may be supposed to) become referring and unambiguously, hence intersubjectively, applicable.

In other words, theoretical terms can be essentially transformed into new observation terms by appropriate theoretical and/or referential induction. Of course, together with earlier or elsewhere accepted observation terms, they can be used for new cases of observational induction. Moreover, with other ones, they will play a crucial role in the separate and comparative evaluation of new theories introducing new theoretical terms, dealing with (partially) new domains, starting another round of the 'empirical cycle'.

11.7 Discussion[19]

In response to Adam Grobler's comments (Grobler 2004), I would like to briefly state the following points.

1) I agree that IBT should not be seen as an autonomous rule of inference characteristic for *the logic of science*, but as a heuristic rule characteristic for *the use of logic in science*. More specifically, I see it as an applied rule of inductive, hence non-monotonic, inference, building upon classical deductive logic, a given domain and vocabulary, and the model-theoretic or structuralist definition of truthlikeness.
2) I thank Grobler for making so explicit that applications of IBT, that is, inferences that the best theory is the closest to the truth, are only plausible, due to a couple of theorems, if it is assumed (not only that the data are correct, but also) that the

[19] That is, the discussion following the presentation of this paper on the 2002-conference on deduction and induction in the sciences.

background knowledge is true. Hence, questioning the background knowledge certainly is a way to avoid the truth approximation conclusion.

3) Grobler's suggestion, in response to my second complaint about IBE, to make its conclusion comparative by replacing it by "[the best theory] is the best candidate (among the alternatives) for the truth" is quite a step in the direction of my favorite IBT for two reasons. First, taking 'the truth' as the point of reference only makes sense if something like my definition of it is accepted, that is, the strongest nomically true theory. Second, assuming that we know that the best theory is false, if the data and the background knowledge are correct, what else could '[being] the best candidate … for the truth' mean than 'being the closest to the truth'?
4) In contrast to Grobler's opinion, my definition of successes and failures includes explanatory successes and failures, at least when they are of a, strictly or approximately, derivational nature. However, it is not immediately clear whether explanatory derivational (general) successes are more functional for truth approximation than non-explanatory ones, assuming that the latter exist at all. The last proviso is made, because 'flagpole' type of examples typically concern the explanation of individual facts and not general facts, such as general successes of theories. Be this as it may, as I concede in my reply to Grobler and Andrzej Wisniewski (2005), in the same volume, that they are certainly right in suggesting that non-derivational types of explanatory successes, in particular successful 'explanations by (causal, functional, and intentional) specification', should be taken into account in the definition of the best theory. As I should have added in my reply there, (correct) explanations by specification amount to specifying special kinds of true statements, hence it follows almost immediately that theories having more of such successes than other ones are likely to be closer to the (nomic or actual) truth.
5) I like to subscribe to Grobler's remark that 'the amount of incorrectness' may matter for the closeness to the truth. It is precisely this intuition that underlies the transition in ICR from the basic theory of truth approximation to the refined one of Chap. 10 (and Chap. 6 of this book), which takes into account that one incorrectly classified possibility may be more similar to a target nomic possibility or impossibility, depending on the relevant claim of the theory, than another.

Let me also briefly discuss the comments made by other participants.

1) Asked for examples of straightforwardly more successful theories, I like to refer to a comparative table of seven electrodynamic theories in view of thirteen experimental results in (Panofsky and Phillips 1962, p. 282, reproduced in ICR, p. 118, and in Kuipers 2001, p. 236).
2) Unlike in this case, in which the special theory of relativity dominates all other theories in success, it may of course be that there is no overall dominant theory, but that there are two or more 'locally dominant' theories, in which case the Rule of Success doesn't prescribe a choice between these theories.
3) Moreover, even in the case of an overall dominant theory, it may well be that this one is still very far from the truth.

4) In general, if we succeed in improving a theory, e.g. in the sense of banishing a counterexample in a non-ad hoc manner, the Comparative Success Hypotheses will predict some novel facts, predictions that may or may not come true.
5) Although Mandelbrot, the founding father of the fractal theory, is certainly right in that 'the truth' about the length of a certain coast does not exist, it does exist as soon as we use a minimal unit of length, that is, a vocabulary, as presupposed throughout.
6) Leaving room for inconsistent theories in the standard sense, as in paraconsistent logic, is not merely a challenge for IBT, it requires an adaptation of the whole theory of truth approximation. This would be a very interesting research project. However, dealing with inconsistency of two-sided theories <M, P>in the sense that M is not a subset of P does not seem very difficult.
7) As suggested in Notes 2 and 9, it is possible to take non-empirical success criteria into account in the definition of 'more successful' (Kuipers 2002, Chap. 9 of this book), and hence in the definition of IBT, of course, as far as not in conflict with the empirical ones and as far as functional for truth approximation. Such criteria may not only depend on the discipline, and its stage of development, but even on the style of scientific reasoning, as the historical evidence for an English and a French style seems to suggest. After all, there is more than one road to Rome. Be this as it may, regarding IBT there seems to be less need for including such non-empirical criteria than in the case of IBE, viz. in view of their rather different conclusions, for in the IBE-case additional arguments are needed for the non-comparative conclusion of being true.

References

Aliseda, A. (1997). *Seeking explanations: Abduction in logic, philosophy of science and artificial intelligence*. Dissertation Stanford. Amsterdam: ILLC Dissertations Series (1997–04).

Aliseda, A. (2006). *Abductive reasoning. Logical investigations into discovery and explanation*. Synthese Library 330. Dordrecht: Springer (Extended and revised version of Aliseda, 1997).

Cartwright, N. (1983). *How the laws of physics lie*. Oxford: Clarendon Press.

Cartwright, N. (1989). *Natural capacities and their measurement*. Oxford: Clarendon Press.

Douven, I. (2005). Empirical equivalence, explanatory force, and the inference to the best theory. In R. Festa, A. Aliseda, & J. Peijnenburg (eds) *Confirmation, empirical progress, and truth approximation. Essays in debate with Theo Kuipers* (Vol. 1, pp. 281–309). Amsterdam: Rodopi (Reply by T. Kuipers, 310–313).

Flach, P., & Kakas, A. (2000). Abductive and inductive reasoning: background and issues. In P. Flach & A. Kakas (Eds.), *Abduction and induction* (pp. 1–27). Dordrecht: Kluwer Academic Publishers.

Grobler, A. (2004). The significance of explanatory considerations. In F. Stadler (Ed.), *Induction and deduction in the sciences*, VCI-yearbook (pp. 53–56). Dordrecht: Kluwer.

Grobler, A., & A. Wisniewski (2005). Explanation and theory-evaluation. In R. Festa, A. Aliseda, & J. Peijnenburg (Eds.), *Cognitive structures in scientific inquiry. Essays in debate with Theo Kuipers* (Vol. 2, pp. 299–310). Amsterdam: Rodopi (Reply by T. Kuipers, 311–314).

Harman, G. (1965). The inference to the best explanation. *Philosophical Review, 74*, 88–95.

Kuipers, T. (1999). Abduction aiming at empirical progress or even truth approximation, leading to a challenge for computational modelling. *Foundations of Science, 4*(3), 307–323.

Kuipers, T. (2000). *From instrumentalism to constructive realism. On some relations between confirmation, empirical progress, and truth approximation*. Synthese Library 287. Dordrecht: Kluwer Academic Publishers.

Kuipers, T. (2001). *Structures in science. Heuristic patterns based on cognitive structures. An advanced textbook in neo-classical philosophy of science*. Synthese Library 301. Dordrecht: Kluwer Academic Publishers.

Kuipers, T. (2002). Beauty, a road to the truth. *Synthese, 131*(3), 291–328.

Lipton, P. (1991). *Inference to the best explanation*. London: Routledge.

Niiniluoto, I. (2005) Abduction and truthlikeness. In R. Festa, A. Aliseda, & J. Peijnenburg (Eds.), *Confirmation, empirical progress, and truth approximation. Essays in debate with Theo Kuipers* (Vol. 1, pp. 255–276). Amsterdam: Rodopi (Reply by T. Kuipers, 276–280).

Panofsky, W., & Phillips, M. (1962). *Classical electricity and magnetism*. London: Addison-Wesley.

Salmon, W., et al. (1999). *Introduction to the philosophy of science*. Indianapolis/Cambridge: Hackett Publishing Cy.

Thagard, P. (1988). *Computational philosophy of science*. Cambridge: MIT Press.

van Fraassen, B. (1980). *The scientific image*. Oxford: Clarendon Press.

van Fraassen, B. (1989). *Laws and symmetry*. Oxford: Clarendon Press.

Chapter 12
Pragmatic Aspects of Truth Approximation

Abstract In this chapter I will draw a very general picture of an overall pragmatic theory of justification from the viewpoint of *truth approximation*. Crucial is the comparative definition of *success* according to which a theory P* is more successful than a theory P if and only if (1) P* shares all general successes of P, that is all empirically confirmed general implications, and has some extra ones, and (2) P shares all empirical counter-examples of P*. The *rule of success* says that if so far P* has been more successful than P, then – for the time being – accept P* as the better one and eliminate P. It is further argued that according to the pragmatic-instrumentalist view of scientific theories, theory elimination should exclusively be guided by the rule of success rather than by Popper's *rule of falsification* – for it is pragmatically reasonable to accept the most successful theory available, even if this theory is known to be false. In the last part of the chapter, it is argued that the connection between pragmatic theory-success and truth approximation is otherwise rather close to Popper: at least 'almost' always, a theory which is closer to the truth will also turn out to be the pragmatically more successful theory. However, in the upshot, 'being false' and hence 'being falsified' are not very important for truth approximation.

Keywords Pragmatic theory of justification · Rule of success · Rule of falsification · Theory elimination · Truth approximation

12.1 Introduction

In (Kuipers 2000), I have argued that there is a coherent set of methodologically characterizable activities, with the hypothetico-deductive (HD-)method as its core, which are so connected that they are functional for truth approximation, viz.,

---

Revised version of: "Pragmatic aspects of truth approximation", in: P. Weingartner, G. Schurz and G. Dorn, *The role of pragmatics in contemporary philosophy*, Hölder-Pichler-Temsky, Vienna, (Proceedings of the 20th International Wittgenstein-Symposium, August 1997), 1998, 288–300.

T. A. F. Kuipers, *Nomic Truth Approximation Revisited*, Synthese Library 399,
https://doi.org/10.1007/978-3-319-98388-2_12

- HD-testing of general test implications of theories
- separate HD-evaluation of theories
- comparative HD-evaluation of theories
- application of the rule of success

The main, highly idealized, train of thoughts is as follows. HD-testing of a general test implication of a theory, to be called a general test, is the elementary form of separate HD-evaluation of a theory. Separate HD-evaluation of a theory may well be continued, by a new general test, even if a former test has generated a counterexample, for the result of the new test further enriches the evaluation report of the theory. This evaluation report is important as such, but also in the comparison with the evaluation reports of other theories. More specifically, if one theory is, at a certain moment, more successful than another, (further) comparative HD-evaluation is guided by the comparative success hypothesis[1] that the first theory will remain more successful than the second. If this hypothesis has sufficiently been confirmed, it is a good reason to provisionally make a choice in favor of the better theory, that is, to take that theory as the new point of departure for further improvement. Finally, and this is the main claim made above, an application of this so-called rule of success (RS) not only marks the establishment of a case of empirical progress, there are good reasons to provisionally conclude that the better theory is closer to the truth than the worse one.

As suggested, this is a highly idealized story, with many hidden hooks and eyes. The main purpose of this chapter is to highlight the pragmatic aspects of the four activities and the provisional truth approximation conclusion. All of them have more or less specific pragmatic aspects. Moreover, there is an overall pragmatic aspect, and, finally, these pragmatic aspects amount together to a strong relativization of falsification and confirmation of theories. I will start with the overall pragmatic aspect, then present the specific pragmatic aspects, and conclude with the resulting relativizations of falsification and confirmation.

12.2 Overall Pragmatism

I begin with a general claim: as far as pragmatism implies the instrumentalist methodology, that is, the view that theory elimination should exclusively be guided by comparative success, rather than primarily by falsification,[2] pragmatist theory selection is pre-eminently functional for truth approximation. In other words, although both the instrumentalist and the falsificationist methodology are effective for truth approximation, the former is more efficient. The reason for the superiority of the

[1] As in the previous chapter, we will keep the distinction in (Kuipers 2000) between the comparative success hypothesis and the rule of success because it makes the further analysis more transparent.

[2] That is, by falsification and only by comparative success as far as non-falsified theories are concerned.

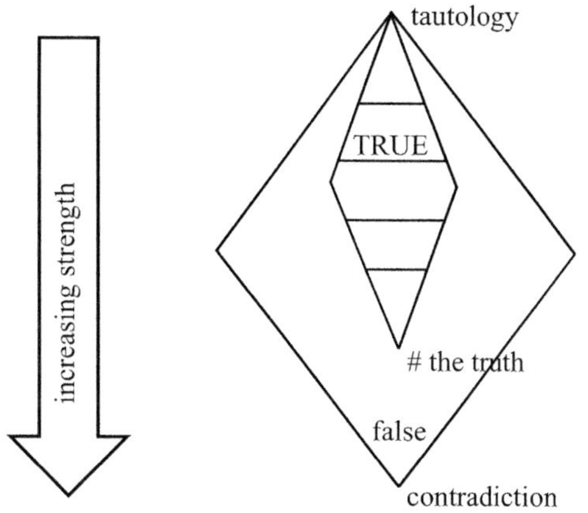

Fig. 12.1 The set of propositions and the set of true propositions (shaded), ordered by strength

instrumentalist methodology is the fact that falsification of a theory, guiding the falsificationist methodology, is not important for its absolute or relative distance to the truth, except that such a theory cannot be equal to the truth itself. A theory may well be false in such a way that it is nevertheless close to the truth, e.g. because it only has a minor mistake. For my purposes even more importantly, a false theory may well be closer to the truth than another one.

That the latter is perfectly possible in an intuitive sense is easy to demonstrate by a picture for a finite propositional language. In Fig. 12.1 the big diamond represents the set of all propositions that can be formulated with such a language. They are vertically ordered by strength. It is easy to check that, starting with the tautology, the number of propositions of equal strength, represented on a horizontal line, first increases, and then decreases. 'The truth' is explicated as the strongest true proposition, indicated by #. All its consequences are weaker and true, hence 'above' #. They may or may not be re-ordered in a small diamond, as it has been done in the figure. Now, assuming any reasonable ternary relation of 'closer to' between propositions, it is clear from the figure that one false theory may be closer to the truth than another one, just as one true theory may be closer to the truth than another one.

Several quantitative and qualitative explications of 'truthlikeness' have been given, that also formally allow such comparative statements. For several reasons, partly fundamental, partly pragmatic, I prefer the qualitative approach, more specifically, the nomic qualitative approach, in which the truth characterizes not only what is conceptually but also what is physically or really possible, and theories are supposed to try to do so as well. Below I will give some more details and specifications.

For the moment, it will be clear that the claim that pragmatist theory selection is pre-eminently functional for truth approximation is only *prima facie* paradoxical. For although pragmatists may not aim at efficient truth approximation, it may well be a consequence of their practice.[3] Before I return to truth approximation, I will indicate the pragmatic aspects of the four methodological activities.

12.3 HD-Testing of General Test Implications of Theories

Testing a general test implications (GTI's) of a theory by deriving and testing individual test implications (ITI's) requires five major types of pragmatic decisions, before a GTI can be rejected or accepted, by an 'inductive jump'. I will briefly elucidate them (for details, see Kuipers 1995b, 2000, Chap. 5), using the terminology that corresponds to experimental tests. For natural tests and past tests the terminology can easily be adapted, wherever necessary. To be sure, there is nothing original in this survey, except perhaps its ordering.

The first two types of decision relate to the fact that a GTI is supposed to be derivable from the theory, by some *logico-mathematical proof*, where a GTI is a general, conditional statement in terms that are treated as observation terms. As a rule, however, the derivation requires *auxiliary hypotheses* of one kind or another. Following the main steps that may occur in the related pattern of explanation of observational laws by theories (Kuipers 1990, 2001), we obtain: specification hypotheses for the application step, statistical hypotheses for an aggregation step, identity hypotheses for an identification step, causal correlation hypotheses for a causal correlation step, and idealization hypotheses for an approximation step. Since logico-mathematical proofs and auxiliary hypotheses may almost always be questioned, we have to accept them explicitly in order to make testing of the GTI meaningful with respect to that theory.

Moreover, we have assumed that the GTI is formulated in observation terms, but in normal scientific practice this means that these terms are nevertheless laden with all kinds of underlying or background theories, called the *observation presuppositions*. Only by accepting these presuppositions, it is possible to derive the relevant 'instrument reading' version, indicated by GTI^{ir}.

In order to test GTI^{ir}, we have to accept, in addition, suitable initial (test) conditions IC^{ir} as true descriptions of the experimental conditions, for only then we may predict that the relevant individual test implication ITI^{ir} is also true. Similarly, we have to accept or reject ITI^{ir} as a true description of the resulting outcome of the experiment. In both cases we need to apply decision criteria of statistical and approximate nature, called *condition and outcome decision criteria*.

The indicated five decisions seem to exhaust the underlying pragmatic decisions for the conclusion that we have obtained a confirming instance (IC & ITI) or a counterexample (IC & not-ITI) of the GTI, and hence of the theory, also called an individual success or an individual problem, respectively. However, although individual problems, that is, counterexamples, form a natural 'problem unit' of evaluation of a

[3] In (Kuipers 1995a) I have, not too seriously, suggested that the effectiveness for truth approximation may be called an illustration of *the cunning of reason*, and that its superior efficiency, relative to the falsificationist methodology, could then be called *the irony* of the cunning of reason.

theory, not individual successes but general successes form the most natural 'success unit' of evaluation. A general success amounts to the acceptance of the GTI itself, and this requires that the GTI is, relative to its initial plausibility, 'sufficiently confirmed'. What this exactly means may be a matter of dispute, but not that some *inductive acceptance criteria* for accepting 'inductive generalizations' are required. Note that a general success may be called an explanatory success in the minimal sense that it can be derived from the theory (with the help of auxiliary hypotheses).

The above survey of decisions underlying the acceptance of a counterexample as such provides at the same time a survey of possible dogmatic strategies. To avoid the conclusion that a favorite theory has been falsified, there are equally many points of impact to question the relevant counterexample as such. Similarly, to avoid the conclusion that a competing theory has obtained a general success, there is even one more point of impact to question that general success as such, viz., the inductive jump.

12.4 Separate HD-Evaluation of Theories

Separate HD-evaluation of a theory amounts to testing several GTI's of the theory, leading to the formation of an *evaluation report* on the basis of these tested GTI's. An evaluation report includes, of course, the identification of previously accepted individual and general facts as counterexamples and general successes, respectively. It should also be mentioned that besides the asymmetric, 'dual' categories of counterexamples and general successes, two other units of evaluation can be used for this purpose, viz., confirming instances, and falsifying general hypotheses. In the same way as a general success summarizes confirming instances, a falsifying general hypothesis summarizes counterexamples. The four categories for evaluation reports are represented in the matrix of Table 12.1. Below I will concentrate on the dual categories of counterexamples and general successes.

Besides the pragmatic decisions involved in each GTI, dealt with above, there are some new ones. To begin with, the above-mentioned identification of previously accepted individual and general facts as counterexamples and general successes, respectively, requires new decisions. Firstly, their acceptance as facts has to be reaffirmed. Secondly, it has to be decided that it concerns facts about the domain of the theory. The second type of decision essentially presupposes one of the general decisions that have to be taken before we can speak of a specific theory.

The general decisions that were assumed in the foregoing without saying are: the choice of a domain vocabulary V_d, a domain D of 'intended applications', well-defined in terms of V_d, an observational vocabulary V_o, and a theoretical vocabulary V_t, where V_d is supposed to be a subset of V_o and, for convenience, V_o of V_t. Theory

Table 12.1 Matrix of evaluation categories

	Problems	Successes
Individual	Counterexamples	Confirming instances
General	Falsifying general hypotheses	General successes

$\underline{P}$,[4] formulated in V_t-terms, is supposed to have observational theory $\underline{Po}$, formulated in V_o-terms, as observational consequence. The choice of the vocabularies V_d, V_o, and V_t, and of the domain D is not only a matter of a decision guided by fundamental and pragmatic interests, it is also a decision which can be revised on the basis of previous findings.

When two (groups of) scientists are *prima facie* dealing with the same theory, they may well disagree about the fine-tuning of the indicated decisions, with the consequence that the evaluation report they give of a theory at a certain moment may differ. However, as far as they agree about the choices and the test fate of the relevant GTI's their evaluation reports will coincide.

12.5 Comparative HD-Evaluation of Theories

Starting from the observation that theory $\underline{P^*}$ is, at a certain moment, more successful than theory $\underline{P}$, comparative HD-evaluation is guided by the

Comparative success hypothesis (CSH or CSH($\underline{P^*}$, $\underline{P}$)):

$\underline{P^*}$ will remain more successful than $\underline{P}$.

The crucial definition presupposed by CSH is that of 'more successful'. That definition has two components, an instantial and an explanatory clause.

Definition: theory $\underline{P^*}$ is *more successful* than theory $\underline{P}$ iff

- explanatory clause: $\underline{P^*}$ shares all general successes of $\underline{P}$, and has some extra ones
- instantial clause: $\underline{P}$ shares all counterexamples of $\underline{P^*}$

For reasons that will become clear below, it is here not assumed that $\underline{P}$ needs to have extra counterexamples. As a consequence, to be more successful, 'extra general successes' is the extra condition added to the obvious definition of 'at least as successful'.[5] The definition is called the basic definition, as opposed to the refined one, in which 'improving models', based on a ternary relation of structurelikeness (one structure is more similar to another than a third) is taken into account. From time to time, I will hint at this refined definition.

CSH($\underline{P^*}$, $\underline{P}$) is a perfect, albeit non-standard, example of a testable hypothesis, even if both theories are known to be false. It amounts to two general comparative test implications:

- all new general successes $\underline{P}$ are also (new) general successes of $\underline{P^*}$
- all new counterexamples of $\underline{P^*}$ are also (new) counterexamples of $\underline{P}$

[4] In line with Chap. 2, '$\underline{P}$' refers informally to a set of postulates, which are supposed to exclude certain possibibilities.

[5] This is a small deviation from previous expositions in which extra counterexamples and extra general successes are treated in a symmetrical way.

Again there are several pragmatic decisions involved. Firstly, if the domains of the two theories are not identical, it has to be agreed what their common domain is for the given definition of 'more successful' obviously presupposes the same domain. It is possible, of course, to generalize that definition in such a way that different domains are taken into account. Secondly, if the vocabularies of both theories are not unproblematically the same, and hence are fundamentally incommensurable, decisions have to be taken to make them pragmatically commensurable. Thirdly, in the refined case, the definition of (observational) structurelikeness has to be agreed upon.

Fourthly, and finally, as soon as we become interested in cases of theories with 'divided success', relative weights have to be assigned to general successes and counterexamples. In this case, even straightforward 'equal treatment' is not as easy as it might seem, because, in particular, general successes can be grouped together or split in different ways. However this may be, since the ultimate target will be to find a third theory which is at least as successful as the two competing theories, the fourth type of pragmatic decisions is less important in the long run.

12.6 Application of the Rule of Success

Testing of CSH may lead to falsification or to persistent confirmation. In the latter case, acceptance may follow. I like to call the corresponding rule of acceptance the rule of success (RS):

> *Rule of Success* (RS): if CSH($\underline{P}^*$,$\underline{P}$) has been 'sufficiently confirmed' to be accepted as true, then eliminate the worse theory, $\underline{P}$, in favor of the better one, $\underline{P}^*$, for the time being.

Application of RS amounts to establishing empirical progress and RS may even be seen as the hall-mark of scientific rationality. However, it requires a comparative inductive jump, again a pragmatic decision, for it does not (yet) seem to be possible to give fully objective criteria. That is, as in the case of accepting GTI's, also here 'sufficiently confirmed' is a debatable notion.

It is important to note that the evidence E on the basis of which the confirmation of CSH has to be judged consists of two components. The first one is the set of observed real possibilities R, and the second one is the set of established observational laws. Their conjunction, the strongest established observational law, will be indicated by S. Of course, S is based on R, in some way or other. R and S, in combination with their relation to the two theories, viz., being a counterexample or not, and being a general success or not, respectively, comprise all decisions that have been mentioned in the foregoing sections.

To be honest, the possibility of a pure application of RS is rare. Some deviations of the conditions for applications may be such that a choice can nevertheless be made. Firstly, there is the possibility of so-called 'Kuhn-loss'. A 'largely better' theory may fail to explain all individual and general facts the worse theory is able to explain, but the exceptions may all be such that there are good reasons to claim either that the task of their explanation should be relegated to a different kind of theory, or that it concerns facts that don't ask for a theoretical explanation but for a historical explanation, guided by relevant theories. One of Kuhn's famous examples of the latter kind is the cosmological-historical explanation (sketch) of the number of planets.

In the rest of the chapter, I will neglect Kuhn-loss, but not the second kind of deviations, viz., the possibility of extra counterexamples of the 'largely better' theory. Below we will see that the realist may have good reasons to be not impressed too much by them as long as the favorite theory is explanatorily more successful.

12.7 Truth Approximation

In the introduction I have claimed that the above described methodological activities are functional for truth approximation. More precisely, the first three activities are clearly functional for applying the Rule of Success (RS), and this is functional for truth approximation. The latter point still has to be elucidated.

The core argument is based on some theorems, and these are based on the following specifications of 'the truth' and definitions of 'more truthlike'. Given a domain D and an observational vocabulary V_o, there is a unique strongest true theory $\mathbf{T_o}$, *the observational truth*, which characterizes the real or nomic possibilities as far as V_o is concerned. Similarly, there is a unique strongest true theory $\mathbf{T_t}$, *the theoretical truth*, which characterizes them in terms of V_t. Of course, $\mathbf{T_o}$ is $\mathbf{T_t}$ restricted to V_o. Note the modal character of $\mathbf{T_o}$ and $\mathbf{T_t}$. Finally, it is possible to define on the basis of $\mathbf{T_t}$ which terms in V_t refer, and which do not, leading to *the referential truth*. The basic idea is that a term in V_t refers, as far as $\mathbf{T_t}$ is concerned, if and only if it plays a role in shaping $\mathbf{T_t}$.

The crucial comparative definitions concern the notions of 'closer to' the theoretical, to the observational, and to the referential truth. Since all three are of a more or less similar nature, I restrict the attention to the first. The basic definition reads as follows:

Definition[6]: $\underline{P^*}$ is *closer to* $\mathbf{T_t}$ than $\underline{P}$ iff

- positive or consequence clause: all rightly excluded possibilities of $\underline{P}$ are excluded by $\underline{P^*}$ and $\underline{P^*}$ rightly excludes some extra possibilities relative to $\underline{P}$ (or, equivalently, $\underline{P^*}$ shares all true consequences of $\underline{P}$, and has some extra ones)
- negative clause: all wrongly excluded possibilities of $\underline{P^*}$ are also excluded by$\underline{P}$

[6]This definition deals only with so-called one-sided theories, more specifically, theories only making an exclusion claim. E.g. $\underline{P}$ is supposed to claim $\mathbf{Tt} \subseteq P$ (or, equivalently, $cP \subseteq c\mathbf{Tt}$). Again there is a small deviation from previous expositions in which extra rightly excluded possibilities and extra wrongly excluded possibilities were treated in a symmetrical way.

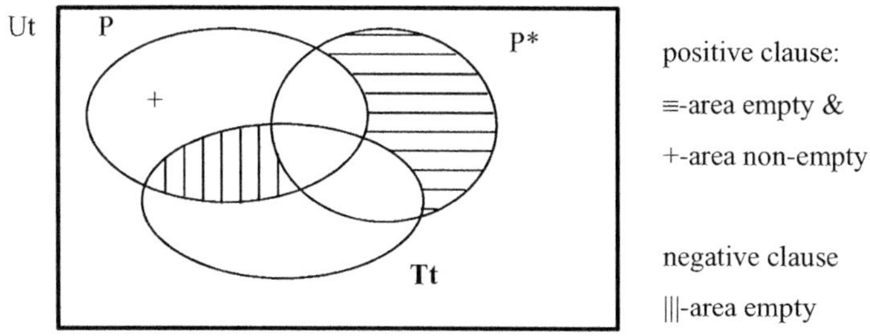

Fig. 12.2 Theoretical truthlikeness: P* is closer (or more similar) to $\mathbf{T}_t$ than P

The situation is depicted in Fig. 12.2, where U_t is the set of conceptual possibilities generated by V_t. It is easy to see that P* may be closer to $\mathbf{T}_t$ than P, even if (the claim of) P* is false, that is, even if $\mathbf{T}_t$–P* is non-empty.

Now, since 'closer to the observational truth' amounts to the truth of the Comparative Success Hypothesis (CSH), the *success theorem* stating that it guarantees 'at least as successful' will be no surprise. The *projection theorem*, however, only states that 'closer to the theoretical truth' *almost* guarantees 'at least as close to the observational truth'. More precisely, it guarantees the relevant consequence clause (P* shares all true observational consequences) without exception, but not the negative clause: if P* is closer to the truth than P on the theoretical level, there still exists the possibility on the theoretical level of a 'lucky miss' of P relative to a mistaken exclusion of both P and P*, leading to a counterexample to the negative part of the claim that P* is at least as close to the truth as P on the observational level.[7]

The two theorems imply together that 'closer to the theoretical truth' *almost* guarantees 'at least as successful'. More precisely, it guarantees the explanatory clause, but not the instantial clause, for P* may have extra observed counterexamples which are no (or at least not yet) observable counterexamples of P. Despite the 'almost' qualification, it is clear that the hypothesis that P* is closer to $\mathbf{T}_t$ than P is a genuine, that is, empirically testable, hypothesis, to be called the theoretical truth approximation hypothesis.

On the basis of the two theorems, and the definition of reference, some conclusions in the other direction may now be drawn. Of course, we may not expect that application of RS guarantees truth approximation. However, RS is nevertheless preeminently functional for it. To be precise, RS is straightforwardly functional for observational truth approximation, the latter is functional for theoretical truth approximation in a rather strong sense, and for referential truth approximation in a weak sense. Following the terminology of Sect. 11.5, let us call 'inference to the best theory as the closest to the observational truth' *observational induction*,

[7]That is, there may be a theoretical (nomic) impossibility x, that is rightly excluded by P*, but not by P (hence, missed by P), however such that there is also a theoretical possibility z, wrongly excluded by both, that has the same observational projection as x. For a figure depicting the situation, see e.g. Sect. 2.3.

'inference to the best theory as the closest to the theoretical truth' *theoretical induction*, and 'inference to the best theory as the closest to the referential truth' *referential induction*. Unlike the case of observational induction, the cases of theoretical and referential induction are not straightforward kinds of (explanatory) abduction, in the sense that an 'abduced' hypothesis (deductively) explains the relevant evidence. Whereas 'closer to the observational truth' straightforwardly explains 'at least as successful', notably theoretical induction requires extra arguments. That is, although 'closer to the theoretical truth' explains 'explanatorily at least as successful', it does not straightforwardly explain 'instantially at least as successful', for, as we have seen, accidentally escaping counterexamples of the worse theory are not excluded. Moreover, strictly speaking, referential induction on its turn does not (deductively) explain something. However, it plays indirectly a crucial role in the foregoing 'theoretical explanation', but it needs considerably extra theoretical and experimental arguments, together amounting to so-called 'triangulation'.

These results have important consequences for the four main epistemological positions, viz., instrumentalism, constructive empiricism, entity or, preferably called, referential realism, and constructive realism, that is, theory realism without essentialism. As far as they are willing to take nomic truth approximation into account, it follows from the success theorem that there is no reason for the instrumentalist not to become a constructive empiricist. In order to give 'deeper' explanations of persistent success differences, it is unavoidable for the latter to become a referential realist. Although this is quite a jump, it may be for good, triangulation reasons. Finally, for the referential realist there is, due to the projection theorem, almost no reason not to become a constructive realist.[8]

It should be stressed that 'the truth' is not something which is only reality and not language dependent. On the contrary, it is always domain and vocabulary dependent, which may be seen as an unavoidable pragmatic aspect. Moreover, truth approximation claims, based on comparative success claims, require additional decisions for pragmatic commensurability. The story becomes even more complicated when we take into account that in actual practice the determination of domains, vocabularies and truths is a matter of 'dialectical interaction'. However, this does not exclude that the resulting theories can be ordered by success and even by their distance from the relevant truth in the qualitative sense.

As in the case of success comparison, there can also be given a refined definition of truthlikeness (see Chaps. 6 and 7). It enables refined versions of the two main theorems and hence it implies, roughly, the same conclusions for the four epistemological positions. Moreover, the basic and the refined story can be told in terms of theories formulated within the same or in different vocabularies and research programs. As far as theories of one research program are concerned, hence with a hard or, preferably, semi-hard core, it follows that truth approximation is possible despite some apparent dogmatism. This holds when one aims at improving one's theory within the program, which distinguishes responsible dogmatism in science from

[8] It may well be that similar arguments may be developed such that 'structural realism' comes in a similar, but parallel, position as referential realism between constructive empiricism and constructive realism. See Chap. 16, Sects. 16.4.2 and 16.5.

pseudoscience. When the theories are embedded in different research programs the pragmatic decisions to be taken for comparison purposes multiply. A case in between has been analyzed in detail in (Hettema and Kuipers 1995): when the theories of Rutherford, Bohr and Sommerfeld are formulated in the general framework of 'the old quantum theory' of Sommerfeld, it can be shown that the mentioned triple of theories is a potential case of refined truth approximation, that is, a case of refined truth approximation assuming that Sommerfeld's theory is the truth.

If one goes from the qualitative perspective on success and truth distance comparison to the quantitative perspective, the number of pragmatic decisions increases even more, for that approach requires a distance function between the relevant conceptual possibilities. As soon as they take two or more types of quantities into account, e.g. length and mass, it requires essentially arbitrary decisions. For this reason, and because such distance functions do not seem to play a role in scientific practice, I prefer the qualitative approach, which does, of course, not exclude certain quantitative comparisons.[9]

12.8 Resulting Relativization of Falsification

From the foregoing exposition it follows that there are several respects in which falsification of a theory can be relativized. The relativizations presented in Sect. 12.3, may be summarized as follows:

(i) The description of a counterexample may be wrong, a possibility which has already been stressed by Hempel and Popper.
(ii) A counterexample may be due to an auxiliary hypothesis or background assumption, a possibility emphasized by Hempel and Lakatos.

Moreover, in Sect. 12.2 a third, rather fundamental relativization was already indicated by Fig. 12.1, and in the previous section it was seen to be a direct consequence of the definition of 'closer to the truth':

(iii) Contra (the standard interpretation of) Popper: 'being false' is not important for the distance to the truth of a theory.

Of course, 'being false' may be important for pragmatic considerations, e.g. regarding the safety of bridges, etc.

Finally, we met in the previous section a fourth, also fundamental relativization of extra counterexamples.

(iv) Realists contra empiricists/instrumentalists/pragmatists: if $\underline{P}^*$ is explanatorily more successful than $\underline{P}$, but instantially less, then $\underline{P}^*$ may still be closer to the theoretical truth than $\underline{P}$.

---

[9] See (Kuipers 1997).

The reason was that an extra counterexample of P* may not (yet) be observable as a counterexample of P. Whereas the third relativization leads to a serious relativization of the falsificationist methodology, the fourth relativization of (extra) counterexamples is a strong point of realism against weaker epistemological positions. For an empiricist, for example, it seems difficult to provide good reasons to prefer an explanatorily superior theory which has nevertheless more counterexamples.

12.9 Resulting Relativization of Confirmation

The foregoing exposition also leads to several qualifications of 'confirmation':

(a) The term 'confirmation' has more or less the connotation that the relevant hypothesis has not yet been falsified. However, regarding theories, it is clear from the comparative success and truth approximation perspective that falsified theories should not be dropped for being false. In other words, instead of speaking of confirmation, the terminology of individual and general successes is much more appropriate.
(b) There can be made relativizations of successes, similar to the first three of the four relativizations of falsification above. For logical reasons, the fourth one, however, has no 'success analogue'.
(c) However, confirmation in the traditional sense of confirmation by successes of not yet falsified hypotheses remains important. More specifically, confirmation of GTI's or, more generally, of general observational hypotheses, of comparative success hypotheses, and of truth approximation hypotheses play an important role in scientific argumentation.

Hence, from (c) it follows that we still need a theory of confirmation or, at least, a coherent language of confirmation. Moreover, rules of acceptance are badly needed. In the latter respect, I have nothing to offer, but they will have many pragmatic aspects, as is clear from e.g. Pollock (1990). For the language game of confirmation, the ratio or relevance theory of confirmation has in my opinion the best qualitative consequences (Kuipers 2000, Chap. 2). Let me summarize the main points, assuming a probability pattern p, hypothesis H and evidence E, and the degree of confirmation r(H,E) defined as the degree of relevance, that is, as p(H/E)/p(H) or p(E/H)/p(E). These ratios are equal when p(H) and p(E) are non-zero, and that if one of the latter is zero the ratio with the other denominator may well be defined. E.g. in the case of deductive confirmation, i.e., when H logically entails E,[10] p(E/H) = 1, and hence r(H,E) is defined, and equal to 1/p(E), even if p(H) is zero, but p(E) is not.

Deductive confirmation satisfies the principle that stronger evidence confirms a hypothesis more than weaker evidence (P.1), whereas a stronger hypothesis is equally confirmed as a weaker one (P.2) (but will nevertheless have a lower posterior probability, due to a lower prior probability).

[10] For simplicity, we here neglect initial conditions and background knowledge.

For confirmation in general, defined by p(E/H) > p(E), and terminologically equating 'probability' and 'plausibility' we have the following qualitative consequences:

- a hypothesis with no plausibility at all may nevertheless be confirmed,
- confirmation of a hypothesis with some plausibility makes it more plausible
- generalized versions of P.1 and P.2 obtain.

Applied to truth approximation hypotheses, we may conclude from the success theorem that the observational truth approximation hypothesis OTAH (<= > CSH) is deductively confirmed by 'being more successful', with r(OTAH,E) = 1/p(<R, S>). From the fact that the projection theorem is almost true, but has exceptions, we may conclude that the theoretical truth approximation hypothesis TTAH is almost as much confirmed as OTAH by 'being more successful'. Of course, the prior and hence the posterior probability of TTAH will be much smaller than that of OTAH.

12.10 Concluding Remarks

It has been argued that truth approximation is not a straightforward matter. Many pragmatic decisions have to be taken in order to qualify a theory transition as a step in the direction of the truth. However, this is not a reason to deteriorate into post-modern *subjective* pragmatism.[11] At least for the natural sciences, where a domain and a vocabulary may be supposed to determine together a stable, unique truth, *objective* pragmatism is the plausible modern or, if you like, post postmodern position.

References

Hettema, H., & Kuipers, T. (1995). Sommerfeld's Atombau: A case study in potential truth approximation. In T. Kuipers & A.-R. Mackor (Eds.), *Cognitive patterns in science and common sense, Poznan Studies* (Vol. 45, pp. 273–297). Amsterdam/Atlanta: Rodopi.

Kuipers, T. (1990). Reduction of laws and concepts. In J. Brzezinski, F. Coniglione, T. Kuipers, & L. Nowak (Eds.), *Idealization I: General problems II: Forms and applications, Poznan Studies* (Vol. 16, pp. 241–276). Amsterdam/Atlanta: Rodopi.

Kuipers, T. (1995a). Falsificationism versus efficient truth approximation. In W. Herfel, W. Krajewski, I. Niiniluoto, & R. Wojcicki (Eds.), *Theories and models in scientific processes, Poznan Studies* (Vol. 44, pp. 359–386). Amsterdam/Atlanta: Rodopi.

Kuipers, T. (1995b). Explicating the falsificationist and the instrumentalist methodology by decomposing the hypothetico-deductive method. In T. Kuipers & A.-R. Mackor (Eds.), *Cognitive patterns in science and common sense, Poznan Studies* (Vol. 45, pp. 165–186). Amsterdam/Atlanta: Rodopi.

[11] See (Rescher 1998) and Schurz (1998) for the important distinction between objective and subjective pragmatism.

Kuipers, T. (1997). Comparative versus quantitative truthlikeness definitions: Reply to Thomas Mormann. *Erkenntnis, 47*(2), 187–192.

Kuipers, T. (2000). *From instrumentalism to constructive realism. On some relations between confirmation, empirical progress, and truth approximation, Synthese Library 287*. Dordrecht: Kluwer Academic Publishers.

Kuipers, T. (2001). *Structures in science. Heuristic patterns based on cognitive structures. An advanced textbook in neo-classical philosophy of science. Synthese library 301*. Dordrecht: Kluwer Academic Publishers.

Pollock, J. (1990). *Nomic probability and the foundations of induction*. Oxford: Oxford University Press.

Rescher, N. (1998). Pragmatism in crisis. In P. Weingartner, G. Schurz, & G. Dorn (Eds.), *The role of pragmatics in contemporary philosophy (Proceedings of the 20th International Wittgenstein -Symposium, August 1997)* (pp. 24–38). Vienna: Hölder-Pichler-Temsky.

Schurz, G. (1998). Kinds of pragmatisms and pragmatic components of knowledge. In P. Weingartner, G. Schurz, & G. Dorn (Eds.), *The role of pragmatics in contemporary philosophy (Proceedings of the 20th International Wittgenstein -Symposium, August 1997)* (pp. 39–57). Vienna: Hölder-Pichler-Temsky.

Chapter 13
Empirical Progress and Nomic Truth Approximation by the 'Hypothetico-*Probabilistic* Method'

Abstract Three related intuitions are explicated in this chapter. The first is the idea that there must be some kind of probabilistic version of the HD-method, a 'Hypothetico-Probabilistic (HP-) method', in terms of something like probabilistic consequences, instead of deductive consequences. According to the second intuition, the comparative application of this method should also be functional for some probabilistic kind of empirical progress, and according to the third intuition this should be functional for something like probabilistic truth approximation. In all three cases, the guiding idea is to explicate these intuitions by explicating the crucial notions as appropriate 'concretizations' of their basic or deductive analogues, being 'idealizations'.

It turns out that the comparative version of the proposed HP-method amounts to a sophisticated likelihood ratio success (SLRS-) method applied to the cumulated evidence. This method turns out to be not only functional for probabilistic empirical progress but also for probabilistic truth approximation. The latter is based on a probabilistic threshold theorem constituting for this reason the analogue of the deductive success theorem.

This chapter is in fact a supplement to Chap. 5. Despite some terminological deviation, there is substantial partial overlap to enable independent reading.

Keywords Hypothetico-probabilistic method · HP-method · Probabilistic empirical progress · Probabilistic truth approximation · Quantitative truth approximation · Sophisticated likelihood method · Probabilistic success theorem · Inference to the best theory

---

Strongly revised version of: "Empirical progress and Truth Approximation by the 'Hypothetico-Probabilistic Method'", *Erkenntnis, 70.3*, 2009, 313–330. *Acknowledgements*: I would like to thank for all the comments I got before and after writing in 2005 the preliminary paper (2007a) when presenting the ideas in Bielefeld (2003), Rotterdam (2004), Konstanz (2004), Trieste (2007), Amsterdam (2007), Leusden (2007) and in the PCCP research group in Groningen (2004, 2007). In particular I want to thank Igor Douven, Dale Jacquette, Jacek Malinowski, Elliott Sober, Ilkka Niiniluoto, Jan-Willem Romeijn, Jeanne Peijnenburg, and David Atkinson. I also like to thank the Netherlands Institute for Advanced Research (NIAS) in Wassenaar for the opportunity to spend in 2005 and 2007 two weeks on this and another paper. Finally, I like to thank two anonymous referees for their constructive questions.

T. A. F. Kuipers, *Nomic Truth Approximation Revisited*, Synthese Library 399,
https://doi.org/10.1007/978-3-319-98388-2_13

13.1 Introduction

In previous work I have argued in detail that the comparative application of the hypothetico-deductive (HD-)method is not only functional for making empirical progress, but even for achieving nomic truth approximation, in both cases in the basic sense (see (Kuipers 2000), for a monograph synthesis, and Chap. 4 of this volume for a generalized, two-sided version). For the last conclusion the so-called (basic or deductive) success theorem is crucial: being closer to the truth entails always being (empirically) at least as successful and it entails almost always becoming more successful in the long run, that is, showing empirical progress.

Since the basic versions of 'more successful' and 'closer to the truth' presented in Chap. 4 can be given a deductive interpretation, see below, and since the quantitative version presented in Chap. 5 is essentially probabilistic, the terms 'deductive' and 'probabilistic', instead of 'basic' and 'quantitative' will mainly be used here for the following reason. It highlights the fact that the probabilistic versions are concretizations of their deductive idealizations, the more so in view of the same relation between the hypothetico-probabilistic method, to be presented, and the hypothetico-deductive method.

Three intuitions form the basis of this chapter. The first is the idea that there must be some kind of probabilistic version of the HD-method, a 'Hypothetico-Probabilistic (HP-) method', in terms of something like probabilistic consequences, instead of deductive consequences. According to the second intuition, the comparative application of this method should also be functional for some probabilistic kind of empirical progress, and according to the third intuition this should be functional for a probabilistic kind of truth approximation. In all three cases, the guiding idea is to explicate these intuitions by explicating the crucial notions as appropriate 'concretizations' of their deductive analogues, being 'idealizations'.

In a previous paper (Kuipers 2007a), the HP-paper for short, the first intuition, already suggested by Ilkka Niiniluoto in (Niiniluoto and Tuomela 1973, p. 204, 209), has already been successfully explicated, and the second as well, at least to a large extent. However, it turned out to be a rather difficult task to explicate the third intuition, for two reasons. First, although 'probabilistically closer to the truth' would have to be explicated as a genuine concretization of 'deductively closer to the truth', this was very difficult to prove for the most plausible *prima facie* concretization. Second, whereas a conceptually interesting, but impractical, probabilistic success theorem in terms of 'expected success' was plausible to think of, finding a probabilistic version of the success theorem with practical value, turned out to be a difficult challenge.

In a forerunner of the present chapter (Kuipers 2009), these problems have been solved, with some surprising aspects. However, from the perspective of the truth- and falsity-content analysis in the present book, the probabilistic versions of 'more successful' and 'closer to the truth' in the 2009-paper are still unsatisfactory. The present chapter revises the paper in harmony with the basic (deductive) and quantitative (probabilistic) explications in Chap. 4 and 5. It integrates brief versions of the explication of the first intuition, that there is something like an HP-method, and a relevant adaptation of the second, that this method is functional for empirical progress.

I like to stress that for this kind of analysis the general (meta-)method of explication (Kuipers 2007b) is indispensable. It goes back to the constructive attitude of Carnap and Hempel to the analysis of 'scientific common language and scientific common sense': it requires no genuine artificial language, but may well require some well-argued revision of concepts and intuitions. More specifically, this project requires the 'conceptual version of idealization and concretization (conceptual-I&C)' (Kuipers 2007b, and more specifically, 2007c), going from the relevant deductive notions to probabilistic ones. In order to be a proper concretization, it should satisfy the *C-test*, which has a formal and material side. The idealizational point of departure should formally appear as an extreme special case of the concretization. Moreover, the latter should be a case of 'conceptual progress' relative to the problems and limitations of the former, which we will obtain in the present case.

13.2 The 'Hypothetico-Probabilistic (HP-) Method' as a Concretization of the 'Hypothetico-Deductive (HD-) Method'

13.2.1 The Primary Concretization

The primary concretization needed is the concretization of the crucial notion of the HD-method, viz. deductive consequences, D-consequences. There have been proposed in the literature quite a number of alternatives. They are briefly presented and compared in the HP-paper (Kuipers 2007a). For our purposes, there are two, related, useful possibilities. Starting from the idealization, for statements H and E:

$H \models E$: *E is a deductive (D-) consequence of H*

and assuming some probability function p(.) the first attractive possibility is:

$p(E/H) \geq p(E)$: *E is a probabilistic (Pr-) consequence of H*

This is also known as the (weak) positive relevance condition in the context of explicating the notion of 'confirmation'. The probability function may be of any kind, in particular, in terms of a classification introduced in (Kuipers 2001, 2006), Popperian (non-inductive), Carnapian (inductive likelihoods), Bayesian (inductive priors), Hintikkian (double inductive). As to the formal C-test, it directly follows from

if $H \models E$, then $1 = p(E/H) \geq p(E)$

that a D-consequence is an extreme special case of a Pr-consequence. Here, and later, it is easy to see how we get conditional versions, that is, assuming conditions such as initial conditions, auxiliary hypotheses, background assumptions, and core principles. Indicating them with C, we get in the present case:

if H, C $\models$ E, then 1 = p(E/H&C)≥p(E&C)

The alternative possibility would be to take "p(E/H) ≥ p(¬E/H)", that is, "p(E/H) ≥ ½", as the defining condition of a Pr-consequence, a condition that is again satisfied in an extreme sense by a D-consequence. Although I have a slight preference for the first possibility, there is not much reason to debate the choice, because the relevance of making a choice disappears as soon as we enter (comparisons of) likelihood ratios of the kind "p(E/H*)/p(E/H)", which will become crucial for our analysis.[1]

13.2.2 From the HD- To the HP-Method of Testing

Although the context of testing of a hypothesis (H) resulting in evidence (E) was suggested by our abbreviations, so far nothing substantial was particularly assumed about such a context. In the context of the HD-method of testing, the crucial notion is that of a (deductive) test implication (E) of a hypothesis (H), which is of course explicated as a (non-trivial) D-consequence of an observational nature. Of course, 'observational' will here be understood in the 'theory-relative' sense that relative to H the non-formal terms occurring in E are supposed to be observational, however theory-laden they may be with underlying or background theories. In sum, we get as the idealized point of departure:

E is a *D-test implication* of H iff
E is an 'observational' D-consequence of H such that not "$\models$E"

The concretization is now almost as plausible:

E is a Pr-*test implication* of H iff
E is an 'observational' Pr-consequence of H such that "p(E/H) > p(E)"

It satisfies the C-test, with a marginal exception, for: "if H $\models$ E and not $\models$ E then p(E/H) = 1 > p(E)" holds for almost all kinds of evidence, but not for 'almost tautological' evidence, that is, non-tautological evidence for which p(E) nevertheless equals 1. Although such evidence is conceptually possible, it is not something to keep taking into account; hence we may conclude that a D-test implication is an extreme case of a Pr-test implication.

Now we are in the position to formulate the basic aim of the HD-method of testing: it aims at deriving D-test implications and finding out, by experiment or mere observation, whether they are true or false, which is called (D-)confirmation and falsification of the hypothesis, respectively. In sum, the HD-method aims at

[1] For both versions it is interesting for various formal reasons to study the set of Pr-consequences of a statement: PCn(A) = $_{def}$ {B/p(B/A) ≥ p(B)}, {B/p(B/A) ≥ p(¬B/A)}, respectively. In our HP-paper we have made a start.

D-confirmation or falsification of D-test implications. Being restricted to D-consequences, it is an idealization. Its suggested concretization, the Hypothetico-Probabilistic (HP-)Method, aims at Pr-confirmation or Pr-disconfirmation of Pr-test implications. Here 'Pr-confirmation' amounts to the coming out to be true of a Pr-test implication, and 'Pr-disconfirmation' its coming out to be false.[2]

13.2.3 Further Challenges

Starting from the deductive explications, taking counterexamples and falsified theories into account, presented in (Kuipers 2000), the main challenges were to give a coherent I&C-explication of probabilistic versions of the deductive ones of (1) confirmation and falsification, of (2) testing, and separate and comparative evaluation, and of (3) empirical progress and truth approximation. The first challenge was already met in (Kuipers 2000, Chap. 3) and improved in (Kuipers 2001, Section 7.1), the second one in the HP-paper (Kuipers 2007a). The 2009-paper builds upon these results and partially redirects them and combines them with meeting the third challenge. The present chapter adapts it to renewed definitions of probabilistically 'more successful' and 'closer to the truth' in line with Chap. 5.

The main claim of the HP-paper, integrated in the 2009-paper, was that the method of likelihood comparisons (the *LC-method*) is *the* concretization of (instrumentalist) comparative HD-evaluation. The present chapter replaces the LC-method by a sophisticated version of it, and adds to this the claims that this method is, like the comparative HD-method, not only functional for empirical progress but also for truth approximation.

More specifically, the HP-paper shows that there is a HP-method as a straightforward concretization of the HD-method. It concerns testing and separate and comparative evaluation[3] and is, depending on the choice of probability function, like the explication of 'probabilistic confirmation', of a Popperian, Carnapian, Bayesian, or Hintikkian nature. The first is non-inductive, by using only Pr-functions without ampliative effect on the re-occurrence of a property, also called 'instantial confirmation'. Carnapian Pr-functions achieve this inductive feature by a likelihood function having this effect, whereas Bayesian Pr-functions achieve it by assuming a prior distribution over a family set of hypotheses. Finally, Hintikkian Pr-functions achieve it by the combination of the two, for which reason these Pr-functions may be called 'double-inductive'.

The relevant aspects of HP-testing and separate HP-evaluation for this chapter have been presented above. The core of the explication of 'comparative HP-evaluation' amounts to comparisons of likelihoods of the kind p(E/H*) / p(E/H) for evidence E and hypotheses or theories H and H*, further on specified and called the Sophisticated Likelihood Ratio Success (SLRS-) method. In Sect. 13.4 we will

[2] See (Kuipers 2001, 7.1.2) for an analysis of probabilistic (dis-)confirmation.

[3] Surprisingly enough, this explication also generates as a side product probabilistic explications of explanation and prediction, also presented in the HP-paper.

present in an integrated form how this method is functional for probabilistic empirical progress and probabilistic truth approximation. It will also provide an adapted summary of the way in which it is argued in (Kuipers 2000) that the comparative HD-method is functional for deductive empirical progress and deductive truth approximation. But first we have to define probabilistic truthlikeness, the core notion of probabilistic truth approximation. The notion of probabilistic more successfulness and hence probabilistic empirical progress will be introduced in the course of Sect. 13.4.

13.3 From Deductive to Probabilistic Truthlikeness

13.3.1 General Assumptions and Restrictions

As in the whole book, the present analysis is also restricted to laws and theories of a deterministic nature dealing with what is nomically possible versus impossible. Such laws and theories can be characterized by subsets of the set U of conceptual possibilities generated by a given vocabulary for the intended domain. In the structuralist approach to theories it has been demonstrated by many realistic examples that it is at least possible to define the type of set-theoretic structures, here called conceptual possibilities, that are involved. See Chap. 12 of (Kuipers 2001) or Sect. 13.2 of (Kuipers 2007d) for a survey.

When subsets A and B of U can be axiomatized, then $A \subseteq B$ amounts to $A \models B$. Note that we do not make a notational distinction between statements and their models, such as in the present case between the possible axiomatization of a subset and the subset itself, that is, the class of models of the axiomatization. It may take some effort to get acquainted with this simplifying convention.

Let (bold) **T** denote the (unknown) subset of U of nomic possibilities. We restrict ourselves to exclusion theories, see Chap. 2 or 4, according to which a theory $\underline{P}$ (with P being a subset of U, e.g. the models of some postulates) makes the claim "$\mathbf{T} \subseteq P$" and theory $\underline{P}$ is true iff this claim is true. If $P^{\#} \subseteq P$ the claim of theory $\underline{P^{\#}}$ is of course stronger than that of theory $\underline{P}$. It directly follows that a characterization of **T**, axiomatic or set-theoretic, indicated by non-bold T, represents 'the strongest true theory', called '*the* (nomic) truth'.

13.3.2 Deductive Truthlikeness

The definition of '(comparative) deductive truthlikeness' can be given in three ways, a model, a (deductive) consequence and a mixed version. Here, Q(P) indicates the set of supersets of P (within U). This set represents in a set-theoretic way the (deductive) consequences of P.

Deductive (D-)definition of 'more truthlikeness':
"P* is ('deductively') at least as close/similar to the (nomic) truth **T** as P"
(or "P* is D-at least-as-close to **T** as P") iff:

(i) $cP \cap c\mathbf{T} \subseteq cP^* \cap c\mathbf{T}$
$(\equiv P^* \cap c\mathbf{T} \subseteq P \cap c\mathbf{T} \quad \equiv P^* \cup \mathbf{T} \subseteq P \cup \mathbf{T} \quad \equiv Q(P) \cap Q(\mathbf{T}) \subseteq Q(P^*) \cap Q(\mathbf{T}))$

(ii) $cP^* - c\mathbf{T} \subseteq cP - c\mathbf{T}$
$(\equiv P \cap \mathbf{T} \subseteq P^* \cap \mathbf{T} \quad \equiv P \cup c\mathbf{T} \subseteq P^* \cup c\mathbf{T} \quad \equiv Q(P^*) \cap Q(c\mathbf{T}) \subseteq Q(P) \cap Q(c\mathbf{T}))$

"P* is D-closer to **T** than P" iff in addition both subset relations are proper subset relations.

The model clauses, first listed, can be interpreted as stating (i) that the truth-content of P is a subset of that of P* and (ii) that the falsity-content of P* is a subset of that of P.[4] The (deductive) consequence clauses, last listed, can be interpreted as stating (i) that all true consequences of P are (true) consequences of P* and (ii) that all 'strongly-false' consequences of P* are (strongly-false) consequences of P.[5] The mixed definition combines of course the consequence reading of (i) with the model reading of (ii).

Note that the definition of 'D-closer to' is strong. Weaker versions (e.g. 'at least one proper subset relation') are also possible. Note also that, in terms of symmetric differences, (i) and (ii) together amount to $\mathbf{T}\Delta P^* \subseteq \mathbf{T}\Delta P$ and in case the theories are axiomatizable also to: $\mathbf{T} \cap P \models P^* \models \mathbf{T} \cup P$. This equivalence makes it even more plausible that the definition is in a crucial sense of a deductive nature.

Figure 13.1 depicts "P* is D-closer to **T** than P" set-theoretically. Note that P* is as it were moving from P to **T**.

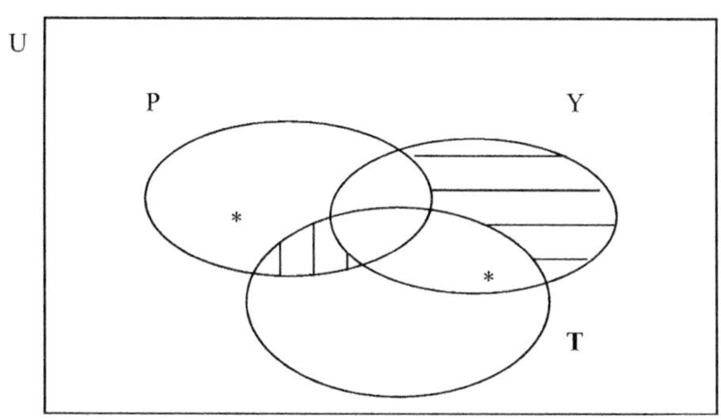

Fig. 13.1 "P* is D-closer to **T** than P": shaded areas empty, *-areas non-empty

[4] Note also that the truth-content of P captures the rightly excluded possibilities and the falsity-content the wrongly excluded possibilities. See further Chap. 4.

[5] A 'strongly-false' consequence is not just not a consequence of **T**, but even a consequence of c**T**. For the consequence and the mixed versions see further (Kuipers 2000, Section 8.1). It should be noted that the versions given there assume maximal theories.

13.3.3 Probabilistic Truthlikeness

Let $m(.)$ represent any probability measure (function) on the set of (measurable) subsets of U. It is important to note that m(P) is no prior in the sense of the probability that (the claim of) theory P is true, i.e., m("**T**⊆P"). For this reason we prefer here 'm' for the probability function, in accordance with Chap. 5, where m(P) is the absolute probability (measure) of P as a subset of the total outcome space U. Hence, $m(.)$ is such that $m(\emptyset) = 0 \leq m(P) \leq m(U) = 1$.

Highly inspired by the deductive definition of 'closer to the truth', see as second listed equivalences in that definition, we define:

Probabilistic (Pr-)definition of 'closer to the truth'
"P* is Pr-at least as close/similar to the (nomic) truth **T** as P" iff
Unconditional (UPr-)version:
(i-p) $m(P^*\cap c\mathbf{T}) \leq m(P\cap c\mathbf{T})$
(ii-p) $m(P\cap\mathbf{T}) \leq m(P^*\cap\mathbf{T})$
Conditional Probabilistic (CPr-)version, assuming $0 < m(\mathbf{T}) < 1$,
(i-p) $m(P^* \mid c\mathbf{T}) \leq m(P \mid c\mathbf{T})$
(ii-p) $m(P \mid \mathbf{T}) \leq m(P^* \mid \mathbf{T})$
Likelihood Ratio (LR-) version, assuming $0 < m(P), m(P^*) < 1$
(i-p) $m(P^*)m(c\mathbf{T}/P^*) \leq m(P)m(c\mathbf{T}/P)$ ($\equiv m(c\mathbf{T}/P^*)/m(c\mathbf{T}/P) \leq m(P)/m(P^*)$)
(ii-p) $m(P)m(\mathbf{T}/P) \leq m(P^*)m(\mathbf{T}/P^*)$ ($\equiv m(P)/m(P^*) \leq m(\mathbf{T}/P^*)/m(\mathbf{T}/P)$)
"P* is Pr-closer to **T** than P" iff in addition the two inequalities are proper inequalities, to be indicated by (i-p$^+$) respectively (ii-p$^+$).

The strong conditional clauses can be interpreted in terms of random sampling: (i-p$^+$) amounts to the claim that the probability of getting a (member of) P* by random sampling in c**T** is smaller than that of getting a P, and (ii-p$^+$) to the claim that the probability to get a (member of) P by random sampling in **T** is smaller than that of getting a P*.

Note that e.g. m(P)m(T/P) is a kind of weighted likelihood. Hence, the strong likelihood clauses may be read as: P* yields a higher weighted likelihood for the nomic possibilities than P (ii-p$^+$) and a lower weighted likelihood for the nomic *im*possibilities (i-p$^+$).

The two likelihood ratio (LR-)clauses can be combined, using the bracketed equivalences, to the following:

Sophisticated Likelihood Ratio Truthlikeness condition (SLRT-condition):
("P* is Pr-at least as close/similar to the (nomic) truth **T** as P" iff)
$m(c\mathbf{T}/P^*)/m(c\mathbf{T}/P) = [1\text{-}m(\mathbf{T}/P^*)]/[1\text{-}m(\mathbf{T}/P)] \leq m(P)/m(P^*) \leq m(\mathbf{T}/P^*)/m(\mathbf{T}/P)$

It is called the Sophisticated Likelihood Ratio Truthlikeness condition, because by leaving out the middle term it reduces to the essentially weaker:

Likelihood *Ratio Truthlikeness condition* (LRT-condition):
("P* is Pr-at least as close/similar to the (nomic) truth **T** as P" entails)
$m(\mathbf{T}/P) \leq m(\mathbf{T}/P^*)$

which may be paraphrased as: P* yields a higher likelihood for the nomic possibilities than P.

According to the SLRT-condition the likelihood-ratio $m(\mathbf{T}/P^*)/m(\mathbf{T}/P)$ has a minimum, i.e., $m(P)/m(P^*)$, and the 'complementary' likelihood ratio $m(c\mathbf{T}/P^*)/m(c\mathbf{T}/P)$ has this value as a maximum. Note that the (bi-directional) threshold $m(P)/m(P^*) =_{def} h$ is essentially known as soon as $m(.)$ has been fixed.

13.3.4 Survey of Successive Concretizations and Generalizations

In order to show that the Pr-definition is, at least in the formal sense, a concretization of the D-definition it is illuminating to schematically present some intermediate definitions and their relations. An arrow indicates 'entails', and '(e)sc' indicates 'is (extreme) special case of'. 'N-closer to' is short for 'Numerically closer to', and is only defined for finite U. 'UPr-/CPr-closer to' is short for 'Unconditional/Conditional Probabilistically closer to'.

	'(i)-conditions'		'(ii)-conditions'
D-closer to	$P^*\cap c\mathbf{T} \subset P\cap c\mathbf{T}$	&	$P\cap\mathbf{T} \subset P^*\cap\mathbf{T}$
	$\Downarrow$	esc	$\Downarrow$
N-closer to	$\lvert P^*\cap c\mathbf{T}\rvert < \lvert P\cap c\mathbf{T}\rvert$	&	$\lvert P\cap\mathbf{T}\rvert < \lvert P^*\cap\mathbf{T}\rvert$
	$\Downarrow$	sc	$\Downarrow$
UPr-closer to	$m(P^*\cap c\mathbf{T}) < m(P\cap c\mathbf{T})$	&	$m(P\cap\mathbf{T}) < m(P^*\cap\mathbf{T})$
	$\Downarrow\Uparrow$	$\Downarrow\Uparrow$	$\Downarrow\Uparrow$
CPr-closer to	$m(P^*/c\mathbf{T}) < m(P/c\mathbf{T})$	&	$m(P/\mathbf{T}) < m(P^*/\mathbf{T})$

In sum, 'D-closer to' is an extreme special case of 'N-closer to', assuming finite U, which is a special case of 'UPr-closer to', which is equivalent to 'CPr-closer to'. Of course, 'D-closer to' can also directly be seen as an extreme special case of 'UPr-closer to'/'CPr-closer to'. Note that all these claims hold already condition-wise and also already for the weak 'at least as close'-definitions.

One important problematic aspect of the definition of 'D-closer to' is that it holds only very rarely because most theories are incomparable according to that definition. Since 'Pr-closer to' is much weaker than 'D-closer to', this naturally suggests the question to what extent the possibilities for more truthlikeness and hence for truth approximation have increased. It is clear that the number of possibilities have increased enormously in any informal sense. In fact, theories $\underline{P}$ and $\underline{P^*}$ are only probabilistically incomparable regarding truthlikeness when e.g. the SLRT-condition holds in neither direction. However, this does not very much influence the extent to which it now covers real cases of 'closer to', that is, cases going beyond the toy level of, for example, theories about simple electric circuits (Chap. 2).

The main problem of the deductive and the probabilistic approach is the fact that in real cases theories $\underline{P}$ and $\underline{P^*}$ will have no overlap with $\mathbf{T}$ at all. It is easy to see from the UPr-version that in this case P* is closer to the truth than P iff $m(P^*) < m(P)$. This is an unsatisfactory situation; it is vulnerable for the probabilistic analogue of the 'child's play objection' to the deductive definition. For in the basic case we have that P* is closer to the truth than P iff $P^* \subset P$, with the consequence that we can come closer to the truth simply by excluding still more possibilities. To meet this 'problem of non-overlap' we have to refine the probabilistic approach by taking structure-likeness into account, as has been done for the basic (deductive) case (Kuipers 2000, Chap. 10; and Chap. 6 in this book). For the probabilistic case a refined version still has to be developed.

13.4 The *Comparative* HP–Method Is Functional for Empirical Progress and Truth Approximation

13.4.1 Empirical Evidence

In order to deal with the way in which the HP-method might be helpful for empirical progress and truth approximation we need a representation of empirical findings. Let R_n represent the set of realized possibilities at time n and S_n the strongest law induced from R_n, that is, the resulting conjunction of all induced laws. In the former we can of course make descriptive mistakes, in the latter inductive mistakes. However, if we assume, by way of strong idealizations, that we have not made such mistakes, we may conclude that R_n is a subset of **T**, for we cannot realize nomic impossibilities, and that S_n is a superset of **T**, for it must then be true as a nomic theory. Moreover, while R_n can only grow, not shrink, in the course of time, S_n can only shrink, that is, become stronger. In sum, we have the following 'cumulative evidence relations':

$R0 = \emptyset \subseteq \ldots R_n \subseteq R_{n+1} \ldots \subseteq \mathbf{T}$, provided no descriptive mistakes
$S0 = U \supseteq \ldots S_n \supseteq S_{n+1} \ldots \supseteq \mathbf{T}$, provided no inductive mistakes
or, equivalently,
$cS0 = \emptyset \subseteq \ldots cS_n \subseteq cS_{n+1} \ldots \subseteq \mathbf{T}$, provided no inductive mistakes

This representation of evidence is useful for both our deductive and probabilistic purposes.

13.4.2 Deductive Success Theorem

We now turn to the introduction of the crucial theorem for the claim that the HD-method is functional for empirical progress and truth approximation. First we need an appropriate definition of (more) successfulness:

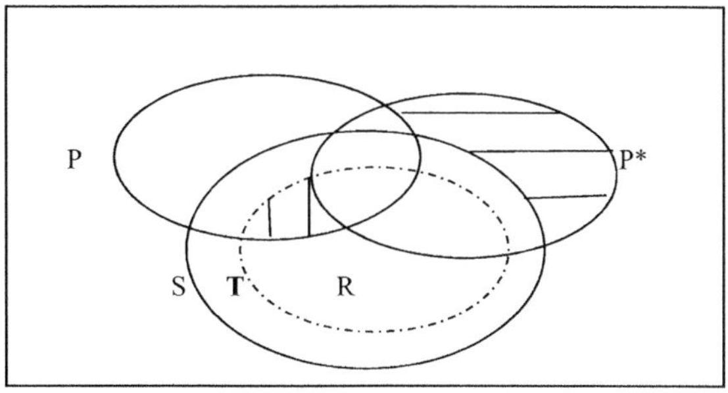

Fig. 13.2 'P* is D-at least as successful as P relative to R/S': shaded areas empty

Definition of 'deductively more successfulness':
"P* is D-at least as successful as P (relative to R/S)", depicted in Fig. 13.2:

(j) $cP \cap cS \subseteq cP^* \cap cS$ ($\equiv P^* \cap cS \subseteq P \cap cS \equiv P^* - S \subseteq P - S \equiv P^* \cup S \subseteq P \cup S$)

(jj) $cP^* - cR \subseteq cP - cR$ ($\equiv R \cap P^* \supseteq R \cap P \equiv R - P^* \subseteq R - P$)

"P* is D-more successful than P (relative to R/S)" iff both subset relations are proper.

The result, "P* is D-more successful than P (relative to R/S)", may well be seen as the ideal result of applying the HD-method, more specifically the method of comparative HD-evaluation to two theories. Let H be a test implication of P*, hence it is a superset of P*. If we get a counterexample of H, that is, if we get a member in R that is excluded by H, hence by P*, condition (jj) (see last mentioned equivalence) requires that it is also a counterexample of P. In other words, according to (jj), all established counterexamples of P* should also be counterexamples of P. On the other hand, if H′ is a test implication of P, hence a superset of P, that turns out not to get counterexamples despite all our attempts, and hence is inductively concluded to be true, that is, to be a superset of S, also called a general success of P, then (j) (see last equivalence) requires that H′ is also a superset of P*, hence a successful test implication of P*. In other words, according to (j), all general successes of P should also be general successes of P*. Of course, P* is more successful (in the strong sense) than P if it drops at least one counterexample of P and counts at least one extra general success.

By the way, the general success reading of (j) and the counterexample reading of (jj) typically reflect the (deductive) consequence interpretation of the given definition of 'more successfulness'.

The rest of the comparative HD-method story in terms of HD-results, i.e., general successes and counterexamples, now goes as follows: Suppose at time n, P* happens to be more successful than P, relative to R_n/S_n. This result need not indicate that P* is really better than P, for it may result from a, relative to P*, happy choice of experiments. Hence, before we draw far reaching conclusions we have to consider and test the Comparative Success Hypothesis (CSH), according to which P* remains more successful than $\underline{P}$. CSH is a genuine empirical hypothesis, albeit of a comparative nature, which can also be tested by the HD-method. One route is trying to derive test implications from $\underline{P}$ that do not follow from $\underline{P^*}$. If they turn out to be a (general) success of $\underline{P}$, they form a kind of counterexample to CSH, more specifically, to the claim that all general successes of $\underline{P}$ are general successes of $\underline{P^*}$. If they turn out to have counterexamples (on the observational level) they confirm CSH, even in a strong sense as far as they are not excluded by $\underline{P^*}$. Similarly, one may derive test implications from $\underline{P^*}$ that do not follow from $\underline{P}$. If they turn out to be (general) successes of $\underline{P^*}$, they confirm CSH in the strong sense, if they are falsified and the relevant counterexamples happen to be allowed by $\underline{P}$, these counterexamples also form a kind of counterexample to CSH, viz. to the claim that all counterexamples of $\underline{P^*}$ should be counterexamples of $\underline{P}$.

As in the context of inductive generalization, one cannot continue testing CSH. After 'sufficiently many and varied' tests, supporting the claim that $\underline{P^*}$ seems so far persistently more successful, one will inductively jump to the conclusion that CSH happens to be true, and hence that we may eliminate $\underline{P}$ in favor of $\underline{P^*}$, for the time being. I have called this the *Instrumentalist Rule of Success* (IRS), rather than a falsificationist rule of success, for it remains to take falsified theories seriously. IRS typically amounts to the conclusion that we have made 'empirical progress' by the transition from theory $\underline{P}$ to theory $\underline{P^*}$, hence the comparative HD-method is functional for achieving empirical progress.

The first question to answer in the sequel is how to extend this story about empirical progress to the comparative HP-method, more specifically, to the SLRS-method, still to be specified. However, first we will introduce the second question.

The next claim is that IRS is also functional for truth approximation. This is based on the *deductive success theorem* (DST), according to which 'D-at least as close to' entails 'D-at least as successful', assuming $R_n \subseteq \mathbf{T} \subseteq S_n$ (i.e., provided no mistakes have been made). The proof is rather easy, as is immediately clear from confronting Fig. 13.1 with Fig. 13.2. Under certain conditions we may even conclude that in the long run: 'D-closer to' leads to 'D-more successfulness'.

The functionality for truth approximation can now be argued as follows. If, as assumed by IRS, $\underline{P^*}$ seems so far persistently more successful than $\underline{P}$, the following holds:

- first, it is possible that $\underline{P^*}$ is D-closer to the truth than $\underline{P}$, a possibility which would, when conceived as hypothesis, according to DST, explain the greater success,
- second, it is impossible that $\underline{P}$ is D-closer to the truth than $\underline{P^*}$, for otherwise, so teaches DST, $\underline{P^*}$ could not be more successful.
- third, it is also possible neither is the case, in which case, however, another specific explanation has to be given for $\underline{P^*}$'s so far seeming persistently more successful than $\underline{P}$.

Hence, since IRS is based on the HD-method, more specifically, the comparative HD-method, the second question to answer in the sequel is how to extend this story about truth approximation to the comparative HP-method, more specifically, the SLRS-method.

13.4.3 Probabilistic Success Theorems

Comparative HP-evaluation may take the following form. Let hypothesis H be a general Pr-test implication of P* and perhaps also, but weaker, of P, in the first sense defined in Sect. 13.2, that is, let H be a general observational statement such that $m(\text{H}) < (m(\text{H/P}) <)\ m(\text{H/P*})$. This may be a good reason to put H to the test, that is, to test whether the claim "$\mathbf{T} \subseteq \text{H}$" is true. Such a test will always lead to realized possibilities and hence enlarge R_n to R_{n+1}, and hence to a stronger claim "$\text{R}_{n+1} \subseteq \mathbf{T}$", and it may or may not lead to a new accepted inductive generalization and, if so, to a transition from S_n to a smaller S_{n+1}, and hence a stronger claim "$\mathbf{T} \subseteq \text{S}_{n+1}$". Note that this description includes the comparative HD-evaluation, viz. the case that $m(\text{H/P*}) = 1$.

There are of course different ways in which R_n and Sn may 'increase', but the crucial question is of course how to compare the successes of two theories and how to relate this to their comparative truthlikeness.

Before we can formulate something like a probabilistic success theorem, we need to specify the probabilistic versions of the cumulative evidence relations and the notion of being 'probabilistically more successful'. Recall the following relations:

$\text{R0} = \varnothing \subseteq \ldots \text{R}_n \subseteq \text{R}_{n+1} \ldots \subseteq \mathbf{T}$, provided no descriptive mistakes
$\text{S0} = \text{U} \supseteq \ldots \text{S}_n \supseteq \text{S}_{n+1} \ldots . \supseteq \mathbf{T}$, provided no inductive mistakes
or, equivalently,
$\text{cS0} = \varnothing \subseteq \ldots \text{cS}_n \subseteq \text{cS}_{n+1} \ldots \subseteq \text{c}\mathbf{T}$, provided no inductive mistakes

hence, assuming no mistakes, in probabilistic terms we have, for any P, $p(\text{P}) > 0$:

$0 = m(\text{R}_0/\text{P}) \leq \ldots . \leq m(\text{R}_n/\text{P}) \leq m(\text{R}_{n+1}/\text{P}) \ldots .. \leq m(\mathbf{T}/\text{P})$
$1 = m(\text{S}_0/\text{P}) \geq \ldots .. \geq m(\text{S}_n/\text{P}) \geq m(\text{S}_{n+1}/\text{P}) \ldots .. \geq m(\mathbf{T}/\text{P})$
or, equivalently,
$0 = m(\text{cS}_0/\text{P}) \leq \ldots . \leq m(\text{cS}_n/\text{P}) \leq m(\text{cS}_{n+1}/\text{P}) \ldots .. \leq m(\text{c}\mathbf{T}/\text{P})$

Assuming in addition that in the end all nomic possibilities can be realized, we may even sharpen these 'probabilistic cumulative evidence relations' to:

$m(\text{R}_n/\text{P})$ is a monotone non-decreasing function approaching $m(\mathbf{T}/\text{P})$
$m(\text{S}_n/\text{P})$ is a monotone non-increasing function approaching $m(\mathbf{T}/\text{P})$
or, equivalently,
$m(\text{cS}_n/\text{P})$ is a monotone non-decreasing function approaching $m(\text{c}\mathbf{T}/\text{P})$

Now we are able to define the probabilistic analogue of 'deductive more successfulness', inspired by the as second listed equivalences in the latter definition:

Probabilistic(Pr-)definition of 'more successful'
"P* is Pr-at least as successful as P, relative to R_n/S_n" iff
Unconditional (UPr-)version
(j-p) $m(P^* \cap cSn) \leq m(P \cap cSn)$
(jj-p) $m(P \cap R_n) \leq m(P^* \cap R_n)$
Conditional Probabilistic (CPr-) version, assuming $0 < m(R_n) < m(S_n) < 1$,
(j-p) $m(P^* \mid cS_n) \leq m(P \mid cS_n)$ $(\equiv m(cS_n/P^*)/m(cS_n/P) \leq m(P)/m(P^*))$
(jj-p) $m(P \mid R_n) \leq m(P^* \mid R_n)$ $(\equiv m(P)/m(P^*) \leq m(R_n/P^*)/m(R_n/P))$
Likelihood Ratio (LR-) version, assuming $0 < m(P), m(P^*) < 1$,
(j-p) $m(P^*)m(cS_n/P^*) \leq m(P)m(cS_n/P)$ $(\equiv m(cS_n/P^*)/m(cS_n/P) \leq m(P)/m(P^*))$
(jj-p) $m(P)m(R_n/P) \leq m(P^*)m(R_n/P^*)$ $(\equiv m(P)/m(P^*) \leq m(R_n/P^*)/m(R_n/P))$
"P* is Pr-more successful than P" iff in addition the two inequalities are proper inequalities, to be indicated by (j-p$^+$) respectively (jj-p$^+$).

The strong conditional clauses can again be interpreted in terms of random sampling: (j-p$^+$) amounts to the claim that the probability of getting a (member of) P by random sampling in cS_n is larger than that of getting a P*, and (jj-p$^+$) to the claim that the probability to get a (member of) P* by random sampling in R_n is larger than that of getting a P.

Noting that R_n represents the *realized* nomic possibilities and cS_n the *induced* nomic impossibilities, the strong likelihood clauses may be read as: P* yields a higher weighted likelihood for the realized nomic possibilities than P (jj-p$^+$) and a lower weighted likelihood for the induced nomic *im*possibilities (j-p$^+$).

The two likelihood ratio (LR-)clauses can be combined, using the bracketed equivalences, to:

Sophisticated Likelihood Ratio Success condition (SLRS-condition):
("P* is Pr-at least as successful as P, relative to R_n/S_n" iff)
$m(cS_n/P^*)/m(cS_n/P) = [1-m(S_n/P^*)]/[1-m(S_n/P)] \leq m(P)/m(P^*) \leq m(R_n/P^*)/m(R_n/P)$

It is called the Sophisticated Likelihood Ratio Success (SLRS-) condition, because by leaving out the middle term it entails the essentially weaker:

Likelihood Ratio Success condition (LRS-condition):
("P* *is* Pr-at least as successful as P, relative to R_n/S_n" entails)
$m(cS_n/P^*)/m(cS_n/P) = [1-m(S_n/P^*)]/[1-m(S_n/P)] \leq m(R_n/P^*)/m(R_n/P)$

which may be paraphrased as: P* leads relative to P to a higher likelihood ratio for realized nomic possibilities than for induced nomic impossibilities.

From the SLRS-condition it follows that the (standard) likelihood-ratio $m(R_n/P^*)/m(R_n/P)$ has the same minimum as $m(\mathbf{T}/P^*)/m(\mathbf{T}/P)$, i.e., $m(P)/m(P^*)$, and that the 'complementary' likelihood ratio $m(cS_n/P^*)/m(cS_n/P)$ has, like $m(c\mathbf{T}/$

P*)/*m*(c**T**/P), this value as a maximum. Together with the fact that the (also here bi-directional) threshold *m*(P)/*m*(P*) $=_{\text{def}}$ h is essentially known as soon as *m*(.) has been fixed, it is a matter of calculation whether the SLRS-condition applies in view of R_n/S_n.

Note that both ratios are likelihood ratios, which shows that the particular kind of comparative HP-method we are proposing is indeed a likelihood comparison method, albeit a sophisticated one, for which reason it is called the Sophisticated Likelihood Ratio Success (SLRS-)method.

13.4.3.1 Expected Success Theorem

Before we present the more general and more relevant 'threshold success theorem', we like to present a success theorem in terms of (mathematical) expectations, that is restricted to the 'numerical probability function' on finite U, according to which *m*(A) = |A| / |U|. It is illuminating because it expresses very well what we might intuitively expect.

Recall that the expectation value E f($\underline{u}$) of real-valued function f($\underline{u}$) of random variable $\underline{u}$ is defined as $\sum$ f(u)p($\underline{u}$ = u), with summation over all the possible values of u. Suppose that the set R, at a certain time n, but now more important, with size r, results from random selection out of the set of all subsets of **T** of size r, denoted by $\mathbf{T}_r$. Hence, $\underline{R}$ is a random variable, with probability p($\underline{R}$ = R) = 1/|$\mathbf{T}_r$|. Similarly, suppose that S, with size s, results from random selection out of the set of all supersets of **T** of size s, denoted by $\mathbf{T}_s$. Hence, $\underline{S}$ is a random variable, with probability p($\underline{S}$=S) = 1/|$\mathbf{T}_s$|.

Expected Success Theorem[6]:
if U is *finite* and 'Pr-closer to the truth' is based on the numerical m, then "$\underline{P^*}$ is Pr-closer to **T** than $\underline{P}$" implies that it is to be expected that "$\underline{P^*}$ is Pr-more successful than $\underline{P}$".

Proof:
We *prove* one of the two claims involved, starting from the unconditional definitions of both notions. The proof of the other claim is similar. Here is, assuming numerical m, the proof of the 'expectation (E-) version' of clause (j-p) assuming clause (i-p) amounts to proving:
if $|(P^*\cap c\mathbf{T})| \leq |(P\cap c\mathbf{T})|$ then E $|P^*\cap cS| <$ E $|P\cap cS|$
Recall, *that* $\mathbf{T}_s$ indicates the set of supersets of **T** of size s. Hence E |P ∩ $\underline{cS}$| amounts to:

$$E\ |P \cap c\underline{S}| = \sum_{\underline{S}=S\in \mathbf{T}s} |P \cap cS|.\ p(\underline{S}=S)$$

[6] In Sect. 5.2.3 a more general, two-sided, version of the theorem is presented. The exemplary proof given there deals with the M-side version of the clause dealing with R. Here we only deal with the P-side, and choose for the clause dealing with S. The proof for the clause dealing with R is essentially similar to the indicated proof in Sect. 5.2.3.

Given p(S̲=S)=1/|$\mathbf{T}_s$|, for supersets S of **T**, and denoting 'truth value' by tv (being 1 or 0), we get that it is equal to:

$(1/|\mathbf{T}_s|) \sum_{\underline{S}=S\in T_s} |P \cap cS| = (1/|\mathbf{T}_s|) \sum_{\underline{S}=S\in T_s} \sum_{x\in P} \text{tv} (\text{"}x \in cS\text{"})$

$=$

$(1/|\mathbf{T}_s|) \sum_{x\in P} \sum_{\underline{S}=S\in T_s} \text{tv} (\text{"}x \in cS\text{"})$.

Note now that the last double summation is proportional to |P ∩ c**T**|, for every x in c**T** is equally many times a member of cS. Hence, E |P ∩ c$\underline{S}$| is proportional to |P ∩ c**T**|. Combining this with the similar proportionality of E |P* ∩ c$\underline{S}$| to |P* ∩ c**T**|, the claim of the theorem directly follows from the relevant defining condition of Pr-closer to, viz. |P* ∩ c**T**| < |P ∩ c**T**|. Q.e.d.

Of course, the expected success theorem has not only a limited and artificial, random sampling, character, but it also has no practical 'detachment' value for a probabilistic analogue of the (instrumentalist) Rule of Success.

13.4.3.2 Threshold Success Theorem

As mentioned in the introduction, finding a sensible success theorem with practical value, to some extent comparable to that of the deductive success theorem, was a difficult task. But it turns out that 'Pr-closer to' entails that both relevant likelihood ratios will in the long run irreversibly pass a certain threshold; however, we never know for certain whether a particular passing is irreversible.

Threshold Success Theorem:

If P̲* is Pr-closer to **T** than P̲, P̲* will in the long run become irreversibly Pr-more successful than P̲.

Proof:

Let '$>_{lr}$' and '$<_{lr}$' indicate 'in the long run larger than' and 'in the long run smaller than', respectively. We will prove it first for one half of the theorem : if m(P)/m(P*) < m(**T**/P*)/m(**T**/P) then in the long run m(P)/m(P*) $<_{lr}$ m(R_n/P*)/ m(R_n/P) will hold. Assume m(P)/m(P*) < m(**T**/P*)/m(**T**/P), that is, [m(P)/m(P*)] m(**T**/P) < m(**T**/P*). According to the probabilistic cumulative evidence relations, m(R_n/P) will not exceed m(**T**/P), whereas m(R_n/P*) will in the long run approach m(**T**/P*) and hence it will pass the threshold [m(P)/m(P*)] m(**T**/P), and hence it will pass [m(P)/m(P*)] m(R_n/P), even earlier, which entails that the ratio m(R_n/P*)/m(R_n/P) will pass [m(P)/m(P*)].

For the other half we have to prove: if m(c**T**/P*)/m(c**T**/P) < m(P)/m(P*) then in the long run m(cS_n/P*)/m(cS_n/P) $<_{lr}$ m(P)/m(P*) will hold. The proof is essentially similar. Q.e.d.

Figure 13.3 illustrates the theorem by way of possible curves of m(R_n/P) and m(R_n/P*). It is assumed that the threshold h = m(P)/m(P*) is larger than 1. Note that when the ratio m(R_n/P*)/m(R_n/P) actually passes h this need not be irreversible. In the beginning it may even be so that this ratio is smaller than 1/h, that is, it may even be that the reverse ratio m(R_n/P)/m(R_n/P*) passes the threshold for P relative to P*.

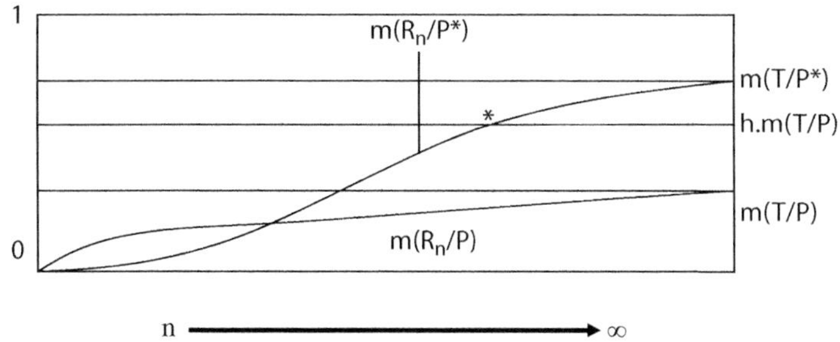

Fig. 13.3 Let h = m(P)/m(P*) > 1. It shows possible curves of $m(R_n/P)$ and $m(R_n/P^*)$, starting unfavorable for P*, with ratio $m(R_n/P^*)/m(R_n/P)$ smaller than 1, hence smaller than h, but later $m(R_n/P^*)$ passes the hidden threshold h.m(**T**/P), indicated by *

Hence we run the risk of concluding too early that the passing of the threshold was irreversible. Note also that if U is denumerably infinite, the irreversible passage of h need in practice not be reached, the more so when U is non-denumerable. Assuming more specific conditions, it will be possible to make 'in the long run' more precise.

13.4.4 The SLRS–Method Is Functional for Empirical Progress and Truth Approximation

Now it is not difficult to extend the functionality story to the probabilistic case. We have defined in general what it means that P* is Pr-more successful (relative to R_n/S_n) than P: the SLRS-condition should be satisfied. Again this may be a matter of lucky choices of experiments in favor of P*. So the probabilistic version of the Comparative Success Hypothesis (PCSH) hypothesizes in this situation that this is from now on irreversibly so. The adapted version of IRS becomes the equally instrumentalist rule:

Probabilistic Instrumentalist Rule of Success (PIRS):
When PCSH is 'sufficiently confirmed', i.e., P*'s being Pr-more successful than P, in the sense of the SLRS-condition, seems irreversible, eliminate P in favor of P*, at least for the time being.

Again we may claim that this is an ideal form of 'empirical progress', now of a probabilistic nature. Since the crucial notion of 'Pr-more successfulness' can be couched in likelihood ratios (relative to cumulated evidence), we may conclude that the SLRS-method is functional for achieving (probabilistic) empirical progress, viz. by applying PIRS whenever this seems justified.

As suggested, the expected success theorem is not useful for a functionality claim. But the Threshold Success Theorem (TST) certainly is. Let P* seem so far irreversibly Pr-more successful than P, i.e., the condition assumed in PIRS. Then there are the following possibilities:

- P* may be Pr-closer to P, which would explain, by TST, its so far seeming irreversibly Pr-more successful, from some time on, for in that case it may have become in fact irreversibly Pr-more successful,
- again on basis of TST, it can't be the reverse, i.e., that P is Pr-closer to **T** than P* and that P* has nevertheless become irreversibly Pr-more successful,
- if neither is Pr-closer to the other, a specific explanation has to be given for the apparent success dominance so far from some time on, i.e., for P*'s so far seeming irreversibly Pr-more successful than P.

Note that there is a difference in the corresponding second clause of the deductive case. In the latter case the simple fact that P* is more successful than P at some time prevents P already from being D-closer to the truth than P*. In the present case P* may, despite all efforts to test PCSH, wrongly seem irreversibly Pr-more successful than P. As a matter of fact, the tables may still turn irreversibly in favor of P, in which case P's being Pr-closer to the truth than P* would be the explanation.

However this may be, the condition of PIRS still provides good reasons to conclude, at least for the time being, that P* is Pr-closer to the truth than P. Since PIRS was based on the SLRS-method, we may conclude that this method is functional for probabilistic nomic truth approximation.

The quest for real cases of 'Pr-closer to' now extends of course to the quest for historical cases of 'seemingly irreversibly Pr-more successful', even of three kinds: cases in which the tables have never turned irreversibly in favor of the other, cases in which this has happened, and cases where the tables have become mixed.

13.4.5 Inference to the Best Theory

Very briefly I will indicate the probabilistic extension of the claim in (Kuipers 2000), and rephrased in terms of abduction in (Kuipers 2004; Chap. 11 of this book), that the (deductive) success theorem, enables the explication of a revised version of so-called Inference to the Best Explanation (IBE).

In the deductive case, the good reasons for extending the instrumentalist conclusion of IRS to the conclusion, for the time being, that P* is D-closer to the truth than P, suggests the following revision of IBT to 'Inference to the Best Theory' (IBT). Whereas the former prescribes to conclude that the best, apparently non-falsified, explanation is true, the latter prescribes to conclude that the best, that is, the persistently D-most successful theory, falsified or not, is the D-closest to the truth. Under even more stringent conditions, including not being falsified, this may justify an inductive jump to the conclusion that the best theory represents the truth, that is, the strongest true theory (Kuipers 2004; Chap. 11 of this book).

In the probabilistic case we have similar good reasons for extending the instrumentalist conclusion of PIRS to the conclusion, for the time being, that P* is

Pr-closer to the truth than P, which now suggests the 'Inference to the Pr-Best Theory' (IPrBT). It prescribes to conclude that the Pr-best, that is, the persistently Pr-most successful theory, falsified or not, is the Pr-closest to the truth.

Assuming some extra conditions this may justify the inductive jump to the conclusion that the Pr-best theory, say P*, is true, that is, that the claim "$\mathbf{T} \subseteq P^*$" is true. Of course, this requires that P* has not been falsified by R_n and S_n, and hence that $R_n \subseteq P^* \subseteq S_n$. Probabilistically, this means that $m(S_n/P^*) = 1$ and that $m(R_n/P^*) = m(R_n)/m(P^*)$.

Again, under more stringent conditions this may even justify the inductive jump to the conclusion that the Pr-best theory represents the truth. In the present case the conditions are of course, first, that we have good reasons to assume that the Pr-best theory, say P*, is true, for which we just have specified the conditions, but, second, we should now also have at least good reasons to assume that $m(P^*) = m(\mathbf{T})$. Of course, if we are only dealing with unstratified theories, we may even have good reasons to assume that S_n captures all possible true inductive generalizations over the nomic possibilities, which would require that S_n equals P* and entail, if true, that P* equals **T**. Although it may be attractive in very special cases to make the indicated inductive jump, it will always be important to do this with some proviso.

13.5 Concluding Remarks: Possible Extensions/ Concretizations

The above probabilistic extension of the deductive story according to which the HD-method is functional for empirical progress and truth approximation is what I wanted to achieve. However, it has a number of abstractions, idealizations, and restrictions asking for specification, concretization and extension.

The main abstraction is that we have left the presupposed probability function *m*(.) further unspecified. Of course, *m*(.) is primarily supposed to be used epistemically, despite occasional use of 'as-if-objective' talk, e.g. interpreting clauses in terms of 'random sampling'. Moreover, the likelihoods, in particular, may well be of an objective nature. However this may be, all claims are relative to the chosen *m*(.). This raises the question whether there are good reasons to restrict probabilistic claims of confirmation, empirical progress and truth approximation to cases such that they hold for all conceivable probability measures functions, or for all members of a certain subclass of such functions, e.g. as distinguished in (Kuipers 2001, 2006), i.e., for all Popperian (non-inductive) or Carnapian (inductive likelihoods) or Bayesian (inductive priors) or Hintikkian (double inductive) probability measure functions. Note that this division makes sense even though e.g. *m*(P) has not the standard meaning of the prior probability that the claim of theory P ($T \subseteq P$) is true, but just the measure of P.

Turning to idealizations, as in the deductive case (Kuipers 2000, Chap. 9; Chap. 7 of this book), a stratified version in terms of the (theory-relative) distinction between observation and theoretical terms would be attractive. However, a simple probabilistic version of the 'projection theorem' seems unlikely, which is, by the way, also absent in the quantitative similarity approach of Ilkka Niiniluoto (1987).

As we have already noted, like in the deductive case (Kuipers 2000, Chap. 10), a refined probabilistic version, taking structurelikeness into account, or distances between structures would be most welcome in order to get realistic cases. In terms of distances, it seems rather plausible to define 'closer to' then in terms of some weighing of the falsity-content cP−c**T** (= **T**−P), for finite U, e.g. as follows:

$$\sum_{z\in \mathbf{T}\text{-}P} m(z)d_{min}(z, P)/d_{max}$$

where d(z, P) is the minimum distance of z to P and d_{max} the maximum distance between two elements of U. Finally, it is assumed that d(x, x) = 0 and $1 \leq d(x, y)$ if $x \neq y$. Assuming that the way of dealing with the truth-content cP ∩ c**T** is kept unchanged. This would lead to the following definition of 'at least as close to':

(i-p) $m(cP \cap c\mathbf{T}) \leq m(cP^* \cap c\mathbf{T})$ $(\equiv m(P^* \cap c\mathbf{T}) \leq m(P \cap c\mathbf{T}))$
&
$(ii\text{-}p)_d$ $\sum_{z\in \mathbf{T}\text{-}P^*} m(z)d_{min}(z, P^*)/d_{max} \leq \sum_{z\in \mathbf{T}\text{-}P} m(z)d_{min}(z, P)/_{dmax}$

Note that, if all distances between different members are 1, $(ii\text{-}p)_d$ reduces to the original condition:

(ii-p) $m(cP^* - c\mathbf{T}) \leq m(cP - c\mathbf{T})$ $(\equiv m(P \cap \mathbf{T}) \leq m(P^* \cap \mathbf{T}))$

Further research is needed for the evaluation of this definition and alternatives.

Regarding restrictions, the laws and theories assumed above were of a deterministic nature, as was the truth. In the context of the (implicit) assumption that the truth is deterministic, laws and theories that are taken into account frequently have nevertheless a statistical or probabilistic nature. It may be possible to reinterpret such use of statistical laws and probabilistic theories as laws and theories that probabilistically deviate from the (deterministic) truth.

Perhaps the most intriguing question is how to extend the analysis to non-deterministic theories, 'the truth' included. One might think of theories and the truth itself as dealing with e.g. objective probabilities, but this is left for future research.

A final concern is the limitations of the SLRS-method in general. Likelihood comparison has come under attack. Forster (2006) has shown by way of detailed counterexamples that there are exceptions to the rule that all the empirical information relevant to the comparison of theories is contained in the likelihoods. Although likelihoods can measure how well a theory accommodates the data, they do not contain the information how well it can predict one part of the data from another. The counterexamples presented by Forster all deal with curve fitting of one kind or another. The deductive and probabilistic approach do not yet cover such 'theories', but refined versions, if based on real-valued distances between structures, certainly do. Hence, a final question for further research is how problematic such counterexamples are for refined probabilistic approaches to empirical progress and truth approximation in the line of the present chapter.

References

Forster, M. (2006). Counterexamples to a likelihood theory of evidence. *Minds and Machines, 16*(3), 319–338.

Kuipers, T. (2000). *From instrumentalism to constructive realism. On some relations between confirmation, empirical progress, and truth approximation, Synthese Library 287*. Dordrecht: Kluwer Academic Publishers.

Kuipers, T. (2001). *Structures in science. Heuristic patterns based on cognitive structures. An advanced textbook in neo-classical philosophy of science, Synthese Library 301*. Dordrecht: Kluwer Academic Publishers.

Kuipers, T. (2004). Inference to the best theory, rather than inference to the best explanation. Kinds of abduction and induction. In F. Stadler (Ed.), *Induction and deduction in the sciences. Proceedings of the ESF-workshop induction and deduction in the sciences, Vienna, July, 2002* (pp. 25–51). Dordrecht: Kluwer Academic Publishers.

Kuipers, T. (2006). Inductive aspects of confirmation, information, and content. In R. E. Auxier & L. E. Hahn (Eds.), *The philosophy of Jaakko Hintikka, The Library of Living Philosophers* (pp. 855–883). Chicago/La Salle: Open Courts.

Kuipers, T. (2007a). The hypothetico-probabilistic (HP-) method as a concretization of the HD-method. In M. Sintonen, S. Pihlström, & P. Raatikainen (Eds.), *Festschrift in honor of Ilkka Niiniluoto* (pp. 179–307). London: King's College.

Kuipers, T. (2007b). Introduction. Explication in philosophy of science. In T. Kuipers (Ed.), *General philosophy of science: Focal issues. Handbook of the philosophy of science* (Vol. 1, pp. vii–xxiii). Amsterdam: Elsevier.

Kuipers, T. (2007c). Empirical and conceptual idealization and concretization. The case of truth approximation. In J. Brzezinski et al. (Eds.), *The courage of doing philosophy. Essays presented to Leszek Nowak* (pp. 75–101). Amsterdam: Rodopi.

Kuipers, T. (2007d). Laws, theories and research programmes. In T. Kuipers (Ed.), *General philosophy of science: Focal issues. Handbook of the philosophy of science* (Vol. 1, pp. 1–95). Amsterdam: Elsevier.

Kuipers, T. (2009). Empirical progress and truth approximation by the 'Hypothetico-Probabilistic Method. *Erkenntnis, 70*, 313–330.

Niiniluoto, I. (1987). *Truthlikeness, Synthese Library 185*. Dordrecht: Reidel.

Niiniluoto, I., & Tuomela, R. (1973). *Theoretical concepts and hypothetico-inductive inference, Synthese Library* (Vol. 53). Dordrecht: Reidel.

Part V
Belief Revision Aiming at Truth Approximation

Chapter 14
Basic and Refined Nomic Truth Approximation by Evidence-Guided Belief Set Revision

Abstract Straightforward theory revision, taking into account as effectively as possible the established nomic possibilities and, on their basis induced empirical laws, is conducive for nomic truth approximation. The question this paper asks is: is it possible to reconstruct the relevant theory revision steps, on the basis of incoming evidence, in the terms of so-called belief revision and, more specifically, in the terms of belief *set* revision, also called the AGM-approach? Assumin g exclusion theories, a positive answer will be given in two rounds, first for the case in which the initial theory is compatible with the established empirical laws, then for the case in which it is incompatible with at least one such a law.

Keywords Theory revision · Belief revision · AGM-revision · Nomic truth approximation · Expansion · Contraction · Revision · Structurelikeness

14.1 Introduction

The AGM-style of belief revision (AGM-BR) grew out of a paper by Carlos Alchourrón, Peter Gärdenfors, and David Makinson (Alchourrón et al. 1985). For an overview of AGM-BR, see (Hansson 1999). It typically aims at coherence optimization between a given set of beliefs and new information in as conservative a way as possible, implicitly taking that new information as true, whatever distance its adherents take to matters of truth. However, as far as aiming at truth approximation at all, AGM-BR seems to be primarily aiming at the truth about the actual world, *actual truth approximation*, in short. Following Grove (1988), Niiniluoto (1999) and Cevolani and Calandra (2009) have been studying the prospects of belief revision for approximation of the actual truth. New and extended attempts focusing on

Strongly revised version of: "Basic and refined nomic truth approximation by evidence-guided belief revision in AGM-terms", *Erkenntnis*, 75, 2, 2011, pp. 223–236. *Acknowledgements*: The author wishes to thank Roberto Festa for the occasion to present this paper in 2009 at a symposium in Trieste, Gustavo Cevolani, Gerhard Schurz and Wlodek Rabinowicz for very useful comments, and the *Netherlands Institute for Advanced Study* (NIAS), Wassenaar, for providing again paradisiacal conditions for research and writing, including linguistic correction by Anne Simpson.

T. A. F. Kuipers, *Nomic Truth Approximation Revisited*, Synthese Library 399,
https://doi.org/10.1007/978-3-319-98388-2_14

this aim have been made at conferences in Trieste (2009) and Amsterdam (2009) by Cevolani, Crupi and Festa, Schurz, Niiniluoto, Baltag and Smets, Zwart and Renardel, and Douven and Kelp (see the special issue of *Erkenntnis* of 2011 75.2)).

However, theorizing and hence theory revision in the natural sciences typically aim at *nomic truth approximation*, that is, an approximation of the truth about what is nomically (e.g. physically) possible and what is not (Kuipers 2000; this book). Nomic truth approximation by theory revision is guided by evidence, where evidence consists of case descriptions and induced empirical laws based on them. In addition to a basic ('content') kind of nomic truth approximation, there is a refined ('likeness') kind, as a concretization of the basic one (see Zwart (2001) for the distinction between content and likeness approaches[1]). Moreover, there exist (observationally-theoretically) stratified variants of both, probabilistic variants and, in principle, all kinds of combinations.

An instructive stratified toy example is the following: let there be a complex, but finite, electric network of switches and bulbs, and a battery. Let the network of (serial and parallel) connections be hidden and the task is to find out the precise structure of this network. The observational nomic truth about the network amounts to a characterization of the physically possible state of the network, as far as positions of the switches (on/off) and the bulbs (lighting or not) are concerned. This nomic truth can be expressed by a propositional formula, of which the disjunctive normal form has as disjuncts the constituents that represent these states. It may be possible to reconstruct from this formula the theoretical nomic truth, that is, the full network, including the hidden connections. However, it may also be that there are empirically equivalent networks, that is, networks generating the same observational nomic truth.

The question asked in this paper is: can something like AGM-BR be helpful for evidence-guided theory revision aiming at (some kind of) nomic truth approximation? In other words, is it possible to reconstruct plausible theory revision steps, on the basis of characteristic evidence, aiming at nomic truth approximation in AGM-terms?

In Sect. 14.2 it will first be argued that, assuming exclusion theories, straightforward basic theory revision, taking into account as effectively as possible the established nomic possibilities and the, on their basis, induced empirical laws, guarantees (unstratified) basic nomic truth approximation. Then it will be shown that this revision can be reconstructed into two AGM-steps, in arbitrary order. One of these is straightforward expansion; the other is an extreme form of contraction, viz. so-called full meet contraction. This revision needs, however, refinement for the difficult but likely case that at least one of the induced laws is incompatible with the original theory.

In Sect. 14.3 it will first be shown that the spheres approach of theory revision developed by Adam Grove (1988) can be used to refine the above indicated two-step theory revision such that it can be used for the hard case, and reduces to the basic case when theory and all induced laws are compatible. Assuming the proper order,

[1] Zwart, however, disagrees about calling the second a concretization of the first.

that is, first a refined kind of revision in the face of the induced laws, viz. a kind of partial meet revision, and then full meet contraction in the face of the remaining counterexamples, the resulting refinement is potentially conducive for *basic* truth approximation. In terms of the likeness foundation of the spheres approach by Wlodek Rabinowicz (1995), based on a four-place similarity relation, it will be shown that even this refinement is potentially conducive for *refined* truth approximation.

In the concluding Sect. 14.4 the main (positive) conclusions, for exclusion theories, will be followed by a number of debunking remarks about the presented AGM-style theory revision from a realist point of view. In the Appendix we will indicate why it is difficult to give a basic and refined AGM-belief revision account for inclusion theories.

14.2 The Basic Account

14.2.1 Basic Definitions and Basic Theory Revision

According to the structuralist theory of truth approximation (Kuipers 2000; this book), nomic truth approximation more specifically aims at the strongest true theory **T** about the set of nomic possibilities within the set of conceptual possibilities U generated by a chosen vocabulary for a chosen domain of phenomena. Nomic truth approximation by evidence-guided *theory* revision requires definitions of 'being closer to the truth' and 'being more successful', or rather primarily their 'at least as'-versions. We will focus on so-called exclusion theories. An exclusion theory $\underline{P}$ amounts to a specified subset P of U with the claim that it is a superset of **T** (**T**⊆P, and hence cP⊆c**T**). The exclusion claim may also be called the *necessity claim*, stating that belonging to P is necessary for being nomically possible.[2]

Informally we can summarize the point of departure as follows: we have a set U of possibilities and every theory concerns a subset of this domain. This applies also to the strongest true theory, **T**. The elements of **T** are the 'real' possibilities, so to speak. All the possibilities outside **T** are not real. The claim concerning a theory $\underline{P}$ is that P does not leave out any real possibilities.

[2] The original paper (Kuipers 2011) was still based on the assumption that theories had to be maximal, i.e., making the exclusion and the inclusion claim, there called the necessity and the sufficiency claim, respectively. For the basic revision, it started by assuming the exclusion claim and then revised the theory in the light of evidence. At the end, the inclusion claim was added ad hoc, which was seen as rather problematic. Although the refined revision story was presented analogously, on closer inspection, the used definition of 'refined closer to' was in fact the refined version for theories making the inclusion claim. For this reason we now not only present the basic revision but also the refined revision in terms of the exclusion claim. In view of the two-sided generalization in this book of basic and refined truth approximation, with one-sided extreme cases, we do not have to add at the end an ad hoc additional claim. In the Appendix we will indicate why it is difficult to give a basic and refined AGM-belief revision account for inclusion theories.

The (qualitative) basic definition of '*P* is at least as close to **T** as P*' amounts to: $P^*\Delta\mathbf{T} \subseteq P\Delta\mathbf{T}$ (where Δ stands for symmetrical difference, i. e. $P\Delta\mathbf{T} = P-\mathbf{T} \cup \mathbf{T}-P$), and hence to:

(i^b) **T**–P* is a subset of **T**–P
(ii^b) P*–**T** is a subset of P–**T**

and '*closer to*' iff, in addition, in at least one case it is a *proper* subset.

Not knowing **T**, we have to try to improve our guesses (theories) of what **T** is on the basis of, or guided by, (new) evidence. Evidence typically comes in by experimentally realizing conceptual possibilities, say R(t) up to time t. They are, of course, nomic possibilities, hence, if we have not made mistakes, R(t) is a subset of **T** ($R(t)\subseteq\mathbf{T}$), whatever **T** is. Neglecting mistakes and forgetfulness, R(t) is an increasing set of established nomic possibilities.

R(t) will grow in particular due to testing general hypotheses, each one with the exclusion claim that all nomic possibilities satisfy it. They may have been derived from the claim of theory P or may have been put to the test in order to test some other theory or for still other reasons. At each point of time we may assume that one or more of them are considered to have been sufficiently established as empirical laws by inductive generalization. Let subset S(t) of U represent at time t the resulting strongest, induced empirical law, which amounts to the claim that S(t) is the smallest induced superset of **T**, whatever **T** is ($\mathbf{T}\subseteq S(t)$). Neglecting mistakes and forgetfulness, S(t) is a decreasing set.

In sum: $R(t)\subseteq\mathbf{T}\subseteq S(t)$, assuming no mistakes. From now on t will be omitted. The following definition is now plausible:

The (qualitative) basic definition of '*P* is at least as successful as P relative to R/S*' amounts to:

(i^b–sf) R–P* is a subset of R–P
(ii^b–sf) P*–S is a subset of P–S ($\equiv P\cup S \subseteq P^*\cup S$)

and '*more successful*' by requiring in addition that in at least one case it is a *proper* subset.

The first clause can be rephrased as: all established counterexamples to P* are counterexamples to P; and the second as: all established laws (represented by supersets of S!) explained by P (i.e., supersets of P) are explained by P*. Note that the above definition implies that a theory P* is maximally successful relative to R/S iff $R\subseteq P^*\subseteq S$. For then, and only then, both R–P* and P*– S are empty sets, which means that P* is, relative to R/S, at least as successful as any theory P.

In general, it is crucial for the proper explication of qualitative notions of more truthlikeness and (corresponding) more successfulness or greater success to be able to prove the following theorem, with or, as in the present basic unstratified case, without further conditions:

Success Theorem: If P* is closer to **T** than P then P* will always be at least as successful as P and become more successful in the long run.

Proof of "$\underline{P^*}$ will always be at least as successful as $\underline{P}$". First clauses, assuming R ⊆ **T**, R−P* ⊆ **T**−P*, and by (i^b), R−P* ⊆ **T**−P. But we also have that R−P* ⊆ R, hence R−P* is a subset of the intersection of R and **T**−P, which equals R−P. Second clauses, assuming **T** ⊆ S, P*−S ⊆ P*−**T**, and by (ii^b), P*−S ⊆ P−**T**. But we have also P*−S ⊆ U−S, hence P*−S is a subset of the intersection of U−S and P−**T**, which equals P−S. Q.e.d.

Proof sketch of "$\underline{P^*}$ will .... become more successful [than $\underline{P}$] in the long run". When (i^b) or (ii^b) can be strengthened to proper subsets, in the long run, in which R approaches **T** by steadily growing and S approaches **T** by steadily shrinking, there will be realized (hence, nomic) possibilities belonging (to **T**) and to P*, but not to P, or there will be laws induced that assign the status of nomic impossibilities (in U−**T**) to conceptual possibilities that are outside P*, but inside P. Of course, a straightforward proof requires precise assumptions about the way in which the experiments are going through **T** and how and when laws are induced. Q.e.d.

This theorem gives good reasons to abduce, under certain conditions and for the time being, that theory $\underline{P^*}$ is closer to the truth than theory $\underline{P}$ when $\underline{P^*}$ is persistently more successful than $\underline{P}$, i.e., when we typically speak of *empirical progress*. Or, conversely: 'truth approximation' provides the *default-explanation* of 'empirical progress'. For the basic case the good reasons are threefold (Kuipers 2000, p. 162), in brief: (1) it is still possible that $\underline{P^*}$ is closer to the truth than $\underline{P}$, which would explain the persistent greater success, (2) it is impossible that $\underline{P}$ is closer to the truth than $\underline{P^*}$, (3) if neither holds, the persistent greater success so far requires a (test history) specific explanation. We can also paraphrase the overall conclusion by saying that persistent greater success is conducive for truth approximation and hence that greater success is *potentially* conducive for truth approximation.

From the above definitions and the theorem we may also draw the preliminary conclusion that, assuming the data are correct, theory revision not only realizes empirical progress but it also makes nomic truth approximation at least formally possible. Moreover, as is easy to check, both are even realistic, in the case of finitely many conceptual possibilities, as in the electric network, and, in general, when a finite propositional language can be used.

It is now easy to show that there is a unique way to revise a theory $\underline{P}$ in the face of evidence R/S such that the revision is, as a rule, not only more successful but even closer to the truth than $\underline{P}$. We call this *basic theory revision of* $\underline{P}$ *by R/S*. The revised theory is $\underline{(P \cap S) \cup R}$ or, equivalently, $\underline{(P \cup R) \cap S}$ and will be indicated by $\underline{P}^b_{R/S}$. Note that $P^b_{R/S}$ equals P when R⊆P⊆S, i.e., when $\underline{P}$ is maximally successful. Moreover, it is easy to see that R⊆ $P^b_{R/S}$ ⊆S, i.e., that $\underline{P}^b_{R/S}$ is maximally successful.

Basic Revision Theorem: Assuming correct data (R⊆**T**⊆S), '*basic theory revision of* $\underline{P}$ *by R/S*', resulting in $\underline{P}^b_{R/S}$, guarantees that $\underline{P}^b_{R/S}$ is (basically) at least as close to **T**, and hence at least as successful as $\underline{P}$ relative to R/S. Moreover, it is even closer to **T**, and more successful, than $\underline{P}$ when $\underline{P}$ is not maximally successful.

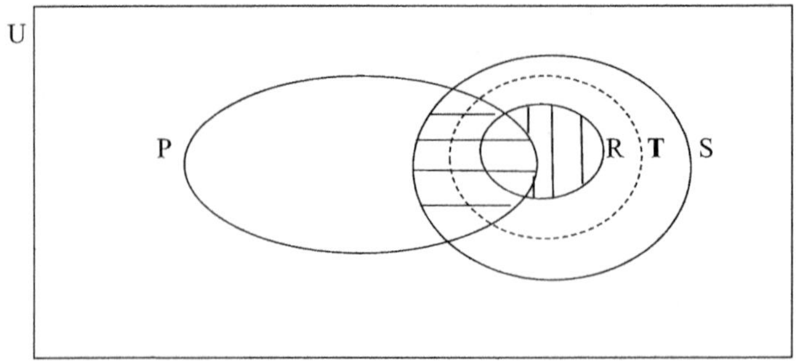

Fig. 14.1 Refined revision when P̲ is compatible with S̲, leading to $\underline{P}^{b}_{R/S} = \underline{(P \cap S) \cup R}$, i.e., the shaded area

Note that the condition that P̲ is not maximally successful amounts to the claim that R is not a subset of P or P is not a subset of S, i.e., R includes counterexamples of P̲ and/or P̲ cannot explain all laws derivable from S, in sum, P̲ is not 'between' R and S (while the revision is!).

The validity of the theorem can easily be checked on the basis of Fig. 14.1, in which the shaded areas together indicate the revised theory $\underline{P}^{b}_{R/S}$ and, in terms of the next subsection, in which the horizontal shading indicates the expansion step and the vertical shading the contraction step.

Note that there are two extreme cases in which the role of P̲ essentially vanishes and which are for that reason of special interest:

If $P \cap S = \emptyset$, then $P^{b}_{R/S} = R$, hence further roles of P and S vanish
If $P \cup R = U$, then $P^{b}_{R/S} = S$, hence further roles of P and R vanish

In particular the first case is of a great interest, for though extreme in some sense, it is certainly not exceptional. It simply amounts to the case in which a theory P̲ is incompatible with at least one established law, and hence with the strongest established law. But first we will deal with the question put in this paper as far as non-extreme cases are concerned.

14.2.2 Basic Theory Revision in Light of AGM-Belief Revision

The main question of the chapter is: is it possible to reproduce the theory revision from P̲ to $\underline{P}^{b}_{R/S}$ by AGM-style belief revision? As is well-known, AGM-*belief revision* centers around three (partially related) operations (Alchourrón et al. 1985), see also e.g. Cevolani and Calandra (2009). A belief set, that is, a deductively closed set of sentences of a given language, is typically excluding states etc. that are incompatible with it. When a belief set is confronted with some 'input sentence' then this sentence should, by minimal further changes of the original belief set, either become a consequence or no longer be a consequence of the revised belief set. For the first case, it makes an important difference whether or not the input sentence is compatible with the belief set. In the first subcase we get so-called *expansion*, viz., the belief set is strengthened to the set of consequences of the union of the belief set and the input sentence. Regarding the input sentence it leads from suspension of judgment about that sentence to its acceptance. In the second subcase the belief set has to be adapted in a more complicated way, satisfying certain axioms. It is called *revision* (in the narrow sense). Regarding the input sentence, revision leads from its rejection to its acceptance, except when the input sentence is inconsistent. Finally, in the second main case, the input sentence is supposed to belong to the belief set, but should no longer belong to the revised set. Hence, now the belief set has to be weakened in a minimal way, again in line with some axioms. It is called *contraction*. Regarding the input sentence it leads from its acceptance to suspension of judgment, except when the input sentence is logically true, in which case it remains accepted after contraction.[3]

The focus in the belief revision program has been the axiomatic characterization of the three indicated operations. Whereas this kind of explication of expansion is relatively simple, it is rather complicated for revision and contraction.

To be sure, we did not present the previous subsection in terms of sentences of a language but in terms of (sets of) conceptual possibilities or structures generated by a language. But we could translate, for example, theory $\underline{P}$ in terms of Th($\underline{P}$), i.e., the (deductively closed) set of sentences that are true of all structures in P. In this way the set of structures P becomes the set of models of Th($\underline{P}$). However, it is characteristic of the structuralist approach to identify a sentence or theory $\underline{P}$ with its set of models and to consider the set of (subsets of U being) supersets of P as representing the set of consequences of $\underline{P}$. In the present context of nomic theories, this essentially model-theoretical notion of consequence is directly transmitted to the exclusion claim of a theory. To be precise, if P* is a superset of P, $P \subseteq P^*$, the exclusion claim of theory $\underline{P^*}$, "$\mathbf{T} \subseteq P^*$", is a consequence of the exclusion claim of theory $\underline{P}$, "$\mathbf{T} \subseteq P$".

In this way we not only get 'model versions' of (sets of) sentences and consequences, but we can also form model versions of the three operations (Hansson 1999, 220–5). For expansion this is almost trivial, for revision and contraction some extreme forms are also rather easy, precisely the ones we need in this section.

Expansion of theory $\underline{P}$ by input 'sentence' $\underline{A}$ amounts to $P \cap A$. The so-called *full meet* (fm-)revision of $\underline{P}$ by $\underline{A}$ amounts to $P \cap A$ when $\underline{P}$ is compatible with $\underline{A}$ ($P \cap A$ is non-empty) and to A when $\underline{P}$ is incompatible with $\underline{A}$. Finally, the so-called *full meet* (fm-)contraction of $\underline{P}$ by $\underline{A}$ amounts to $P \cup cA$ when $\underline{P}$ entails $\underline{A}$ and to $\underline{P}$ when it does not. Note that fm-revision of $\underline{P}$ by $\underline{A}$ not only entails $\underline{A}$, as informally required of revision, but also coincides with expansion of $\underline{P}$ by $\underline{A}$ when $\underline{P}$ and $\underline{A}$ are compatible and fully jumps to $\underline{A}$ when they are incompatible. In the latter case it is an extreme form of (AGM-)revision. Note also that fm-contraction of $\underline{P}$ by $\underline{A}$ no longer entails

[3]The formal definition also leaves room for the case in which the input sentence does not belong to the original belief set. Then the outcome of contraction is simply the original belief set, i.e., judgment about the input sentence was and remains suspended.

$\underline{A}$, when $\underline{P}$ does entail $\underline{A}$, as informally required of contraction, but that it then fully allows *all* possibilities in cA. In this sense it is an extreme form of (AGM-)contraction. Note, moreover, that it remains $\underline{P}$ when $\underline{P}$ does not entail $\underline{A}$.

Since the AGM-operations typically deal with the consequences of the relevant belief set, it is plausible that we have focused on exclusion theories and hence that our leading question now is the way in which the exclusion claim of theory $\underline{P}$ has to be adapted. Recall that we exclude in this subsection two extreme cases, of which the most important one is that $\underline{S}$ and $\underline{P}$ are incompatible.

From the indicated perspective it is immediately clear that the first step in the basic revision, the transition from P to P∩S, is a clear case of (the model version of) *expansion* of $\underline{P}$ by $\underline{S}$. Similarly, the transition from P∪R to (P∪R)∩S is a case of expansion of $\underline{P \cup R}$ by $\underline{S}$. Recall, for later purposes, that expansion and fm-revision of $\underline{P}$ by $\underline{S}$ coincide when they are compatible, which we are assuming.

Regarding the transition from P to P∪R, or from P∩S to (P∩S)∪R, the situation is a bit more complicated. Focusing on the transition from P to P∪R, that is, from P to P∪(R−P), of which R−P amounts to the set of realized counterexamples of $\underline{P}$, we see that whereas P is a subset of, and hence entails, c(R−P), P∪R does no longer entail this consequence. It even allows all possibilities in R−P. This amounts to fm-contraction of $\underline{P}$ by $\underline{c(R-P)}$. Similarly, the transition from P∩S to (P∩S)∪R amounts to fm-contraction of $\underline{P \cap S}$ by $\underline{c(R-(P \cap S))}$.

In sum, we may now conclude that basic theory revision of $\underline{P}$ by R/S, leading to $\underline{(P \cap S) \cup R}$ or, equivalently, $\underline{(P \cup R) \cap S}$, can be seen as the successive application of expansion and fm-contraction (or vice versa) on the exclusion claim.

From our perspective the above analysis completes our task of an AGM-presentation of basic theory revision for the non-extreme cases in which $\underline{P}$ and $\underline{S}$ are compatible (P∩S ≠ ∅) and in which P and R do not exhaust U (P ∪ R ≠ U), respectively. As noted before, in both extreme cases the role of $\underline{P}$ essentially vanishes. Whereas the second extreme case (P ∪ R = U) seems rather rare, the first extreme case (P∩S = ∅) certainly is not: it merely assumes that $\underline{P}$ is incompatible with at least one induced empirical law. Hence, the main remaining task is to refine the expansion of $\underline{P}$ by $\underline{S}$ in some way for the case in which $\underline{P}$ and $\underline{S}$ are incompatible.

One might suggest that another aspect of the transition from P to P∪R (or from P∩S to (P∩S)∪R) may require refinement. Instead of fm-contraction of $\underline{P}$ by $\underline{c(R-P)}$ one might think of so-called partial meet contraction of $\underline{P}$ by $\underline{c(R-P)}$, in which case not all possibilities in R−P are allowed. This would require some kind of degree of trustworthiness of the various experimentally realized conceptual possibilities. A similar kind of refinement of the transition from P to P∩S arises when we would assume a degree of trustworthiness of the induced laws, in which case S is no longer taken for granted. However, these kinds of refinement, which amount to weakening of the correct data assumption, go beyond the scope of the present chapter.

Finally, here, and for later purposes, it is interesting to see what would have resulted when we would have defined basic theory revision of $\underline{P}$ by R/S in terms of fm-revision, instead of expansion, of $\underline{P}$ by $\underline{S}$ followed by the relevant fm-contraction, or vice versa. By the indicated alternative definition we would obtain $\underline{(P \cap S) \cup R}$ when P∩S is non-empty, and $\underline{S \cup R}$, that is, S, when P∩S is empty, hence deviating from our primary definition in, and only in, the second case. By the 'vice versa' definition, fm-contraction by $\underline{c(R-P)}$, followed by fm-revision, instead of expansion, by $\underline{S}$, we would obtain $\underline{(P\cup R)\cap S}$, which reduces to $\underline{R}$ when $P\cap S$ is empty. Hence, in this case the result would not differ from our primary definition. Let us call the deviating alternative definition the *fm-definition* of basic theory revision of $\underline{P}$ by R/S.

14.3 The Refined Account

The main problem of basic revision of $\underline{P}$ by R/S is that it reduces to $\underline{R}$ when P and S do not overlap, i.e., when $\underline{P}$ and $\underline{S}$ contradict each other. Expansion of $\underline{P}$ by $\underline{S}$ then gives the empty set, the subsequent weakening with R, i.e., fm-contraction by $\underline{c(R)}$, just amounts to $\underline{R}$, hence a result that in no way reminds us of $\underline{P}$. It is easy to check that the other order leads to the same result. For this route it is crucial to keep in mind that R is a subset of S. The plausible direction for refinement is to try to concretize basic revision of $\underline{P}$ by R/S in terms of a likeness approach that reduces to the basic (content) approach under the appropriate idealization conditions (IC-test), i.e. when $\underline{P}$ and $\underline{S}$ are compatible.

From the AGM-BR-perspective and our structuralist view of theories the spheres approach of Adam Grove (1988) is highly plausible. The spheres may seem to fall rather out of the air, but later we will see that they can be given a plausible 'similarity foundation' which, moreover, enables us to connect the spheres approach more specifically to the (structuralist) likeness approach of qualitative truth approximation.

The basic idea of Grove is to postulate nested spheres around P, satisfying a number of conditions, notably, and plausibly, that P is the smallest and U is the largest sphere. Consider the smallest sphere $c_P(S)$ around P overlapping with S. It is now plausible to define (refined) theory revision of $\underline{P}$ by $\underline{S}$ ($\underline{P}^r_S$) as resulting from the intersection of S and $c_P(S)$, i.e., as the expansion of $\underline{c_P(S)}$ by $\underline{S}$, leading to $\underline{c_P(S)\cap S}$. It is easy to check (IC-test!) that, when $\underline{P}$ and $\underline{S}$ are compatible, $c_P(S) = P$ and hence $P^r_S = P\cap S$ $(=P^b_S)$. Grove has proven that P^r_S satisfies the original AGM-axioms of belief revision presented in (Alchourrón et al. 1985). This corresponds to what later has been called 'transitively relational partial meet revision' (Hansson 1999, p. 223), which we will simply abbreviate by 'pm-revision'.

In sum, $P^r_S = c_P(S)\cap S$ is the most straightforward AGM-way to deal with the revision of $\underline{P}$ by S, but how to take R into account now? Recall that R is a non-empty subset of S. Recall also that the transition from P to $P\cup R$ amounted to fm-contraction of $\underline{P}$ by $\underline{c(R-P)}$ and the transition from $P\cap S$ to $(P\cap S)\cup R$ to fm-contraction of $\underline{P\cap S}$ by $\underline{c(R-(P\cap S))}$. As Hansson (1999, 224–5) describes, contraction can also get a spheres interpretation, giving rise to partial meet (pm-)contraction. However, this would mean that we have to make a selection of members of $R-P$ or of $R-(P\cap S)$, respectively. In the present context there seems no reason for this kind of refinement.

Therefore, the only question that remains is the order in which the refined revision by $\underline{S}$ and the basic revision by R should take place. According to a first alternative one can first apply fm-contraction of $\underline{P}$ by $\underline{c(R-P)}$, followed by pm-

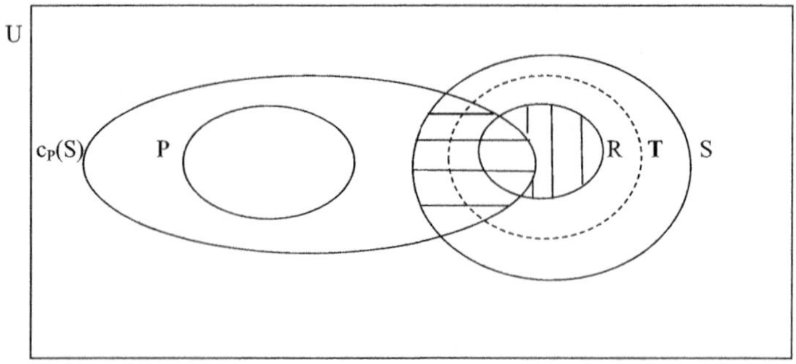

Fig. 14.2 Refined revision when P is incompatible with S, leading to $\underline{P}^r_{R(S)} = \underline{(c_P(S) \cap S) \cup R}$, i.e., the shaded area

revision of the result $\underline{(P \cup R)}$ by $\underline{S}$, leading to $P^r_{S(R)} =_{def} (P \cup R)^r_S = (P \cup R) \cap S = P^b_{R/S}$, since $P \cup R$ itself is the smallest sphere around $P \cup R$ overlapping with S. Hence, $P^r_{S(R)} = P^b_{R/S}$, i.e., basic revision of $\underline{P}$ by R/S, even if $P \cap S = \emptyset$. Hence, the first alternative is no solution of the main problem.

The second alternative fares better: first pm-revision of $\underline{P}$ by $\underline{S}$, followed by fm-contraction of the result $\underline{c_P(S) \cap S}$ by $\underline{c(R-(c_P(S) \cap S))}$. In this way we get: $P^r_{R(S)} =_{def} P^r_S \cup R = (c_P(S) \cap S) \cup R$. Note that (IC-test) $P^r_{R(S)}$ reduces to $(P \cap S) \cup R = P^b_{R/S}$ when P and S overlap. However, $P^r_{R(S)} \neq P^b_{R/S} = \emptyset$ when P and S do not overlap. Hence, the order matters a lot, and we will opt for this second alternative. In Fig. 14.2 the shaded areas together indicate the revised theory $\underline{P}^r_{R(S)}$. Of the suggested overlaps, the only required overlap is that of $c_P(S)$ with S, not that with **T**, let alone that with R. The horizontally shaded area indicates the result of the pm-revision step and the vertically shaded area the 'addition' by the contraction step.

Let us now evaluate refined revision first in terms of basic truthlikeness and basic successfulness. Let us begin by the latter. It is not difficult to check that $\underline{P}^r_{R(S)}$ is basically at least as successful as $\underline{P}$. It is even maximally successful, for it holds that $R \subseteq P^r_{R(S)} \subseteq S$, hence, $\underline{P}^r_{R(S)}$ has no established counterexamples and it explains the strongest established law, hence it explains all established laws. However, already in view of being basically at least as successful, the proposed revision is, due to the (basic) success theorem, potentially conducive for basic truth approximation, even if $\underline{P}$ is incompatible with $\underline{S}$. But in this extreme case, the proposed revision is not basically at least as close to the truth, except in a very extreme, lucky case. This is typically grist to the mill of refined truth approximation. Hence, the question is how the revision fares in terms of refined truthlikeness and corresponding refined successfulness.

Refined truth approximation, as presented in (Kuipers 2000, Chap. 10) and adapted for inclusion and exclusion theories in this book (Chap. 6), is a qualitative likeness approach to truth approximation. It is based on a three-place 'structure-likeness' relation on the set of structures:

s(x,y,z) y is at least as similar (close) to z as x

When s(x,y,z) holds, y is also said to be, qua kind of structure, *between* x and z. It is supposed to satisfy some plausible minimal (s-)conditions.[4] Moreover, we need not assume that all pairs of structures are comparable in the sense of being related by some intermediate structure. Hence we define: x and z are *related*, r(x,z), iff $\exists$y s(x,y,z). Finally, we say that s is trivial if: for all x, y, and z s(x,y,z) iff x = y = z.

Before we introduce further definitions, let us introduce the likeness foundation of spheres and indicate the connection with the likeness approach to truth approximation. Not all of Grove's sphere axioms are very plausible. However, Wlodek Rabinowicz (1995) provided plausible foundations in terms of a four-place similarity relation:

sim(x,y;u,v) x is at least as close (similar) to y as u is to v

satisfying four plausible conditions and one Limit Assumption (see below). Given a set of structures P, Rabinowicz now defines a binary relation between structures

$x \leq_P y$ iff $\forall y' \in P\ \exists x' \in P\ \mathrm{sim}(x',x;y',y)$

This relation might be paraphrased by: P has at least as similar representatives of x as of y.

The relation enables the definition of a sphere (Rabinowicz 1995, p. 92):

P′ is a *sphere* around P iff (i) if $P \neq \emptyset$ then $P' \neq \emptyset$
(ii) $\forall x\ \forall y \in P'$ if $x \leq_P y$ then $x \in P'$

It is not difficult to check that this definition satisfies Grove's four axioms, among them that P and U are the smallest and the largest sphere, respectively.

Recall that $c_P(S)$ was the 'smallest' sphere around P that overlaps with S and that $P^r_S = c_P(S) \cap S$ was defined as the refined revision of P by S. Rabinowicz proved that $P^r_S = \{x' \in S \mid \exists x \in P\ \forall y \in P\ \forall y' \in S\ \mathrm{sim}(x',x;y',y)\}$, where the latter set corresponds to Rabinowicz' specification of P^r_S. The idea behind this version is that it forms "the set of S-worlds that are as similar to some worlds in P as possible, as compared with other worlds in P" (Rabinowicz (1995, p. 82, P substituted for X and S for Y). The Limit Assumption that now is needed instead of a, here not presented, very arbitrary assumption of Grove is not at all that arbitrary: if P and S are non-empty then P^r_S is non-empty.

Now we can turn to the connection between s and sim. Assuming that z in s(x,y,z) is a kind of target the most plausible one certainly is:

s(x,y,z) iff sim (y,z;x,z) i.e., y is at least as similar to z as x (is to z)

---

[4] They are: centered, centering and conditionally left and right reflexive. Here s is centered iff s(x,x,x) and centering iff s(x,y,x) implies x = y. s is conditionally left/right reflexive if s(x,y,z) implies all kinds of left and right reflexivity, i.e., s(x,x,y), s(x,x,z), s(y,y,z) and s(x,y,y), s(x,z,z), s(y,z,z), respectively. Note that this conditional form leaves room for incomparable structures (see text), which otherwise would not be the case.

With this connection in mind we now arrive at the crucial definition of refined truth approximation for exclusion theories (Chap. 6), including a decomposition of the first clause, based on some of the minimal s-conditions.[5]

Definition: *P* is refined at least as truthlike as P* iff

(i^r) $\forall x \in cP\ z \in c\mathbf{T}\ r(x,z) \rightarrow \exists y \in cP^*\ s(x,y,z)$

with the decomposition

(i^b) $cP \cap c\mathbf{T} \subseteq cP^* \cap c\mathbf{T}$

&

$\forall x \in cP - c\mathbf{T}\ z \in c\mathbf{T}\ r(x,z) \rightarrow \exists y \in cP^*\ s(x,y,z)$

$\equiv \forall x \in \mathbf{T} - P\ z \in c\mathbf{T}\ r(x,z) \rightarrow \exists y \in cP^*\ s(x,y,z)$

(ii^r) $\forall y \in cP^* - (cP \cup c\mathbf{T})\ \exists x \in cP - c\mathbf{T}\ \exists z \in c\mathbf{T} - cP\ s(x,y,z)$

$\equiv \forall y \in (cP^* - c\mathbf{T}) - (cP - c\mathbf{T})\ \exists x \in \mathbf{T} - P\ \exists z \in P - \mathbf{T}\ s(x,y,z)$

$\equiv \forall y \in (P \cap \mathbf{T}) - P^*\ \exists x \in \mathbf{T} - P\ \exists z \in P - \mathbf{T}\ s(x,y,z)$

The decomposition of (i^r) shows that it is a strengthening of (i^b) of the basic definition and the equivalent versions of (ii^r) that this is a weakening of (ii^b), for it is easy to check that the latter states that the domain of the first quantifier is empty. (i^r) roughly says that every comparable pair of structures, one of cP and one of c**T**, has an 'intermediate' in cP*. In other words, P* characterizes, by cP*, what is to be excluded, i.e., c**T**, at least as precisely as P by cP. (ii^r) states that, if $cP^* - (cP \cup c\mathbf{T})$ is at all non-empty, which is excluded in the basic case, these structures are 'useful' in realizing (i^r). As is easy to check, the definition reduces to the basic one when s is trivial.

It is interesting to see in detail what the difference is between the basic and refined definition in the light of the truth- and falsity-content analysis for exclusion theories as presented in e.g. Chap. 6, according to which TC(P) = $cP \cap c\mathbf{T}$ and FC(P) = $cP - c\mathbf{T} = \mathbf{T} - P$. Let us start with (ii^r) and (ii^b). Regarding (ii^r), as indicated, the domain of the first quantifier $cP^* - (cP \cup c\mathbf{T})$ is equal to $(cP^* - c\mathbf{T}) - (cP - c\mathbf{T})$ = FC(P*) − FC(P), that is, the extra falsity-content of P* relative to that of P. According to (ii^b), this is empty. Hence, the refined definition leaves room for increasing the falsity-content, provided it satisfies the condition specified in the rest of the clause, of course for a purpose. Let us now first turn to this purpose, as specified in clause (i^r). The first, basic component, amounts to the claim that TC(P) ⊆ TC(P*), and hence that P has no extra truth-content relative to P*.

The second component ($\forall x \in \mathbf{T} - P\ z \in c\mathbf{T}\ r(x,z) \rightarrow \exists y \in cP^*\ s(x,y,z)$) can be paraphrased as stating that every mistakenly excluded possibility by P (hence, a member of **T**−P, i.e., FC(P)) has, relative to every comparable nomic impossibility (hence, to be excluded), a companion that is (rightly or wrongly) excluded by P* that is at least as similar to the latter. In other words, P* equals or mitigates every mistake of P. It is easy to check that the extra falsity-content of P* relative to P suggested by (ii^r) has to serve this purpose according to the rest of this clause.

---

[5] It must be noted that the set-theoretic part of the refined approach sounds more intuitive when we take the original and refined theories as inclusion or representation theories, see the Appendix. However, it does not seem plausible to give a belief revision interpretation in AGM-terms of them.

Whereas the basic revision $\underline{P}^{b}_{R(S)}$ was easily seen to be basically at least as truthlike as $\underline{P}$, the refined revision $\underline{P}^{r}_{R(S)}$ is now not necessarily at least as truthlike as $\underline{P}$ in the refined sense. Hence, there is now even more reason to turn to successfulness.

Definition: *$\underline{P}$* is refined at least as successful as $\underline{P}$, relative to R/S*, iff

(i^r-sf) $\forall x \in cP\ z \in cS\ r(x,z) \rightarrow \exists y \in cP^*\ s(x,y,z)$

(ii^r-sf) $\forall y \in cP^* - (cP \cup cR)\ \exists x \in cP - cS\ \exists z \in cR - cP\ s(x,y,z)$

The Refined Success Theorem tells now that, assuming correct data, 'refined at least as truthlike' entails 'refined at least as successful'. Again the proof is not difficult. Similar to the basic case, the consequence of the theorem is that being persistently more successful in the refined sense is conducive for refined truth approximation.

The final and crucial question now is whether the (AGM-interpretable) refined revision $\underline{P}^{r}_{R(S)}$ of $\underline{P}$ by R/S is at least as successful as $\underline{P}$ in the refined sense. In that case it would be potentially conducive for truth approximation for it may become persistently more successful in the refined sense and hence conducive for refined truth approximation. This happens to be the case according to the following:

Main Theorem: $\underline{P}^{r}_{R(S)}$ is refined at least as successful as $\underline{P}$, relative to R/S.

Note first that $\underline{P}^{r}_{R(S)}$ is maximally successful in the basic sense, whereas $\underline{P}$ certainly is not. So, we may expect a trivial proof. Let us look at the first claim to be proven:

(i^r-sf-wrt $P^{r}_{R(S)}$) $\forall x \in cP\ z \in cS\ r(x,z) \rightarrow \exists y \in cP^{r}_{R(S)}\ s(x,y,z)$

This is trivial, for cS is a subset of $cP^{r}_{R(S)}$ and $r(x,z) \rightarrow s(x,z,z)$ is a (plausible) minimal s-condition. The second claim is:

(ii^r-sf-wrt $P^{r}_{R(S)}$) $\forall y \in cP^{r}_{R(S)} - (cP \cup cR)\ \exists x \in cP - cS\ \exists z \in cR - cP\ s(x,y,z)$

This is also trivial, for $cP^{r}_{R(S)}$ is a subset of cR, hence $cP^{r}_{R(S)} - (cP \cup cR)$ is empty.

In view of the Refined Success Theorem we may now abductively conclude that if $\underline{P}^{r}_{R(S)}$, if kept fixed, remains persistently more successful than $\underline{P}$ in the refined sense in the light of new evidence, this is due to being closer to the truth than $\underline{P}$ in the refined sense. Of course, we will then even be more interested in the revision taking also all later evidence into account.

14.4 Conclusions

The main conclusions of this paper are:

First, basic revision of theory $\underline{P}$ in light of evidence R/S, assuming $\underline{P}$ and $\underline{S}$ compatible, based on expansion by $\underline{S}$, leading to $\underline{P \cap S}$, followed by fm-contraction by $\underline{c(R-(P \cap S))}$, leading to $\underline{(P \cap S) \cup R}$, is basically at least as successful as $\underline{P}$ and even basically at least as close to **T** as $\underline{P}$ in the nomic sense.

Second, refined theory revision in light of evidence R/S, assuming $\underline{P}$ and $\underline{S}$ incompatible, based on pm-revision by $\underline{S}$, along Grove-Rabinowicz lines, leading to $\underline{c_P(S) \cap S}$, followed by fm-contraction by $\underline{c(R-(c_P(S) \cap S))}$, leading to $\underline{(c_P(S) \cap S) \cup R}$, is at least as successful as $\underline{P}$ in the refined sense, and hence potentially conducive for refined nomic truth approximation.

At this point a number of debunking remarks are in order:

(a) Both revisions are rather ad hoc. However, as in general for ad hoc changes in a theory, the crucial question is whether they can be put to new (HD-)tests, and this is evidently the case. After all, it could even be the case that all further tests indicate that no new ad hoc maneuvers have to be made.
(b) Both revisions are rather diehard empiricist or instrumentalist. The 'instrument' $\underline{P}$ is precisely so adapted that it just saves the phenomena, not only with respect to R but also with respect to S.
(c) If there is something like well-formed theories, there do not seem to be good reasons to expect that the two revisions will satisfy the criteria, even if R and S satisfy some derived criteria.
(d) Last, but not least, what remains of the idea behind $\underline{P}$? A proper theory, even if it is without theoretical terms, in some sophisticated sense, is usually based on one or two ideas. It is difficult to imagine that such ideas do not become 'mutilated' by the revision.

Be this as it may, the two results may stimulate the interaction between truth approximation and belief revision approaches for they fundamentally show that AGM belief revision provides means for nomic truth approximation.

Appendix

It must be noted that the set-theoretic part of the refined approach sounds more intuitive when we take the original and refined theories as inclusion or representation theories. However, it does not seem plausible to give a belief revision interpretation in AGM-terms of them. In the next chapter we show, in a very general way, that the belief base approach to belief revision (due to Sven Owe Hansson) leaves perfectly room for at least a basic account for two-sided theories, and hence for the extreme cases of purely exclusion and purely inclusion theories.

For completeness, we like to tell here the story that results when we transform (the formulations of) the last part of Sect. 14.3 in terms of inclusion theories, indicated by changing symbols from P, P* and $P^r_{R(S)}$ to M, M* and $M^r_{R(S)}$. Recall that an inclusion theory M makes the claim "$M \subseteq \mathbf{T}$", but we will not make a notational difference between the set and the theory. The definition of refined truth approximation for (pure) inclusion theories reads then (see Chap. 6).

Definition: M* is refined at least as truthlike as M iff

(ir) $\forall x \in M\ z \in T\ r(x,z) \rightarrow \exists y \in M^*\ s(x,y,z)$

(iir) $\forall y \in M^* - (M \cup T)\ \exists x \in M - \mathbf{T}\ \exists z \in \mathbf{T} - M\ s(x,y,z)$

It is again easy to check that (i^r) is a strengthening of (i^b) of the corresponding basic definition and that (ii^r) is a weakening of (ii^b). (i^r) roughly says that every comparable pair of structures, one of M and one of **T**, has an 'intermediate' in M*. In other words, M* represents what is to be included, i.e., **T**, at least as precisely as M. It is this clause that seems intuitive more plausible than the corresponding clause for exclusion theories. (ii^r) states that if $M^* - (M \cup \mathbf{T})$ is at all non-empty, which is excluded in the basic case, these structures are 'useful'. The definition reduces to the basic one when s is trivial.

Whereas the corresponding basic revision $M^b_{R(S)}$ is again easily seen to be basically at least as truthlike as <u>M</u>, the refined revision $M^r_{R(S)}$ is again not necessarily at least as truthlike as <u>M</u> in the refined sense. Hence, there is again even more reason to turn to successfulness.

Definition: *M* is refined at least as successful as* <u>*M*</u>*, relative to R/S*, iff
(i^r-sf) $\forall x \in M\ z \in R\ r(x,z) \rightarrow \exists y \in M^*\ s(x,y,z)$
(ii^r-sf) $\forall y \in M^* - (M \cup S)\ \exists x \in M - R\ \exists z \in S - M\ s(x,y,z)$

The Refined Success Theorem tells now again that, assuming correct data, 'refined at least as truthlike' entails 'refined at least as successful'. Again the proof is not difficult. Similar to the basic case, the consequence of the theorem is that being persistently more successful in the refined sense is conducive for refined truth approximation.

The final crucial question now is whether the refined revision $M^r_{R(S)}$ of <u>M</u> by R/S is at least as successful as <u>M</u> in the refined sense. In that case it would be potentially conducive for refined truth approximation for it may become persistently more successful in the refined sense and hence conducive for refined truth approximation. This happens to be the case according to the following:

<u>Main Theorem (inclusion)</u>: $M^r_{R(S)}$ is refined at least as successful as M, relative to R/S.

Let us look at the specific claims:

(i^r-sf-wrt $M^r_{R(S)}$) $\forall x \in M\ z \in R\ r(x,z) \rightarrow \exists y \in M^r_{R(S)}\ s(x,y,z)$

This is trivial, for R is a subset of $M^r_{R(S)}$ and $r(x,z) \rightarrow s(x,z,z)$ is a (plausible) minimal s-condition.

(ii^r-sf-wrt $M^r_{R(S)}$) $\forall y \in M^r_{R(S)} - (M \cup S)\ \exists x \in M - R\ \exists z \in S - M\ s(x,y,z)$

This is also trivial, for $M^r_{R(S)}$ is a subset of S, hence $M^r_{R(S)} - (M \cup S)$ is empty.

In view of the Refined Success Theorem we may now abductively conclude that if $M^r_{R(S)}$, if kept fixed, remains persistently more successful than M in the refined sense in the light of new evidence, this is due to being closer to the truth than M in the refined sense. Of course, we will then even be more interested in the revision taking also all later evidence into account.

In sum, as far as set-theoretic aspects are concerned, the refined revision story for inclusion theories is easy to tell, and it is at least as intuitively plausible as that for exclusion theories, but how to give an AGM-interpretation of the a basic and refined account for such theories?

References

Alchourrón, C., Gärdenfors, P., & Makinson, D. (1985). On the logic of theory revision: partial meet contraction and revision functions. *Journal of Symbol Logic, 50*(2), 510–530.

Cevolani, G., & Calandra, F. (2009). Approaching the truth via belief change in propositional languages. In M. Suárez, M. Dorato, & M. Rèdei (Eds.), *Launch of the European philosophy of science association, EPSA Epistemology and Methodology of Science* (Vol. 1, pp. 47–62). Dordrecht: Springer.

Grove, A. (1988). Two modellings for theory change. *Journal of Philosophical Logic, 17*, 157–170.

Hansson, S. O. (1999). *A textbook of belief dynamics*. Dordrecht: Kluwer Academic Publishers.

Kuipers, T. (2000). *From instrumentalism to constructive realism. On some relations between confirmation, empirical progress, and truth approximation, Synthese Library 287*. Dordrecht: Kluwer Academic Publishers.

Niiniluoto, I. (1999). Belief revision and truthlikeness. In S. Halldén, B. Hansson, W. Rabinowicz, & N-E. Sahlin (Ed.), *Spinning ideas, Electronic essays dedicated to Peter Gärdenfors on His Fiftieth Birthday*. http://www.lucs.lu.se/spinning/

Rabinowicz, W. (1995). Global belief revision based on similarities between possible worlds. In S. O. Hansson, & W. Rabinowicz (Ed.), *Logic for a change. Essays dedicated to Sten Lindström on the occasion of his fiftieth birthday* (pp. 80–105). Uppsala Prints and Preprints in Philosophy 1995–1999. Uppsala: Department of Philosophy, Uppsala University.

Kuipers, T. (2011). Basic and refined nomic truth approximation by evidence-guided belief revision in AGMterms. *Erkenntnis, 75*(2), 223–236.

Zwart, S. (2001). *Refined verisimilitude, Synthese Library 307*. Dordrecht: Kluwer Academic Publishers.

Chapter 15
Dovetailing Belief Base Revision with Truth Approximation

Abstract Gustavo Cevolani et al. (Erkenntnis 75(2):183–202, 2011) have shown that their account of verisimilitude of 'conjunctive theories' of a finite propositional language can be nicely linked to a variant of AGM belief *set* revision, viz. belief *base* revision, in the sense that the latter kind of revision is functional for truth approximation according to the conjunctive account. In the present chapter I offer a generalization of these ideas to the case of approaching any divide of a (finite or infinite) universe, allowing several interpretations, besides true (false) atomic propositions, notably nomic states (not) in equilibrium, nomic (im-)possibilities, (non-) instantiated 'Q-predicates' of a monadic language. It shows how and why approximation of 'the true boundary' takes place by belief base revision guided by evidence.

In the nomic (im-)possibilities interpretation this chapter essentially deals with a belief *base* revision perspective on *basic* and *quantitative* nomic truth approximation of two-sided theories in the sense of Chaps. 4 and 5. The previous chapter dealt with a belief *set* revision (i.e., AGM-) perspective on *basic and refined* nomic truth approximation of (one-sided) exclusion theories. This chapter will leave the challenge open of a belief base revision perspective on refined nomic truth approximation of two-sided theories, and, more generally, a belief base perspective on refined approximation of 'the true boundary' belonging to whatever interpretation.

Keywords Belief base revision · Conjunctive theories · Generalization · Approaching a divide · Expansion · Contraction · Revision · Truth approximation · Truthlikeness · Nomic interpretation · Propositional interpretation · Partition interpretation · Monadic interpretation

---

Revised version of: "Dovetailing belief base revision with (basic) truth approximation", in E. Weber, D. Wouters, and J. Meheus, *Logic, Reasoning, and Rationality*, proceedings-conference in Ghent, 2010, Springer, 2014, pp. 77–93. *Acknowledgements*: I am very grateful for the stimulating interaction with Gustavo Cevolani and Roberto Festa, and for the comments I got at conferences in Gent (Logic, Reasoning, and Rationality) and Padua (Science, Truth, and Ontology) in September 2010. Finally, I am grateful to the two anonymous referees for their scrupulous remarks and a number of interesting suggestions.

T. A. F. Kuipers, *Nomic Truth Approximation Revisited*, Synthese Library 399,
https://doi.org/10.1007/978-3-319-98388-2_15

15.1 Introduction

Roberto Festa (2007), Gustavo Cevolani and Francesco Calandra (2009) have developed a qualitative and quantitative account of verisimilitude of 'conjunctive theories' of a finite propositional language. The qualitative version of this 'conjunctive' account turned out to be formally equivalent to the definition of 'descriptive verisimilitude' in (Kuipers 1982). Cevolani et al. (2011) managed to show that this account can be nicely linked to a variant of AGM belief *set* revision, viz. belief *base* revision, in the sense that the latter kind of revision is functional for truth approximation according to the conjunctive account. Following an idea of Festa, it is shown in Cevolani et al. (2013) that all this can be transformed for the purposes of basic theoretical or nomic verisimilitude, as initiated in (Kuipers 1982) and elaborated in (Kuipers 2000), and corresponding nomic truth approximation by belief base revision.

In the present chapter I offer a generalization of these ideas to the case of approaching any divide of a (finite or infinite) universe allowing several interpretations, besides true (false) atomic propositions, notably nomic states (not) in equilibrium, nomic (im-)possibilities, and (non-)instantiated 'Q-predicates' of a monadic language. To be sure, the chapter is not very original relative to Cevolani et al. (2011). They convincingly show (in Sect. 15.4) that the restriction to conjunctive (propositional) theories not only enables the application of the belief *base* revision operations, but even in such a way that they, as mentioned, become functional for truth approximation. This is a great improvement relative to the problematic relation between belief *set* revision and truth approximation, as already pointed out by Niiniluoto (1999). Moreover, they claim already in the final section (p. 198) that their "approach can be generalized to any language characterized by a suitable notion of constituent […]. In such languages, in fact, a c[onjunctive]-theory can be conveniently defined as a 'fragment' of a constituent" and they refer more specifically to first order, nomic and even to statistical languages. The main purpose of the present chapter is to offer a transparent formalization of such a general account, though restricted to non-statistical languages, which may well function as a start for extensions and refinements.

Point of departure is a, possibly infinite, universe of discourse (U), and a vocabulary (V) in which subsets of U, e.g. X^+, Y^-, R^+, are characterized. $\underline{\underline{\mathrm{T}}}$ = «$\mathbf{T}^+$,$\mathbf{T}^-$»,[1] note the bold characters, is a divide (formally: a bivalent partition) of U, that is, $\mathbf{T}^+$ and $\mathbf{T}^-$ are non-overlapping subsets of U that together exhaust U. Crucial is the assumption that the divide is not (yet) given in terms of V. The *target of research* is supposed to be the identification, if possible, of the boundary of the two sets in V-terms.

---

[1] We use the double symbols in $\underline{\underline{\mathbf{T}}}$ = «$\mathbf{T}^+$,$\mathbf{T}^-$» and later in e.g. $\underline{\underline{\mathrm{X}}}$ = «X^+, X^-» and $\underline{\underline{\mathrm{R}}}$ = «R^+, R^-» to clearly distinguish these ordered pairs from those in previous chapters, where the second set would correspond to its complement in the present notation, and vice versa. Hence, e.g. «X^+, X^-» and «R^+, R^-», typical of this chapter, formally correspond to <X^+, cX^-> and <R^+, cR^->, respectively. Conversely, e.g. <M, P> and <R, S>, typical in previous chapters, formally correspond in this chapter to «M, cP» and «R, cS», respectively.

Such a characterization, if it exists, will be denoted by $\underline{\underline{\mathrm{T}}} = «\mathrm{T}^+,\mathrm{T}^-»$, with non-bold $\underline{\underline{\mathrm{T}}}$, T^+ and T^-. This target will be more specifically called 'the true theory/boundary', or 'the truth' which is searched for.

In general, U amounts to a set of objects or items and $\mathbf{T}^+$ ($\mathbf{T}^-$) to the subset of objects to which a certain property F does (not) apply. In the propositional interpretation, U and V essentially coincide by being a (finite) set of atomic propositions and (non-)F corresponds to true (false). In the 'partition-interpretation' (Kuipers 1982), U is a set of conceptually possible states generated by V and (non-)F corresponds to (not) being in a so-and-so state, for example, (not) in equilibrium or the bulb is (not) lighting. As far as these interpretations are concerned both alternatives can be verified, they will be called symmetric interpretations, leading to symmetric sets of data. A typically asymmetric interpretation is the nomic interpretation, leading to asymmetric sets of data. Here U is a set of conceptual possibilities generated by V and (non-)F corresponds to physically or, more general, nomically (im-)possible. Whereas we may be able to verify by a single experiment that a conceptual possibility is in fact physically possible, it is not possible in some straightforward way to show that it is in fact not physically possible. It is easy to check that the same holds for the monadic-existential interpretation, in which U is a set of Q-predicates, i.e., a set of mutually exclusive and together exhaustive predicates generated by a vocabulary of primitive predicates. Here (non-)F corresponds to (not) being instantiated. It is plausible to assume that there will be other essentially different interpretations of symmetric or asymmetric nature.

Our program is as follows. In Sect. 15.2 I will define a suitable notion of theories, their truth- and falsity-content, and the comparative expression "theory $\underline{\underline{\mathrm{Y}}}$ is at least as close to the truth/the true boundary as theory $\underline{\underline{\mathrm{X}}}$". In Sect. 15.3 I will tailor the belief (base) revision operations, viz., expansion, contraction, and revision, to the present context. More specifically, I will define the revision, $\underline{\underline{\mathrm{X}}}*\underline{\underline{\mathrm{A}}}$, of theory $\underline{\underline{\mathrm{X}}}$ in the light of an input theory $\underline{\underline{\mathrm{A}}}$. In Sect. 15.4 I will introduce the distinction between symmetric and asymmetric data and prove for (a) data (theory) $\underline{\underline{\mathrm{R}}}$ of either type that the revision ($\underline{\underline{\mathrm{X}}}*\underline{\underline{\mathrm{R}}}$) is closer to the truth than $\underline{\underline{\mathrm{X}}}$ if we assume that the data are true/correct. In Sect. 15.5 I will present quantitative versions of the relevant notions; notably, a quantitative measure for the closeness to the truth of a theory, i.e., its truthlikeness or verisimilitude (cf. Sect. 15.2) and the increase of verisimilitude by belief base revision by true data (cf. Sect. 15.4). I will end with some questions for further research and concluding remarks in Sect. 15.6.

15.2 A Set-Theoretic Framework for Kinds of Basic Truth Approximation

In any context in which the target of research is supposed to be the identification of the boundary of a divide $\underline{\underline{\mathbf{T}}} = «\mathbf{T}^+,\mathbf{T}^-»$, hence with $\mathbf{T}^-=\mathrm{c}\mathbf{T}^+$, of U in terms of a given vocabulary V, it is plausible to assume that a theory is a tentative proposal for objects or items that belong to one of the two subsets.

Formally this can be represented in set-theoretic fashion as follows. A theory $\underline{\underline{X}}$ is a tuple «X^+, X^-» of non-overlapping subsets X^+, X^- of U, defined in terms of V, with the claim "$X^+ \subseteq \mathbf{T}^+$" associated to X^+ and the claim "$X^- \subseteq \mathbf{T}^-$" associated to X^-. Note that these claims are universally quantified conjunctive claims, that is:

$$X^+ \subseteq \mathbf{T}^+ \equiv \forall_{u \in X^+} u \in \mathbf{T}^+ \text{ and } X^- \subseteq \mathbf{T}^- \equiv \forall_{u \in X^-} u \in \mathbf{T}^-$$

Each conjunct will be called a *basic (b-)claim*[2, 3]. All possible 'positive' or 'negative' b-claims, for different u in U, are assumed to be logically independent. X^+ is called the *positive range* of $\underline{\underline{X}}$, X^- its *negative range*, and $X^+ \cup X^-$ its *(total) range*. It is important to stress that «X^+, X^-» is supposed to represent theory $\underline{\underline{X}}$, including the associated claims. Note that $\underline{\underline{\mathbf{T}}}$=«$\mathbf{T}^+$,$\mathbf{T}^-$» is an improper theory: the constitutive sets are not defined in V-terms and its two claims are trivially true. However, the target $\underline{\underline{T}}$=«$T^+$,$T^-$» is a proper theory.

It is not easy to evaluate the assumption of the logical independence of the b-claims. At first sight it might seem a very strong restriction. However, it should be realized that the assumption does not preclude that the formal characterization of a theory $\underline{\underline{X}}$ fulfills criteria of being well-formed, such as formalizability in a certain way and forms of generality. Moreover, it may well be based on a fundamental idea. The more important restriction is that to conjunctive theories in the sense indicated by the universally quantified form of its claims. In the final section we will hint at the generalization of the present approach to theories in general, where the truth-value of individual b-claims may depend on that of other, not on logical grounds, but because the theory requires so.

In Table 15.1, we present both the logical and the set-theoretic representation of one example, first in general terms and then in terms of the four interpretations that were mentioned in the introduction.

Some general notions will be helpful below. Theory $\underline{\underline{X}}$ is *maximal* (or a constituent theory) iff its range is maximal, i.e., $X^+ \cup X^- = U$. In this case $X^- = cX^+$ (where cX^+ denotes the complement of X^+, i.e., $U - X^+$). Note that $\underline{\underline{\mathbf{T}}}$ = «$\mathbf{T}^+$,$\mathbf{T}^-$» is maximal (but, recall, an improper theory). Finally, for theory $\underline{\underline{X}}$ = «X^+, X^-», theory «X^-, X^+» is called the *specular* of $\underline{\underline{X}}$, indicated by sp($\underline{\underline{X}}$). Hence the claims of sp($\underline{\underline{X}}$) are "$X^- \subseteq \mathbf{T}^+$" and "$X^+ \subseteq \mathbf{T}^-$", respectively.

Now we can start to introduce the crucial notions of truth- and falsity-content and, in these terms, the notion of basic qualitative truthlikeness. The *positive truth-content* of $\underline{\underline{X}}$ (or the set of true positives) is $t^+(\underline{\underline{X}}) = X^+ \cap \mathbf{T}^+$. The *negative truth-content* of $\underline{\underline{X}}$ (or the set of true negatives) is $t^-(\underline{\underline{X}}) = X^- \cap \mathbf{T}^-$. The *truth (t-)content* of $\underline{\underline{X}}$ is the union of these sets: $t(\underline{\underline{X}}) = t^+(\underline{\underline{X}}) \cup t^-(\underline{\underline{X}})$. Similarly, the *positive falsity-content* of $\underline{\underline{X}}$ (the set of false positives) is $f^+(\underline{\underline{X}}) = X^+ \cap \mathbf{T}^-$ (= $X^+ - \mathbf{T}^+$), the *negative falsity-content* of $\underline{\underline{X}}$ (the set of false negatives) is $f^-(\underline{\underline{X}}) = X^- \cap \mathbf{T}^+$ (= $X^- - \mathbf{T}^-$), and

[2] Note that the two claims "$X^+ \subseteq \mathbf{T}^+$" and "$X^- \subseteq \mathbf{T}^-$" are compatible conjunctive claims, which means that the conjunction of these claims can be seen as a 'conjunctive theory' in the sense of Cevolani et al. (2011).

[3] As suggested before, note that the nomic interpretation, as represented in notably Chap. 4, with theory <M, P>, inclusion claim "M$\subseteq$**T**" and exclusion claim "**T**$\subseteq$P", that is "cP$\subseteq$c**T**", corresponds in the present representation to the theory <<M, cP>>and its claims, assuming $\mathbf{T} = \mathbf{T}^+ = c\mathbf{T}^-$.

Table 15.1 Interpretations and representations of an example

Interpretation	Logical representation	Set-theoretical representation $\underline{\underline{X}}$ = «X^+ “$X^+ \subseteq \mathbf{T}^+$”	,X^-» “$X^- \subseteq \mathbf{T}^-$”
General	Fa_1 & Fa_2 & $\neg Fa_3$ $(\neg)Fa_i$ F applies (does not apply) to a_i	«$\{a_1,a_2\}$ “$\{a_1,a_2\} \subseteq \mathbf{T}^+$”	,$\{a_3\}$» “$\{a_3\} \subseteq \mathbf{T}^-$”
Propositional	p_1 & p_2 & $\neg p_3$ $(\neg)p_i$ p_i is true (false)	«$\{p_1, p_2\}$ “$\{p_1, p_2\} \subseteq \mathbf{T}^+$”	,$\{p_3\}$» “$\{p_3\} \subseteq \mathbf{T}^-$”
Partition	Es_1 & Es_2 & $\neg Es_3$ $(\neg)Es_i$ s_i is a (non) equilibrium state	«$\{s_1,s_2\}$ “$\{s_1,s_2\} \subseteq \mathbf{T}^+$”	,$\{s_3\}$» “$\{s_3\} \subseteq \mathbf{T}^-$”
Nomic	Nu_1 & Nu_2 & $\neg Nu_3$ $(\neg)Nu_i$ u_i is a nomic (im-)possibility	«$\{u_1,u_2\}$ “$\{u_1,u_2\} \subseteq \mathbf{T}^+$”	,$\{u_3\}$» “$\{u_3\} \subseteq \mathbf{T}^-$”
Monadic-existential	$\exists xQ_1(x)$ & $\exists xQ_2(x)$ & $\neg\exists xQ_3(x)$ $(\neg)\exists xQ_i(x)$ Q_i is (not) instantiated	«$\{Q_1,Q_2\}$ “$\{Q_1,Q_2\} \subseteq \mathbf{T}^+$”	,$\{Q_3\}$» “$\{Q_3\} \subseteq \mathbf{T}^-$”

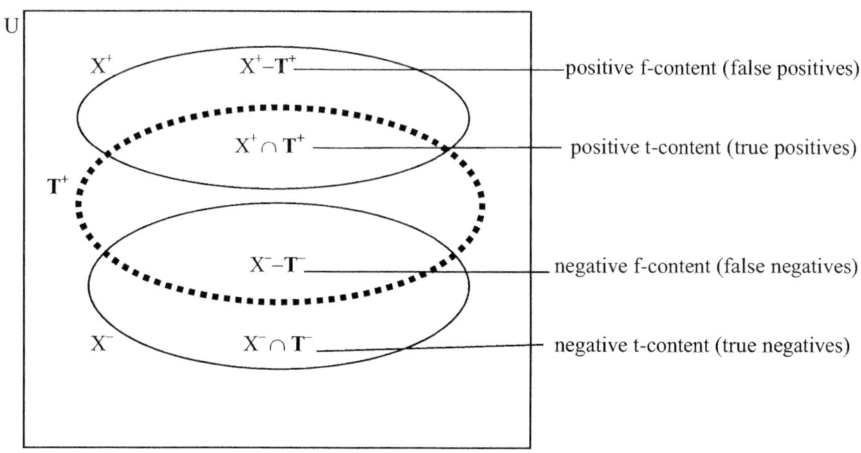

Fig. 15.1 The four content parts of a theory

the *falsity (f-)content* of $\underline{\underline{X}}$ is the union of both: $f(\underline{\underline{X}}) = f^+(\underline{\underline{X}}) \cup f^-(\underline{\underline{X}})$. It is easy to check that $\underline{\underline{\mathbf{T}}}$ itself has extreme values for these six notions for trivial reasons: $t^+(\underline{\underline{\mathbf{T}}}) = \mathbf{T}^+$, $t^-(\underline{\underline{\mathbf{T}}}) = \mathbf{T}^-$, $t(\underline{\underline{\mathbf{T}}}) = U$, and $f^+(\underline{\underline{\mathbf{T}}}) = f^-(\underline{\underline{\mathbf{T}}}) = f(\underline{\underline{\mathbf{T}}}) = \varnothing$. However, the target $\underline{\underline{T}}$ = «T^+,T^-» has them non-trivially.

Note also that the union of the truth-content and the falsity-content of theory $\underline{\underline{X}}$ exhausts its total range, formally: $t(\underline{\underline{X}}) \cup f(\underline{\underline{X}}) = X^+ \cup X^-$. Moreover, the union of the positive truth-content and the positive falsity-content $\underline{\underline{X}}$ equals its positive range, formally: $t^+(\underline{\underline{X}}) \cup f^+(\underline{\underline{X}}) = X^+$. Similarly, the union of the negative truth-content and the negative falsity-content equals the negative range, formally, $t^-(\underline{\underline{X}}) \cup f^-(\underline{\underline{X}}) = X^-$. The four basic notions are depicted in Fig. 15.1, where $\mathbf{T}^-$, i.e., the complement of the indicated T^+, is left implicit.

In my previous work on nomic truth approximation (Kuipers 2000), I have always restricted the attention to maximal theories. To connect the present unrestricted (and 'interpretation-free') analysis with it, I will from time to time present the results for maximal theories. To begin with, if $\underline{\underline{X}}$ is maximal, the falsity-content $f(\underline{\underline{X}})$ equals the symmetric difference of X^+ and $\mathbf{T}^+$, that is, $\Delta(X^+,\mathbf{T}^+) =_{\text{def}} (X^+-\mathbf{T}^+) \cup (\mathbf{T}^+-X^+)$, which equals of course also the symmetric difference of their complements: $\Delta(X^-,\mathbf{T}^-)$. Moreover, the truth-content is the complement of the falsity-content, formally: $t(\underline{\underline{X}}) = c(f(\underline{\underline{X}}))$.

We can now also give plausible definitions of theories being true or false: $\underline{\underline{X}}$ is *true* iff $f(\underline{\underline{X}}) = \emptyset$, and *false* otherwise. As an aside, $\underline{\underline{X}}$ is said to be *completely false* iff $t(\underline{\underline{X}}) = \emptyset$, explicating the idea that the theory does not contain a grain of truth.

For the idea that there is something like 'the truth' to be approached, it is now crucial to note, and easy to check, that there is at most one true maximal (proper) theory, viz. $\underline{\underline{T}} = «T^+,T^-»$, the target of research!

We can now state the definition of *basic truthlikeness* or *basic verisimilitude* in general terms.

Definition 1 Theory $\underline{\underline{Y}}$ is (qualitatively) at least as close to theory $\underline{\underline{\mathbf{T}}}$, or to the truth, or verisimilar, as theory $\underline{\underline{X}}$, $\underline{\underline{X}} \leq_{vs} \underline{\underline{Y}}$, iff

$t(\underline{\underline{X}}) \subseteq t(\underline{\underline{Y}})$	&	$f(\underline{\underline{Y}}) \subseteq f(\underline{\underline{X}})$
or, in decomposed form:		
$t^+(\underline{\underline{X}}) \subseteq t^+(\underline{\underline{Y}})$ & $t^-(\underline{\underline{X}}) \subseteq t^-(\underline{\underline{Y}})$	&	$f^+(\underline{\underline{Y}}) \subseteq f^+(\underline{\underline{X}})$ & $f^-(\underline{\underline{Y}}) \subseteq f^-(\underline{\underline{X}})$

By requiring at least once a *proper* subset relation we get the definition of "Theory $\underline{\underline{Y}}$ is (qualitatively) closer to theory $\underline{\underline{\mathbf{T}}}$, or to the truth, than theory $\underline{\underline{X}}$", but I will focus on the notion of 'at least as close to'. In the case of maximal theories the definition 'at least as close to' can be reduced due to the following theorem.

Theorem 1 If theories $\underline{\underline{X}}$ and $\underline{\underline{Y}}$ are maximal, $t(\underline{\underline{X}}) = c(f(\underline{\underline{X}}))$ and, vice versa, $f(\underline{\underline{X}}) = c(t(\underline{\underline{X}}))$, and hence $t(\underline{\underline{X}}) \subseteq t(\underline{\underline{Y}})$ iff $f(\underline{\underline{Y}}) \subseteq f(\underline{\underline{X}})$.

Consequently, for maximal theories one of the two conditions is enough for $\underline{\underline{Y}}$ being at least as close to the truth as $\underline{\underline{X}}$. As an aside, the 'f-condition', $f(\underline{\underline{Y}}) \subseteq f(\underline{\underline{X}})$, corresponds to the (basic) definition in the nomic interpretation, restricted to maximal theories, viz. $\Delta(Y^+,\mathbf{T}^+) \subseteq \Delta(X^+,\mathbf{T}^+)$, (Kuipers 2000).

Figure 15.2 depicts the situation in general that $\underline{\underline{Y}}$ is at least as close to the truth as $\underline{\underline{X}}$. The horizontally shaded areas are empty due to the fact that the f-content of $\underline{\underline{Y}}$ is supposed to be a subset of the f-content of $\underline{\underline{X}}$, whereas the vertically shaded areas are empty due to the fact that the t-content of $\underline{\underline{X}}$ is supposed to be a subset of the t-content of $\underline{\underline{Y}}$. Note that there are two double-shaded areas.

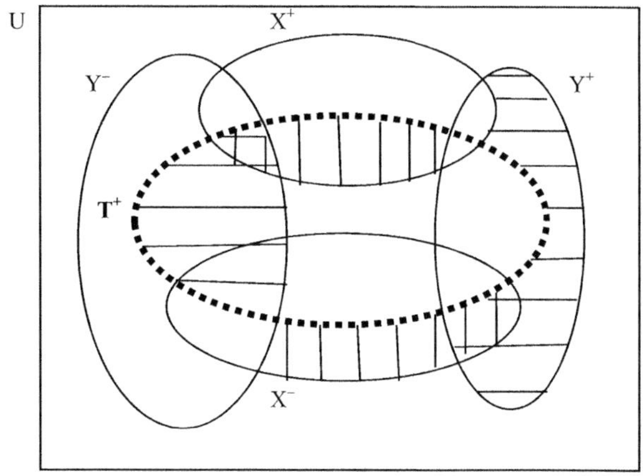

Fig. 15.2 Theory $\underline{\underline{Y}}$ is at least as close to the truth as theory $\underline{\underline{X}}$

15.3 Basic Belief Base Revision, Set-Theoretically Presented

As is well-known, *AGM-belief revision* centers around three (partially related) operations (Alchourrón et al. 1985), see also e.g. Hansson (1999, 2006), and Cevolani and Calandra (2009). A belief set, that is, a deductively closed set of sentences of a given language, is confronted with some 'input sentence' that, by minimal changes of the original belief set, either should become a consequence or no longer be a consequence of the revised belief set. For the first case, it makes an important difference whether or not the input sentence is compatible with the belief set. In the first subcase we get so-called *expansion*, viz., the belief set is strengthened to the set of consequences of the union of the belief set and the input sentence. Regarding the input sentence, assuming it was not already included, it leads from suspension of judgment about that sentence to its acceptance. In the second subcase the belief set has to be adapted in a more complicated way, satisfying certain axioms (see e.g. Alchourrón et al. 1985, p. 513; Hansson 1999, Chap. 3^+). This operation is called *revision* (in the narrow sense). Regarding the input sentence revision leads from its rejection to its acceptance, except when the input sentence is inconsistent. Finally, in the second main case, the input sentence is supposed to belong to the belief set, but should no longer belong to the revised set. Hence, now the belief set has to be weakened in a minimal way, again in line with some axioms (see e.g. Alchourrón et al. 1985, p. 513; Hansson 1999, Chap. 2^+). This operation is called *contraction*. Regarding the input sentence contraction leads from its acceptance to suspension of judgment, except when the input sentence is logically true, in which case it remains accepted after contraction. The three operations are related by the so-called

Levi-identity, according to which the result of the revision of a belief set by an input sentence is or should be identical to the contraction of the belief set by the negation of the input sentence, followed by the expansion of the resulting belief set by the input sentence. The focus in the belief revision program has been on the link between, on the one hand, axiomatic characterizations of revision and contraction, and on the other hand, specific definitions of these operations, sometimes resulting in representation theorems.

As a variant of belief *set* revision, Hansson (1999, 2006) has developed so-called belief *base* revision, in which the operations are primarily defined for and applied to a base of sentences, taking its consequences only into account when relevant. This variant fits best with our set-theoretic set-up. The (positive/negative) base of a theory $\underline{\underline{X}}$ corresponds to its set of b-claims (associated with the positive range X^+ and with the negative range X^-, respectively). New input corresponds to the base of another theory $\underline{\underline{A}}$, which may represent empirical data or some other information that is given priority over $\underline{\underline{X}}$.

Before we introduce the relevant definitions it will be useful to pictorially show the overlapping (o), conflicting (c) and excess (e) parts of theories $\underline{\underline{X}}$ and $\underline{\underline{A}}$, where three of them are named by way of example, see Fig. 15.3. That the pairs X^+ and X^-, and A^+ and A^- are non-overlapping is essential. However, that the components of both pairs (non-overlappingly) touch each other, even at the same place, is inessential and only for simplification of the pictures. The general picture is followed by Fig. 15.4 in which the following *propositional example* has been inserted. $\underline{\underline{X}}$ corresponds to $(p_1\&p_2\&\neg p_3\&p_4)$ and $\underline{\underline{A}}$ to $(p_2\&p_3\&\neg p_5)$. Hence, we get $\underline{\underline{X}} = «X^+,X^-» = «\{p_1,p_2,p_4\}, \{p_3\}»$ and $\underline{\underline{A}} = «A^+,A^-» = «\{p_2,p_3\}, \{p_5\}»$. This example will be used for further illustrations.

It is of course possible to set-theoretically characterize all parts. The ordered pairs of corresponding positive and negative parts of the one relative to the other form theories themselves. The *overlapping* part of $\underline{\underline{X}}$ relative to $\underline{\underline{A}}$ is « $X^+\cap A^+$, $X^-\cap A^-$ », which amounts to «$\{p_2\}$, ∅» in the example. The *conflicting* part of $\underline{\underline{X}}$ relative to $\underline{\underline{A}}$ is «$X^+\cap A^-$, $X^-\cap A^+$», that is, «∅, $\{p_3\}$» in the example. Finally, the

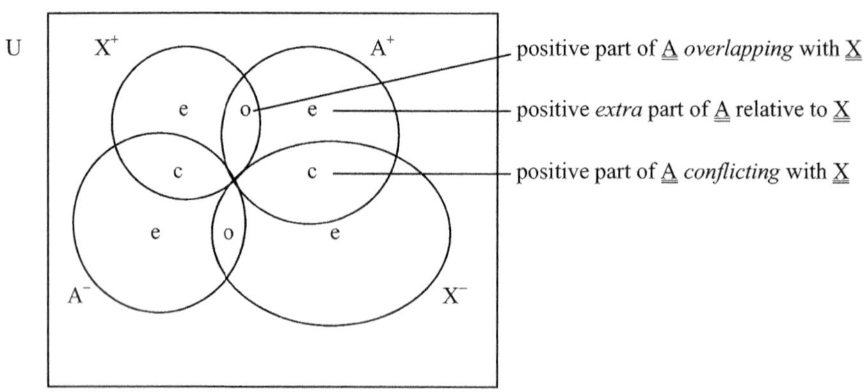

Fig. 15.3 The three positive parts of theory $\underline{\underline{A}}$ relative to theory $\underline{\underline{X}}$

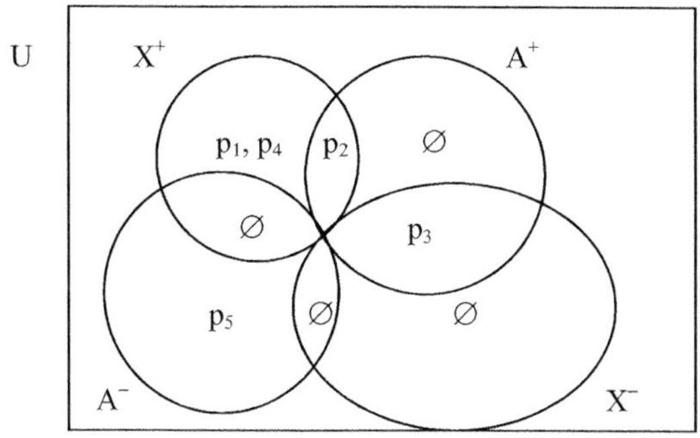

Fig. 15.4 All parts of $\underline{\underline{X}}$ = «{p_1, p_2, p_4}, {p_5}» relative to $\underline{\underline{A}}$ = «{p_2, p_3}, {p_5}» and vice versa

excess part of $\underline{\underline{X}}$ relative to $\underline{\underline{A}}$ is «$X^+ - (A^+ \cup A^-)$, $X^- - (A^+ \cup A^-)$», which can be rewritten as « $(X^+ - A^+) - A^-$, $(X^- - A^-) - A^+$», that is «{p_1, p_4}, ∅» in the example. The overlapping part of $\underline{\underline{A}}$ relative to $\underline{\underline{X}}$ is of course the same as that of $\underline{\underline{X}}$ relative to $\underline{\underline{A}}$. The conflicting part of $\underline{\underline{A}}$ relative to $\underline{\underline{X}}$ is «$A^+ \cap X^-$, $A^- \cap X^+$», which is the specular of that of $\underline{\underline{X}}$ relative to $\underline{\underline{A}}$, that is, sp(«$X^- \cap A^+$, $X^+ \cap A^-$»). The excess part of $\underline{\underline{A}}$ relative to $\underline{\underline{X}}$ is «$A^+ - (X^+ \cup X^-)$, $A^- - (X^+ \cup X^-)$» = «$(A^+ - X^+) - X^-$, $(A^- - X^-) - X^+$».

See Cevolani et al. (2011) for a detailed analysis of these parts, in the propositional interpretation, and for the way in which the three revision operations can be defined in terms of these parts. This can easily be generalized in set-theoretic terms. However, here I prefer another, straightforward, way of defining the operations.

We start with expansion, which is defined only for the case in which $\underline{\underline{X}}$ and $\underline{\underline{A}}$ are compatible, which is of course defined by the condition that the mutually conflicting parts are empty: $A^+ \cap X^- = \varnothing$ and $A^- \cap X^+ = \varnothing$, i.e., the c-areas in the general figure are empty. This amounts to the assumption that none of the b-claims of $\underline{\underline{X}}$ and $\underline{\underline{A}}$ are in conflict. Note that $\underline{\underline{X}}$ and $\underline{\underline{A}}$ in the example are incompatible, hence, expansion cannot be illustrated with it.

In terms of b-claims, the idea of expansion of $\underline{\underline{X}}$ by a compatible $\underline{\underline{A}}$ is that the b-claims of $\underline{\underline{A}}$ are added to those of $\underline{\underline{X}}$. Formally: *expansion* of $\underline{\underline{X}}$ by a *compatible* $\underline{\underline{A}}$, denoted by $\underline{\underline{X}} + \underline{\underline{A}}$, is defined as « $X^+ \cup A^+$, $X^- \cup A^-$ ». It is easy to check that $\underline{\underline{X}} + \underline{\underline{A}}$ is equal to the expansion of $\underline{\underline{X}}$ by the excess part of $\underline{\underline{A}}$ relative to $\underline{\underline{X}}$. In Fig. 15.5 the positive and the negative range of the resulting theory have been indicated by plus- and minus signs, respectively, representing in total all b-claims of $\underline{\underline{X}}$ and the excess b-claims of $\underline{\underline{A}}$ relative to $\underline{\underline{X}}$.

In terms of b-claims, the idea of contraction of $\underline{\underline{X}}$ by $\underline{\underline{A}}$ is that all relevant b-claims of $\underline{\underline{A}}$ are deleted from those of $\underline{\underline{X}}$. Formally: *contraction* of $\underline{\underline{X}}$ by $\underline{\underline{A}}$, denoted by $\underline{\underline{X}} - \underline{\underline{A}}$, is defined as « $X^+ - A^+$, $X^- - A^-$». It is now easy to check that $\underline{\underline{X}} - \underline{\underline{A}}$ is equal to the contraction of $\underline{\underline{X}}$ by the overlapping part of $\underline{\underline{A}}$ relative to $\underline{\underline{X}}$. It is also equal to the expansion of the excess part of $\underline{\underline{X}}$ relative to $\underline{\underline{A}}$ by the conflicting part of $\underline{\underline{X}}$ relative to $\underline{\underline{A}}$. Again, in Fig. 15.6, the positive and the negative range of the resulting theory have been indicated, representing in total all conflicting b-claims of $\underline{\underline{X}}$ rela-

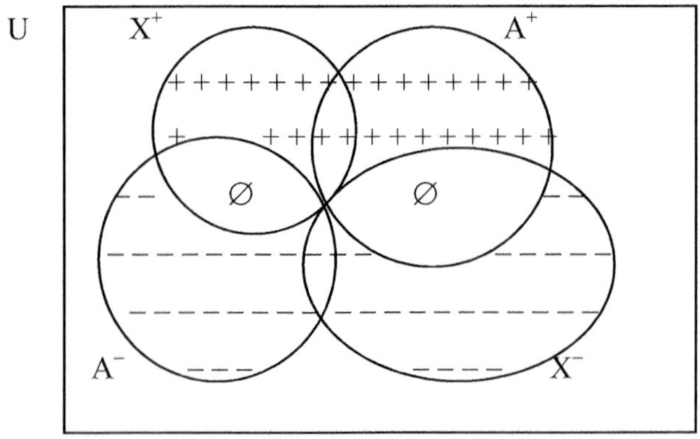

Fig. 15.5 The expansion of X by (compatible) A

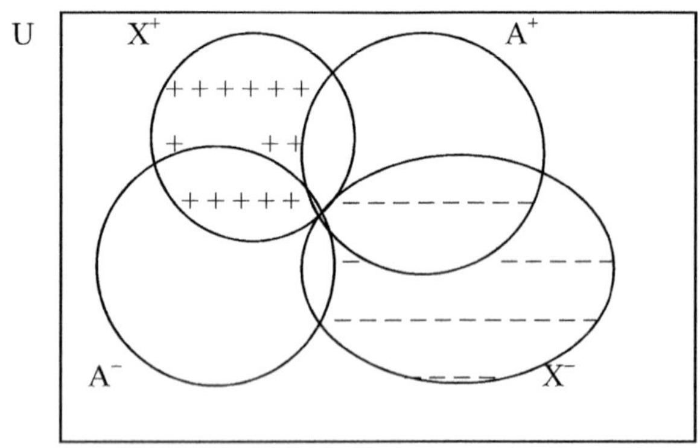

Fig. 15.6 The contraction of X by A

tive to A and all excess b-claims of X relative to A, that is, all b-claims of X that are non-overlapping with those of A. In the example X – A amounts to «$\{p_1,p_4\}$, $\{p_3\}$».

As an aside, I would like to note that from the AGM-perspective the present kind of contraction is rather elementary. AGM-contraction is based on the so-called *remainder set*. In terms of sentences, the remainder set of a belief set or belief base after contraction by a sentence is the set of maximal subsets that do not imply that sentence. Such subsets are called *remainders* (Hansson 1999, 2006). Note that for belief base contraction it is not sufficient that these subsets do not contain that sentence, for they might still (jointly) imply it. Belief contraction becomes interesting (and complicated) when the remainder set contains more than one remainder. In this case the general form of so-called *partial meet contraction* is defined as the intersection of the sets in some subset of the remainder set, to be selected by some selection function. In our present context of contraction the relevant remainder set is already unique, due to the fact that we are assuming that all possible b-claims are

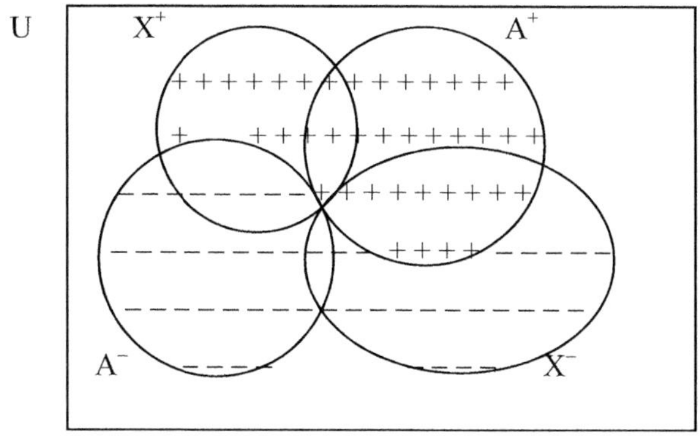

Fig. 15.7 The revision of $\underline{\underline{X}}$ by $\underline{\underline{A}}$

logically independent and that we are only considering contraction by (sets of) b-claims.

Finally, we turn to (belief base) revision (in the narrow sense). As announced, we base our definition on the relevant form of Levi's identity. That is, in the revision of $\underline{\underline{X}}$ by $\underline{\underline{A}}$ there are two steps: first the b-claims of $\underline{\underline{X}}$ that are in conflict with those of $\underline{\underline{A}}$ are deleted, and second all b-claims of $\underline{\underline{A}}$ are added. The first step amounts to the contraction of $\underline{\underline{X}}$ by the specular of $\underline{\underline{A}}$, sp($\underline{\underline{A}}$), after which expansion of the result, $\underline{\underline{X}}$ − sp($\underline{\underline{A}}$), by $\underline{\underline{A}}$ follows. Formally: *revision* of $\underline{\underline{X}}$ by $\underline{\underline{A}}$, denoted by $\underline{\underline{X}} * \underline{\underline{A}}$, is defined by [$\underline{\underline{X}}$ − sp($\underline{\underline{A}}$)] + $\underline{\underline{A}}$. It is now easy to check that $\underline{\underline{X}} * \underline{\underline{A}}$ is equal to the expansion of the excess part of $\underline{\underline{X}}$ relative to $\underline{\underline{A}}$ by $\underline{\underline{A}}$ or, equivalently, the expansion of $\underline{\underline{A}}$ by the excess part of $\underline{\underline{X}}$ relative to $\underline{\underline{A}}$. Again, in Fig. 15.7, the positive and the negative range of the resulting theory have been indicated, representing in total all b-claims of $\underline{\underline{A}}$ and all excess b-claims of $\underline{\underline{X}}$ relative to $\underline{\underline{A}}$.

Recall the example $\underline{\underline{X}}$ = «{p_1,p_2,p_4}, {p_3}» and $\underline{\underline{A}}$ = «{p_2,p_3}, {p_5}». Hence, sp($\underline{\underline{A}}$) = «{$p_5$}, {$p_2$, p_3}», which leads to $\underline{\underline{X}}$ − sp($\underline{\underline{A}}$) = «{$p_1$,$p_2$,$p_4$}, ∅» and hence to [$\underline{\underline{X}}$ − sp($\underline{\underline{A}}$)] + $\underline{\underline{A}}$ = «{p_1,p_2,p_3,p_4}, p_5}».

For the same reason as in the case of contraction, the present form of revision is a rather special case from the AGM-perspective. Due to the uniqueness of contraction and the rest of the Levi-identity form of our definition, revision is also unique. In line with calling truth approximation in terms of (comparing) sets of basic claims (Kuipers 2000), it is plausible to call belief base (expansion, contraction and) revision in terms of sets of basic claims also *basic*.

15.4 Truth Approximation by Data Guided Theory Revision

Now we can start to dovetail theory revision with truth approximation. First we have to represent empirical data as they may come in by experiments or otherwise. It is important to distinguish two types of possible data, depending on the relevant interpretation as already alluded to in the introduction. In both cases we can represent data as theories.

Symmetric data can come in when, for example, in the propositional interpretation it is possible to verify whether some p is true or not, depending on what is in fact the case. A specific example of this arises in the partition interpretation, for example, when it can be verified whether a state is in equilibrium or not. In general, we will speak of (sets of) realized examples of equilibrium, denoted by R^+, and (sets of) realized non-examples, denoted by R^-. In combination we get a *symmetric data-theory* $\underline{\underline{R}} = «R^+,R^-»$, assuming that (examples and) non-examples can be realized. In the following we will assume that the data are correct. For later purposes we introduce the *Correct Data (CD-)hypothesis*, i.e., all b-claims of $\underline{\underline{R}}$ are true or, simply, $\underline{\underline{R}}$ is true, which amounts in the symmetric case to: $R^+ \subseteq \mathbf{T}^+$, $R^- \subseteq \mathbf{T}^-$, i.e., $R^+ \subseteq \mathbf{T}^+ \subseteq cR^-$ (recall: $\mathbf{T}^- = c\mathbf{T}^+$).

Asymmetric data arise when non-examples cannot be realized in some direct sense by the nature of the interpretation, such as physical impossibilities, non-existence claims, etc. In this case we may get in due course a set of realized examples R and on their basis we may make inductive jumps to laws, claiming, for the time being, that certain conceptual possibilities cannot exist by the nature of physical reality or that certain types of individuals do not occur in the relevant universe of discourse. Note that it is plausible to assume that R arises, at least partly, by testing general hypotheses and hence that the inductive steps presuppose this kind of serious testing. Having induced laws implies having an indirectly induced strongest law, to be obtained by conjunction. It is easy to see that a law, and hence the strongest one, can set-theoretically be represented by a subset S of U with the claim that $\mathbf{T}^+$ is a subset of S or, equivalently, that cS is a subset of $\mathbf{T}^-$. In combination we now get an *asymmetric data-theory* $\underline{\underline{R}} = «R,cS»$, in particular for nomic and monadic-existential interpretations. In the asymmetric case the Correct Data (CD-)hypothesis, i.e., $\underline{\underline{R}}$ is true, amounts to: $R \subseteq \mathbf{T}^+$, $cS \subseteq \mathbf{T}^-$, i.e., $R \subseteq \mathbf{T}^+ \subseteq S$. Of course, whereas symmetric data may contain descriptive mistakes, asymmetric data may contain descriptive as well as inductive mistakes. Hence, in the asymmetric case the CD-hypothesis is much more demanding than in the symmetric case.

We start with the revision of theories by symmetric data. The revision of theory $\underline{\underline{X}}$ by symmetric data theory $\underline{\underline{R}} = «R^+,R^-»$, $\underline{\underline{X}} * \underline{\underline{R}}$, is by definition: $[\underline{\underline{X}} - sp(\underline{\underline{R}})] + \underline{\underline{R}}$, that is, $[« X^+, X^-» - «R^-, R^+»] + «R^+,R^-»$, which leads via $«X^+ - R^-, X^- - R^+» + «R^+,R^-»$ to $«(X^+ - R^-) \cup R^+, (X^- - R^+) \cup R^-»$. Now it is easy to prove the following:

Theorem 2 Assuming the CD-hypothesis for symmetric data $\underline{\underline{R}}$, $\underline{\underline{X}} * \underline{\underline{R}}$ is (qualitatively) at least as close to the truth as $\underline{\underline{X}}$, i.e., $\underline{\underline{X}} \leq_{vs} \underline{\underline{X}} * \underline{\underline{R}}$.

For the proof we will first compare $«(X^+ - R^-) \cup R^+, (X^- - R^+) \cup R^-»$ with $«X^+, X^-»$. In this comparison it is easy to see that all established false b-claims of $\underline{\underline{X}}$ are deleted and that the established (excess) true b-claims are added. In more detail, without 'established': the false positive b-claims of $\underline{\underline{X}}$ are deleted and true positive b-claims of $\underline{\underline{R}}$ are added and the false negative b-claims of $\underline{\underline{X}}$ are deleted and true negative b-claims of $\underline{\underline{R}}$ are added. For the proof in the strict sense it is only important to observe that only false b-claims of $\underline{\underline{X}}$, if any, are deleted and that only true b-claims, if any, are added. In other words, that the falsity-content of the result does not increase and the truth-content of the result does not decrease, in accordance with the definition of ‘at least as close to’.

Let us now turn to the revision of theories by asymmetric data. Of course, formally we can formulate and prove the relevant theorem by just replacing above ‘R^+’ by ‘R’ and ‘R^-’ by ‘cS’. However, for the sake of transparency, I will also present the resulting asymmetric story. The revision of theory $\underline{\underline{X}}$ by asymmetric data theory $\underline{\underline{R}}$ = «R,cS», $\underline{\underline{X}} * \underline{\underline{R}}$, is by definition: [$\underline{\underline{X}}$ − sp($\underline{\underline{R}}$)] + $\underline{\underline{R}}$, that is, [« X^+, X^-» − «cS, R»] + «R,cS», which leads via «$X^+ - cS$, $X^- - R$» + «R, cS» to «$(X^+ - cS) \cup R$, $(X^- - R) \cup cS$». Again it is easy to prove the following theorem:

Theorem 3 Assuming the CD-hypothesis for asymmetric data $\underline{\underline{R}}$, $\underline{\underline{X}} * \underline{\underline{R}}$ is (qualitatively) at least as close to the truth as $\underline{\underline{X}}$, i.e., $\underline{\underline{X}} \leq_{vs} \underline{\underline{X}} * \underline{\underline{R}}$.

By comparing «$(X^+ - cS) \cup R$, $(X^- - R) \cup cS$» with «X^+, X^-» it is again easy to see that only, even all, established false b-claims are deleted and that only, even all, established (excess) true b-claims are added, where the two times ‘only’, even if restricted to ‘only, if any’, are already sufficient to prove the ‘at least as close to’-claim. A slight transformation of the resulting revision, viz. «$(X^+ \cap S) \cup R$, $(X^- \cap cR) \cup cS$», enables to paraphrase the comparison with «X^+, X^-» by: the positive b-claims are restricted to those compatible with S and extended with those about R, and the negative b-claims are restricted to those compatible with the complement of R and extended with those implied by S, that is, those compatible with the complement of S. Note, that the resulting revision «$(X^+ \cap S) \cup R$, $(X^- \cap cR) \cup cS$» can be further transformed (using $R \subseteq S$) in terms of the laws claimed by $\underline{\underline{X}}$, viz. “$\mathbf{T}^+ \subseteq cX^-$”, leading to «$(X^+ \cap S) \cup R$, $(ccX^- \cap cR) \cup cS$» = «$(X^+ \cap S) \cup R$, $c(cX^- \cup R) \cup cS$» = «$(X^+ \cap S) \cup R$, $c((cX^- \cup R) \cap S)$» = «$(X^+ \cup R) \cap S$, $c((cX^- \cap S) \cup R)$». Hence, the law-claim of the resulting revision amounts to “$\mathbf{T}^+ \subseteq (cX^- \cap S) \cup R$”.

Finally, it is easy to see that the two theorems can be strengthened to the claims that the revised theory is even closer to the truth, if we assume that $\underline{\underline{R}}$ is not just a subtheory of $\underline{\underline{X}}$ in the sense that $R^+ \subseteq X^+$ and $R^- \subseteq X^-$ in the symmetric case and $R \subseteq X^-$ and $cS \subseteq X^-$ in the asymmetric case, respectively. For by the contraction step some false b-claims of $\underline{\underline{X}}$ will be deleted and/or by the final expansion the revised theory will get extra true b-claims relative to $\underline{\underline{X}}$.

15.5 Generalized Feature-Contrast Measure of Verisimilitude

It is not difficult to present a quantitative version of the main points of the previous sections. We will first define a measure for the verisimilitude of a theory and then show that, and to what extent, the verisimilitude of the revision by data, assumed to be true, has increased.

Let m be a so-called normalized countably additive real-valued measure function on the set of (measurable) subsets of U, that is, for all subsets V, W of U: $m(\emptyset) = 0 \leq m(V) \leq 1 = m(U)$ and $m(V \cup W) = m(V) + m(W) - m(V \cap W)$.

We define the following degrees of positive/negative truth-/falsity-content:

Degree of positive truth-content	$\mathrm{Cont}^{+}_{t}(\underline{\underline{X}}) = m(t^{+}(X)) = m(X^{+} \cap \mathbf{T}^{+})$
Degree of negative truth-content	$\mathrm{Cont}^{-}_{t}(\underline{\underline{X}}) = m(t^{-}(X)) = m(X^{-} \cap \mathbf{T}^{-})$
Degree of positive falsity-content	$\mathrm{Cont}^{+}_{f}(\underline{\underline{X}}) = m(f^{+}(X)) = m(X^{+} \cap \mathbf{T}^{-}) = m(X^{+} - \mathbf{T}^{+})$
Degree of negative falsity-content	$\mathrm{Cont}^{-}_{f}(\underline{\underline{X}}) = m(f^{-}(X)) = m(X^{-} \cap \mathbf{T}^{+}) = m(X^{-} - \mathbf{T}^{-})$

In order to take different absolute weights of truth and falsity and relative weights of true (false) positives and negatives into account, we introduce three positive real-valued parameters, viz. φ,γ, and δ, all positive, and define.

Definition 2 The degree of verisimilitude of $\underline{\underline{X}}$: $\mathrm{Vs}_{\varphi}^{\gamma,\delta}(\underline{\underline{X}})$

$$=_{\mathrm{def}} [\mathrm{cont}^{+}_{t}(\underline{\underline{X}}) + \gamma\, \mathrm{cont}^{-}_{t}(\underline{\underline{X}})] - \varphi\, [\mathrm{cont}^{+}_{f}(\underline{\underline{X}}) + \delta\, \mathrm{cont}^{-}_{f}(\underline{\underline{X}})]$$
$$= [m(X^{+} \cap \mathbf{T}^{+}) + \gamma\, m(X^{-} \cap \mathbf{T}^{-})] - \varphi\, [m(X^{+} - \mathbf{T}^{+}) + \delta\, m(X^{-} - \mathbf{T}^{-})]$$

Here φ represents the relative weight of truth and falsity and γ and δ represent the relative weight of true positives and negatives and false positives and negatives, respectively. The parameters γ, δ may or may not depend on $m(\mathbf{T}^{+})$ and $m(\mathbf{T}^{-})$, respectively, or they may be related in some other way. The reason to claim that they may depend on $m(\mathbf{T}^{+})$ and $m(\mathbf{T}^{-})$ is that, for example, if (one may assume that) $m(\mathbf{T}^{+})$ is much smaller than $m(\mathbf{T}^{-})$ it seems reasonable that true (false) positives are valued higher than true (false) negatives. For this reason the parameters could even both be set equal to $m(\mathbf{T}^{+})/m(\mathbf{T}^{-})$. To be sure, by making the weights dependent on $m(\mathbf{T}^{+})$ and $m(\mathbf{T}^{-})$, it becomes impossible to calculate the increase of verisimilitude of a revised theory (see below) in the standard situation in which we don't know $\underline{\underline{\mathbf{T}}} = «\mathbf{T}^{+}, \mathbf{T}^{-}»$.

Other kinds of special cases we get by putting φ=1 or by putting γ=δ or even further γ=δ=1, the latter specification leading to

$$\mathrm{Vs}_{\varphi}^{1,1}\left(\underline{\underline{X}}\right) =_{\mathrm{def}} \mathrm{Vs}_{\varphi}\left(\underline{\underline{X}}\right) = m\left(\mathrm{t}\left(\underline{\underline{X}}\right)\right) - \varphi m\left(\mathrm{f}\left(\underline{\underline{X}}\right)\right)$$

which is the definition in Cevolani et al. (2011), for which reason our general definition is called a 'generalized' measure. The authors call $\mathrm{Vs}_{\varphi}^{1,1}$ a 'feature-contrast' measure, because they relate it to a standard statistical practice of feature-contrast representation. Feature-contrast measures of verisimilitude have been introduced (without using this name) by Cevolani and Calandra (2009). Cevolani and Festa (manuscript) gives an overview of feature-contrast measures. See also Cevolani and Festa (to appear) as the most recent presentation, including a significant extension to the case of arbitrary theories, notably disjunctive theories.

Of course, the definition of the quantitative measure would be inadequate when it would not cohere with the qualitative definition in Sect. 15.2. However, it is not difficult to prove

Theorem 4 if $\underline{\underline{X}} \leq_{vs} \underline{\underline{Y}}$ then for all $\varphi, \gamma, \delta > 0$ $Vs_{\varphi}^{\gamma,\delta}(\underline{\underline{X}}) \leq Vs_{\varphi}^{\gamma,\delta}(\underline{\underline{Y}})$

This theorem automatically implies that the verisimilitude of the revision of $\underline{\underline{X}}$ by true data $\underline{\underline{R}}$, viz. $Vs_{\varphi}^{\gamma,\delta}(\underline{\underline{X}}*\underline{\underline{R}})$, is higher (or at least equal to) than that of $\underline{\underline{X}}$, viz. $Vs_{\varphi}^{\gamma,\delta}(\underline{\underline{X}})$, because the qualitative relation holds.

The increase of verisimilitude of the revision of $\underline{\underline{X}}$ by true data $\underline{\underline{R}}$ can also be expressed. Recall:

$$Vs_{\varphi}^{\gamma,\delta}\left(\underline{\underline{X}}\right) = \left[m\left(X^{+} \cap \mathbf{T}^{+}\right) + \gamma m\left(X^{-} \cap \mathbf{T}^{-}\right)\right] - \varphi\left[m\left(X^{+} - \mathbf{T}^{+}\right) + \delta m\left(X^{-} - \mathbf{T}^{-}\right)\right]$$

Hence, in the symmetrical representation, $Vs_{\varphi}^{\gamma,\delta}(\underline{\underline{X}}*\underline{\underline{R}}) =$

$$\left[m\left(\left(X^{+} - R^{-}\right) \cup R^{+}\right) \cap \mathbf{T}^{+}\right) + \gamma m\left(\left(X^{-} - R^{+}\right) \cup R^{-}\right) \cap \mathbf{T}^{-}\right)\right]$$
$$-\varphi\left[m\left(\left(X^{+} - R^{-}\right) \cup R^{+}\right) - \mathbf{T}^{+}\right) + \delta m\left(\left(X^{-} - R^{+}\right) \cup R^{-}\right) - \mathbf{T}^{-}\right)\right]$$

It is not difficult to sort out the components of the four measure expressions

$$m\left(\left(X^{+} - R^{-}\right) \cup R^{+}\right) \cap \mathbf{T}^{+}\right) = m\left(X^{+} \cap \mathbf{T}^{+}\right) + m\left(R^{+} - X^{+}\right)$$
$$m\left(\left(X^{-} - R^{+}\right) \cup R^{-}\right) \cap \mathbf{T}^{-}\right) = m\left(X^{-} \cap \mathbf{T}^{-}\right) + m\left(R^{-} - X^{-}\right)$$
$$m\left(\left(X^{+} - R^{-}\right) \cup R^{+}\right) - \mathbf{T}^{+}\right) = m\left(X^{+} - \mathbf{T}^{+}\right) - m\left(X^{+} - R^{-}\right)$$
$$m\left(\left(X^{-} - R^{+}\right) \cup R^{-}\right) - \mathbf{T}^{-}\right) = m\left(X^{-} - \mathbf{T}^{-}\right) - m\left(X^{-} - R^{+}\right)$$

Now it is easy to express the increase of verisimilitude, in the symmetric as well as in the asymmetric representation (the latter again by just replacing 'R^{+}' by 'R' and 'R^{-}' by 'cS'), that is, it is easy to prove the following:

Theorem 5 $Vs_{\varphi}^{\gamma,\delta}(\underline{\underline{X}}*\underline{\underline{R}}) - Vs_{\varphi}^{\gamma,\delta}(\underline{\underline{X}}) =$
symmetric: $[m(R^{+} \cap X^{+}) + \gamma\, m(R^{-} - X^{-})] - \varphi\, [m(X^{+} - R^{-}) + \delta\, m(X^{-} - R^{+})]$
asymmetric: $[m(R \cap X^{+}) + \gamma\, m(cS - X^{-})] - \varphi\, [m(X^{+} - cS) + \delta\, m(X^{-} - R)]$

Assuming that the parameters do not depend on **T**, it is interesting to note that the increase of verisimilitude can be calculated on the basis of $\underline{\underline{X}}$ and $\underline{\underline{R}}$ alone.

15.6 Concluding Remarks

We have presented a very general approach, with several interpretations, to the true boundary within some universe of discourse by belief base revision, guided by input data of symmetric or asymmetric nature. Sect. 15.2 provided the qualitative definition for the statement "theory $\underline{\underline{Y}}$ is at least as close to the truth/the true boundary as theory $\underline{\underline{X}}$", and Sect. 15.3 the definition of the statement "the revision $\underline{\underline{X}}*\underline{\underline{A}}$ of theory $\underline{\underline{X}}$ in the light of input theory $\underline{\underline{A}}$". In Sect. 15.4 followed the theorem for any input data (theory) $\underline{\underline{R}}$ of symmetric or asymmetric nature: "if $\underline{\underline{R}}$ is true/correct, then $\underline{\underline{X}}*\underline{\underline{R}}$ is closer to the truth than $\underline{\underline{X}}$". In Sect. 15.5 the main lines of a quantitative version of this account were given.

There are some plausible challenges for further research. To begin with, the generalization from a divide, i.e., a bivalent partition, to a (finite) partition in general does not seem to be difficult in principle, although it seems technically rather complicated. More interesting is the following challenge. Cevolani and Festa (manuscript) deal with quantitative measures of verisimilitude for constituent languages as well as for quantitative languages, where the latter incorporate quantitative, statistical and tendency hypotheses. Whereas our qualitative account essentially covers constituent languages of all kinds, the interesting question is whether this account, or its quantitative version, can be adapted to quantitative languages.

Another challenge to cope with is the problem in the basic qualitative account in Sect. 15.4 that, for example, in the nomic interpretation one or more of the established laws and hence the strongest law S, may not only be incompatible with the nomic theory «M, cP» in the weak sense that P is not a subset of S, but even in the strong sense that S does not overlap with P at all, that is, $P \subseteq cS$, in which case the (belief base) revision of «M, cP» by $\underline{\underline{R}}$ = «R, cS», i.e., «$(M \cap S) \cup R$, $(cP \cap cR) \cup cS$» = «$(M \cap S) \cup R$, $c[(P \cap S) \cup R]$», is just «R, cR». Hence, any feature or trace of «M, cP» and of S has disappeared.

In a previous attempt (Kuipers 2011, strongly revised in the previous chapter) to dovetail belief revision and truth approximation, restricted to the nomic interpretation, I succeeded in overcoming the corresponding problem by taking refined forms of belief revision into account, notably partial meet revision. That dovetail attempt was initially dealing with 'maximal' theories leading to unsatisfactory ad hoc maneuvers, but this problem has been removed in the previous chapter in such a way that it could focus on exclusion theories. The result is, as is not difficult to check, that the basic belief *set* revision operations of the previous chapter and the basic belief *base* revision operations of this chapter are equivalent in their set-theoretic representation. The consequence is that the exclusion side of the revision of «M, cP» = <M, P> by $\underline{\underline{R}}$ = «R, cS» = <R, S>, i.e., $c[(P \cap S) \cup R]$, corresponds to the belief set revision of the nomic exclusion theory $\underline{P}$ (underlining with the exclusion interpretation) by evidence R/S of the previous chapter, viz. $\underline{(P \cap S) \cup R}$. Hence, in the extreme case that P and S do not overlap, this leads to the set R, with no trace of P, and not even of S. The question for further research is whether the present (basic) belief base approach can be refined in a similar way as the belief set approach was refined in the previous chapter in order to deal with this extreme case. This can of course be done in the specific setting of a nomic theory «M, cP» = <M, P > or in the general setting of a theory in this chapter: «X^+,X^-».

A further challenge is to generalize the present approach from conjunctive theories to theories in general, where the latter can be seen as disjunctions of constituents of the relevant kind, e.g. propositional constituents. In this sense, theories in general are disjunctive theories. As a matter of fact, conjunctive theories are disjunctive theories of a special kind. In terms of propositional constituents, for convenience's sake, a conjunctive theory is the disjunction of all constituents that fully agree for a certain subset of atomic propositions, but differ in all possible ways about all the other. In the generalization to (propositional) theories in general, the truth-value of some atomic propositions may depend on that of other, not on logical grounds, but because the theory requires so. It may well be that such a generalization is only possible in quantitative terms. More specifically, it seems plausible to try to use the recent generalization of Cevolani and Festa (to appear) of their quantitative approach of conjunctive theories to arbitrary theories, notably disjunctive theories.

I ended the previous chapter, dealing with 'basic and refined nomic truth approximation by evidence-guided belief *set* revision' with a number of debunking remarks. I am afraid they are valid for this chapter as well, at least as far as the nomic interpretation is concerned and hence 'basic nomic truth approximation by evidence-guided belief *base* revision'. So I repeat them in adapted form for this chapter, restricted to the case of (two-sided) nomic theories and asymmetric data, in which interpretation they seem the most relevant:

(a) The revision is rather ad hoc. However, as in general for ad hoc changes in a theory, the crucial question is whether they can be put to new (HD-)tests, and this is evidently the case. After all, it could even be the case that all further tests indicate that no new ad hoc maneuvers have to be made.
(b) The revision is rather diehard empiricist or instrumentalist. The 'instrument' «M, cP» is precisely so adapted that it just saves the phenomena, not only with respect to individual cases (R), but also with respect to empirical generalizations (S).
(c) If there is something like well-formed theories, satisfying certain criteria, e.g. of formalizability and generality, there do not seem to be good reasons to expect that the revision of a theory that satisfies them, will also satisfy these criteria, even if R and S satisfy some derived criteria.
(d) Last, but not least, what remains of the idea behind «M, cP»? A proper theory, in some sophisticated sense even if without theoretical terms, is usually based on one or two ideas. It is difficult to imagine that such ideas do not become 'mutilated' by the revision.

Again, be this as it may, the result may stimulate the interaction between truth approximation and the belief base revision approach for it fundamentally shows that belief base revision provides means for truth approximation in general and nomic truth approximation in particular.

References

Alchourrón, C., Gärdenfors, P., & Makinson, D. (1985). On the logic of theory revision: Partial meet contraction and revision functions. *Journal of Symbol Logic, 50*(2), 510–530.

Cevolani, G., & Calandra, F. (2009). Approaching the truth via belief change in propositional languages. In M. Suárez, M. Dorato, & M. Rèdei (Eds.), *Launch of the European philosophy of science association* (EPSA epistemology and methodology of science, Vol. 1, pp. 47–62). Dordrecht: Springer.

Cevolani, G., & Festa, R. Manuscript. *Features of verisimilitude*. Unpublished.

Cevolani, G., & Festa, R. To appear. *A partial consequence account of truthlikeness*. To appear in a special issue of Synthese in honor of Gerhard Schurz.

Cevolani, G., Crupi, V., & Festa, R. (2011). Verisimilitude and belief change for conjunctive theories. *Erkenntnis, 75*(2), 183–202.

Cevolani, G., Festa, R., & Kuipers, T. (2013). Verisimilitude and belief change for nomic conjunctive theories. *Synthese, 190*, 3307–3324.

Festa, R. (2007). Verisimilitude, qualitative theories, and statistical inferences. In S. Philström, P. Raatikainen, & M. Sintonen (Eds.), *Approaching truth: Essays in honour of Ilkka Niiniluoto* (pp. 143–178). London: College Publications.

Hansson, S. O. (1999). *A textbook of belief dynamics*. Dordrecht: Kluwer Academic Publishers.

Hansson, S. O. (2006). *Logic of belief revision*. http://plato.stanford.edu/entries/logic-belief-revision. Accessed 11 July 2017.

Kuipers, T. (1982). Approaching descriptive and theoretical truth. *Erkenntnis, 18*(3), 343–378.

Kuipers, T. (2000). *From instrumentalism to constructive realism. On some relations between confirmation, empirical progress, and truth approximation* (Synthese library 287). Dordrecht: Kluwer Academic Publishers.

Kuipers, T. (2011). Basic and refined nomic truth approximation by evidence-guided belief revision in AGM-terms. *Erkenntnis, 75*(2), 223–236.

Niiniluoto, I. (1999). Belief revision and truthlikeness. In S. Halldén, B. Hansson, W. Rabinowicz, & N. -E. Sahlin (Eds.), *Spinning ideas, electronic essays dedicated to Peter Gärdenfors on his fiftieth birthday*. http://www.lucs.lu.se/spinning/.

Part VI
Conclusion: Comparative Constructive Nomic Realism

Chapter 16
Comparative Realism as the Best Response to Antirealism

Abstract Arguments for and against scientific realism usually presuppose as the main epistemic claim about theories that we may have good reasons to conclude that they are true, or at least approximately true. The antirealist charges against this claim are not easy to counter. In this chapter it is argued that the defense of realism is much easier if we relativize its epistemic claim in the light of (theories of) truth approximation. From this *comparative realist* perspective the main epistemic claim becomes that we may have good reasons to conclude that successor theories are closer to the truth than their predecessors. For example, although Einstein's theory of general relativity may still be false, and not even approximately true, we have good reasons to assume that it is closer to the truth than Newton's theory of gravitation. A similar relativization of claims that theoretical terms refer to things in the world is argued for in terms of 'being closer to the referential truth'. For both purposes it is also plausible to relativize 'being empirically successful' to the comparative notion of 'being (persistently) empirically more successful'.

Comparative realism hence is realism guided by the comparative perspective on success and on truth approximation, that is, the notions of 'more successful' and 'closer to the (observational, referential, and theoretical) truth', and their mutual relations. This approach is defended against the antirealist charges and compared with the main other realist responses. The main positive claim of comparative realism is that (theoretical) truth approximation provides the (stratified) default explanation and prediction of empirical progress between non-empirically equivalent theories and of 'aesthetic progress' between empirically equivalent theories. Here 'aesthetic progress' is based on the comparative notion of 'aesthetically more

Revised version of: "Comparative realism as the best response to antirealism". In Logic, Methodology and Philosophy of Science. Proceedings of the Thirteenth International Congress (August 9–15, 2007, Beijing), eds. Clark Glymour, Wang Wei and Dag Westerstahl. College Publications, King's College, London, 2009, pp. 221–250. *Acknowledgements*: I like to thank a number of people for various useful comments received (at the try-out) in Groningen and in Beijing: David Atkinson, Luis-Alberto Cordero-Lecca, Richard Dawid, Lars-Göran Johansson, James Joyce, Barteld Kooi, Ulianov Montano, Ilkka Niiniluoto, Robert Nola, Jeanne Peijnenburg, Isabelle Peschard, Gerhard Schurz, Jan-Willem Romeijn Allard Tamminga, Sjoerd Zwart. Finally, I like to thank the Netherlands Institute for Advanced Research (NIAS) in Wassenaar for the opportunity to spend two weeks to complete this and another paper.

T. A. F. Kuipers, *Nomic Truth Approximation Revisited*, Synthese Library 399,
https://doi.org/10.1007/978-3-319-98388-2_16

successful' and specified in terms of the prevailing 'aesthetic canon' in the relevant field and period, that is, the prevailing non-empirical virtues of theories.

Keywords Realism · Antirealism · Comparative realism · Empirical progress · Aesthetic progress · Truth approximation · No-miracles argument · Inference to the best explanation · Referential realism · Constructive realism

16.1 Introduction

Arguments for and against scientific realism usually presuppose as the main epistemic claim about theories that we may have good reasons to conclude that they are true, or at least approximately true. The antirealist charges against this claim are not easy to counter. In this chapter it will be argued that the defense of realism is much easier if we relativize its epistemic claim in the light of (theories of) truth approximation. From this *comparative realist* perspective the main epistemic claim becomes that we may have good reasons to conclude that successor theories are closer to the truth than their predecessors. For example, although Einstein's theory of general relativity may still be false, and not even approximately true, we have good reasons to assume that it is closer to the truth than Newton's theory of gravitation. A similar relativization of claims that theoretical terms refer to things in the world will be argued for in terms of 'being closer to the referential truth'. For both purposes it is also plausible to relativize 'being empirically successful' to the comparative notion of 'being (persistently) empirically more successful'.

Comparative realism hence is realism guided by the comparative perspective on success and on truth approximation, that is, the notions of 'more successful' and 'closer to the (observational, referential, and theoretical) truth', and their mutual relations. This approach will be defended against the antirealist charges and compared with the main other realist responses.

In Sect. 16.2 I will draw heavily upon the survey of (non-comparative) arguments pro and contra realism presented by James Ladyman in his textbook of 2002 and his handbook exposition (2007) of ontological, epistemological, and methodological positions. In particular, the *pessimistic meta-induction*, the *no-miracles argument* and *inference to the best explanation* and their counter arguments will be analyzed in detail and will lead to a number of conditions of adequacy for explicating the crucial expressions.

There are at least two ways to specify the truth approximation perspective, the quantitative approach of Niiniluoto (1987, 1999) and my own qualitative, comparative approach (Kuipers 2000; this book). For the present purposes the latter, in a sense more strict, and hence more cautious, approach is sufficient. More specifically, in Sect. 16.3 I will sketch some of its main lines and in Sect. 16.4 I will show to what extent this approach fulfils the conditions of adequacy that have been gathered in Sect. 16.2. I will conclude, in Sect. 16.5, with a final comparison of comparative realism with the main kinds of non-comparative realism that have been suggested to meet the antirealist charges. It concerns in particular the restriction of realism to entities, structures, 'mature' theories, theories with 'novel' predictive success, and 'essential' parts/aspects of theories, or stretching realism by stretching the underlying (causal) theory of reference.

The core of *comparative realism* becomes that the (comparative) phenomenon that one theory persistently is empirically more successful than another provides a good reason for the claim that the first theory is closer to the (theoretical and referential) truth than the second; a good, analytical, reason being that this very claim amounts to the generic *default explanation* for that comparative empirical phenomenon. Among empirically equivalent theories, it may occur that one theory persistently is more successful than another according to some non-empirical, notably aesthetic criteria. In this case there may be empirically justified reasons, however weak, for the truth approximation claim and the corresponding default explanation.

I conclude this introduction with a puzzling observation about the realism-antirealism debate. One important success of the first decades of (constructive) analytical philosophy was the discovery, notably by Russell, Carnap, Hempel, Beth and Barth, of asymmetric relations as a means to solve age-old problems by refined concept explication. This concerned not least the asymmetric relations, constitutive of comparative concepts, such as 'longer than', 'being caused by', etc. Of course, conceptions of 'better than', hence of improvement or progress, are comparative.

Now it is very surprising that in the supposedly analytic realism – antirealism debate there is almost no sign of awareness of the possible relevance of this insight. One continues to talk about true and false theories and reference claims versus empirically adequate and inadequate theories, that is, observationally true and false theories, respectively. The retreat to 'approximately true' is of no help, for that remains basically non-comparative and hence it cannot capture progress. Moreover, it requires a necessarily arbitrary threshold. On the other hand, from the point of view of relations, it is highly plausible to think in terms of comparative notions, such as 'closer to the truth' and 'more successful than'. To be sure, in the comparative approach theories will frequently not be straightforwardly comparable; but for 'mixed' cases the 'principle of dialectics' (that is, try to improve both) and the quantitative approach are plausible concretizations.

16.2 Antirealist Arguments to Be Met

In this section I will heavily draw upon Ladyman (2002, 2007) in looking for the main arguments that have been put forward against realism. I'll start with a brief preview of the arguments and the responses that naturally follow from the comparative realist point of view. Recall, that this point of view conceives realism guided by the comparative perspective on success and on truth approximation, that is, the notions of 'more successful' and 'closer to the truth', and their mutual relations.

Laudan's so-called pessimistic meta-induction is rebutted by the claim that it is not at all a problem that previously the best theories turned out to be false (and even did not refer). For the main question is whether they remained, at least as a rule, more successful than their predecessor theories and hence can still be argued to be closer to the truth than (and refer at least as well as) their predecessors. Hence, similarly for of our currently the best theories, they will presumably turn out to be false, but the interesting question remains whether they will continue to be more successful than their predecessors, so that we can still hold them as closer to the truth.

Regarding the arguments against Putnam's so-called no-miracles argument in favor of realism, the comparative response is that the former will miss their point when the latter is replaced by the observation that the success of science would be miraculous on anything but a scientific realist view in the sense that specific occurrences of (persistent) empirical progress would be miraculous if not, as a rule, due to truth approximation. This, notwithstanding the aptness of Van Fraassen's Darwinian analogy that, as a rule, the best theories survive because they are selected for that reason.

Finally, the arguments (notably of Van Fraassen) against inference to the best explanation (IBE) are rebutted by the claim that IBE has so far not been adequately explicated. The proposed comparative explication reads: inference to the best explanation as the closest to the truth among the available explanations or theories, even among empirically equivalent theories.

In the course of the presentation, there will be derived a number of conditions of adequacy an acceptable explication of comparative realism, including IBE, will have to satisfy. In Sect. 16.4, it will be investigated to what extent they can be satisfied by the theory of truth approximation to be sketched in Sect. 16.3 and elaborated in Sect. 16.4.

16.2.1 Arguments from Theory Change

16.2.1.1 The Pessimistic Meta-induction

One of the main arguments against realism is known as 'the pessimistic meta-induction' and is due to Larry Laudan. Ladyman (2002, pp. 236–7, 2007, p. 345, as quoted) presents it as follows (my italics):

(i) There have been *many empirically successful* theories in the history of science which have subsequently been rejected and whose central theoretical terms do not refer according to our best current theories.
(ii) Our best current theories are no different in kind from those discarded theories and so we have no reason to think they will not ultimately be replaced as well.

> So, by induction we have positive reason to expect that our best current theories will be replaced by new theories according to which some of the central theoretical terms of our best current theories do not refer, *and hence, we should not believe in the approximate truth or the successful reference of the theoretical terms of our best current theories.*

According to Ladyman the most common response is that the realist claims should be restricted to mature theories and/or theories having novel predictive success, for which (i) would not apply, and hence the inductive argument would be blocked. This retreat has its own problems, see below. Another way to block this argument is to liberalize the realist elements in the argument according to the comparative perspective:

(iCR) There have been (some, several, many[1]) theories in the history of science *that were and still are empirically more successful than their predecessor theories*, which have *nevertheless* subsequently been rejected and whose central theoretical terms do not *all* refer according to our best current theories.

(iiCR) (= (ii)) Our best current theories are no different in kind from those discarded theories and so we have no reason to think they will not ultimately be replaced as well.

So, by induction we have positive reason to expect that our best current theories will be replaced by new theories according to which some of the central theoretical terms of our best current theories do not refer, *however we may still* (instead of: and hence, we should not) *believe that our best current theories are closer to the truth than their predecessor theories or that their theoretical terms more successfully refer than those of their predecessor theories.*

Hence, it is not plausible anymore to think that the second part of the suggested, partially bracketed conclusion of the CR-version of the inductive argument, starting with "(and hence, we should not) believe that…." is still 'valid'. By consequence, it seems plausible to think that, starting from the basic explications to be given in Sect. 16.3, when making a distinction between (relatively) observational and theoretical terms, refined explications can be given of the crucial phrases, viz. 'empirically more successful', 'closer to the truth' and 'more successfully refer', such that the following conditions of adequacy are satisfied, in which we add 'novel predictive success' in the second one for later purposes.

CA1: The explications of the crucial phrases should leave room for the possibility that false theories, even with some non-referring theoretical terms, are not only (persistently) empirically more successful but even closer to the (theoretical) truth and refer at least as well as other theories.

CA2: The explications should be such that 'being (persistently) empirically more successful' is, *as a rule*, due to 'being closer to the theoretical truth', which on its turn entails, as a rule, 'referring at least as well' and 'novel predictive success'.

Or, in a plausible 'progress version'

CA2: The explications should be such that empirical progress is, *as a rule*, due to theoretical truth approximation, which on its turn entails, as a rule, referential truth preservation and novel predictive success.

In Sect. 16.3 we will indicate some basic explications that allow 'false theories closer to the truth than other ones' and therefore 'being at least as successful', however, not yet with the distinction between observational and theoretical terms. In

---

[1] Depending on whether we will use qualitative or quantitative versions of 'more successful' and 'closer to the truth'.

Sect. 16.4 we will try to meet these conditions while taking this distinction into account.

16.2.1.2 The Antirealist Attack on the No-Miracles Argument

As mentioned, according to Ladyman, the usual realist response to the pessimistic meta-induction is the retreat to mature theories and/or theories having novel predictive success. The weakness of both, *prima facie* rather ad hoc, retreats is, of course, that they require precise non-ad hoc definitions of 'mature theories' and 'novel predictive success', both of which tasks have so far not generated a proposal that is agreed upon. Note that comparative realism remains to take all theories seriously.

Besides such technical problems with these rejoinders there is a more general problem with such retreats. The antirealist counter argument to such a retreat is that one does not need the (unrestricted) inductive argument to undermine the so-called no-miracles argument in favor of realism, that is, "the idea that the success of science would be miraculous on anything but a scientific realist view" (Ladyman 2007, p. 342). Illustrated by the ether theory of light and the caloric theory of heat one may claim, according to Ladyman (2002, p. 244, 2007, p. 346, as quoted; bracket insertions [] and deletion () are mine):

(a) Successful reference of its central theoretical terms is a necessary condition for the approximate truth of a theory.
(b) There are [at least two] examples of theories that (were mature and) had novel predictive success but which are not approximately true.
(c) [Hence,] approximate truth and successful reference of central theoretical terms is not a necessary condition for the novel-predictive success of scientific theories.

> So, the no-miracles argument is undermined since, if approximate truth and successful reference are not available to be part of the explanation of some theories' novel predictive success, there is no reason to think that the novel predictive success of other theories has to be explained by realism.

I have some reservation against this argument as such. Although (a), (b), and (c) may be granted,[2] the conclusion drawn from them seems too strong. Why would maturity and novel predictive success not be possible without approximate truth and successful reference? One could imagine that there are logical gaps leaving room for such cases. Of course, in such cases the novel predictive success would have to be explained in another, specific way. However, the suggestion that such cases prevent the use of approximate truth and successful reference in other specific cases for the explanation of novel predictive success seems unjustified or, at least no such justification is reported by Ladyman. In other words, why is it excluded that *prima facie* similar cases have different explanations?

---

[2] However, even with respect to (a), there may be room for doubt, for it very much depends on the definition of 'approximate truth' whether there is the logical possibility of an approximately true theory without reference of (all of) its theoretical terms.

However this may be, the current realist responses to the antirealist rejoinder are twofold according to Ladyman (2002, p. 245–8, 2007, p. 346–8). (I) Stretch realism by stretching the (causal) theory of reference such that the relevant abandoned theoretical terms refer after all or (II) Restrict realism to those theoretical claims about unobservables that feature in an *essential* way in the derivation of novel predictions. For example, Stathis Psillos (1999) uses (I) to save the ether theory as referring after all, viz. to the electromagnetic field, and (II) to explain the success of the caloric theory without 'caloric' as a referring term.

The comparative perspective does not need such interventions, although even within this perspective there may be *other* good reasons for such moves. However, in the present context the point is whether the abandoned theories were and are closer to the truth than their respective predecessor theories. And this may well be the case in several historical examples, notably not only for the ether theory but also for the caloric theory.

Hence, let us turn to the CR-versions of the antirealist premises (a) – (c) enabled by CA1&2. This will lead to a defensible CR-version of the no-miracles argument.

(aCR) Being closer to the truth entails, as a rule, referring at least as well, and novel predictive success ('novel', now in whatever favorite sense).
(bCR) Exceptionally there (may) have been examples of theories that (were mature and) had novel predictive success relative to their predecessor theories but which were not closer to the truth and even not referring at least as well.
(cCR) Being (persistently) more successful is, as a rule, due to being closer to the truth, even more so when novel predictive success is involved

Note that (aCR) and (cCR) are taken care of by CA2. Note also that the caloric and the ether theory may well fit in (aCR)&(cCR) relative to their predecessors. But, by the insertion *'as a rule'*, (aCR) and (cCR) leave room for other exceptions (bCR) and suggest a cautious CR-version of the no-miracles argument.

CR-no-miracles argument:

- As a rule, (persistently) empirically more successful theories are closer to the truth than, and refer at least as well as their predecessors.
- If this were not the case, the regular occurrence of persistently empirically more successful theories, including some novel predictive success, would be miraculous.
- Occasionally, other, case-specific, explanations of persistent empirically more successfulness, novel or not, may be appropriate.

This cautious version of the CR-no-miracles argument suggests that truth approximation provides a kind of *default explanation and prediction* of (persistent) empirical more successfulness, including some novel predictive success. It is attractive to reformulate the CR-version in terms of progress.

CR-no-miracles argument, progress version:

- As a rule, (persistent) empirical progress is due to theoretical truth approximation and referential truth preservation, if not referential truth approximation.

- If this were not the case, the regular occurrence of (novel) empirical progress would be miraculous.
- Occasionally, other, case-specific, explanations of (novel) empirical progress may be appropriate.

And, hence, in this form the argument suggests that (theoretical) truth approximation provides a kind of *default explanation and prediction of (novel) empirical progress*.

For standard realism the ether theory of light and the caloric theory of heat give rise to problems because in both cases the crucial theoretical terms do not refer in a straightforward sense according to our present lights, whereas the theories were mature and had quite some novel predictive success. It surely is an intriguing question how these theories can combine both features. However, they may not be so very relevant for the realism debate since these problems do not arise for comparative realism. What matters in this perspective are only comparative questions, that is, questions in comparison with previous and successor theories. For example, in the case of the ether theory the focus in the debate is on (Young and) Fresnel's revised version of Huygens' wave theory of light in ether, by replacing his longitudinal waves by transversal waves. Instead of trying to answer, as a defense of realism – however interesting it may be – the question as to how Fresnel's theory could have impressive novel predictive success, by our present lights, without reference of the crucial theoretical term 'ether', some of the plausible CR-questions are:

> Was Fresnel's theory empirically more successful than its predecessor theory, viz. that of Huygens, and its competitor theory, the particle emission theory of Newton?

As is well known, at least with respect to Foucault's quasi-crucial velocity of light experiment, the result was in favor of Fresnel's theory relative to both Huygens' version of the wave theory and Newton's particle alternative. Hence, the comparative realist questions are:

> If so, could this be explained by the hypothesis that Fresnel's theory is closer to the truth than, and hence, presumably, refers at least as well as, those of Huygens and Newton?

Not knowing the truth, such questions can only be answered in the abstract. However, assuming that the relativistic version of Maxwell's electromagnetic theory, that is, Einstein's special theory of relativity encompassing Maxwell's theory, is the truth (or at least a close approximation to the truth), we can form potential answers to these questions.

> Can the greater empirical success of Fresnel's theory be explained by the fact that Fresnel's theory is closer to the theory of Maxwell (-Einstein) than those of Huygens and Newton, and hence, presumably, refers at least as well?

Assuming precise enough explications, these questions can be answered. It is at least plausible to think that Fresnel's theory is closer to Maxwell's theory than Huygens' theory. Moreover, it may well be that the sequence of theories of Huygens, Fresnel, Maxwell, Maxwell-Einstein is a matter of truth approximation. However, it does not seem plausible that the sequence of theories of Newton, (Huygens),

Fresnel, Maxwell (-Einstein) can be reconstructed as such. Newton's theory is indeed incommensurable with versions of the wave theory, at least without the leap to the first versions of quantum physics, suggesting that the photon theory of light is closer to the truth than both Newton's theory and (non-quantum) versions of the wave theory.[3] I leave it as a challenge to raise similar questions about the caloric theory of heat.

16.2.2 Inference to the Best Explanation

So-called inference to the best explanation (IBE) plays an important role in the defense of standard realism. Ladyman (2002, p. 209, 2007, p. 341 as quoted) describes this (putative) rule of inference as follows:

> ...where we have a range of competing hypotheses all of which are empirically adequate to the phenomena in some domain we should infer the truth of the hypothesis which gives us the best explanation of those phenomena.

Here it is understood that one hypothesis provides a better explanation than another when either the first explains more of these phenomena than the second or when they explain the same phenomena but the first satisfies some additional non-empirical evidential criteria, or epistemic virtues, whereas the second does not satisfy all of them or to a lesser degree. Ladyman (p. 340) lists as such non-empirical criteria simplicity, non-adhocness, novel predictive power, elegance, and explanatory power; Douven (2005) mentions simplicity, elegance, inner beauty, fertility, coherence with background theories and/or metaphysical suppositions.[4] Following McAllister (1996), I will call such non-empirical criteria *aesthetic criteria* when they are put forward by scientists for evaluative purposes. McAllister deals in particular with: symmetry, simplicity, and visualizability. Weinberg (1993) deals in his chapter (6), entitled "Beautiful theories", not only with simplicity and symmetry, but also with inevitability or rigidity.

According to the (standard) reading of IBE suggested by Ladyman, the competing hypotheses are not only assumed to be empirically adequate to the phenomena in the relevant domain in the sense that they are compatible with them, that is, they have not been falsified by one of them, but all of these hypotheses are even assumed to explain *all* phenomena in the domain. In this case, the best explanation is that one among those hypotheses that satisfies the non-empirical criteria the best.

---

[3] Several more encompassing truth approximation questions are suggested by the set of success comparisons of electrodynamic theories that is provided by a table in (Panofsky and Phillips 1962², p. 282, reprinted and commented upon in Kuipers 2000, p. 118, and 2001, p. 236). It represents the records in the face of 13 general experimental facts of the special theory of relativity (STR) and six alternative electrodynamic theories that were largely developed and defended in the nineteenth century, viz., three versions of the ether theory and three emission theories.

[4] See (Niiniluoto 2007) for a discussion of ten empirical and non-empirical virtues of theories.

IBE is of course not deductively valid, but it is used as a kind of inductive or abductive argument at two levels. At the local level it is used to escape in specific cases of empirically equivalent theories from the so-called '(strong) underdetermination by empirical data'. It is supposed to enable specific theoretical truth and reference claims. At the global level it is used to defend realism as the best explanation of the overall success of scientific theorizing. (According to the strong version of the no-miracles argument it is even the only explanation.) The global use can be defended against Van Fraassen's Darwinian analogy (Van Fraassen 1980, p. 40) that, as a rule, the best theories survive because they are selected for that reason, by noting that realism predicts and explains the survival so far but also the continuation of success. Against the charge of circularity the global use has been defended by David Papineau and Stathis Psillos in a similar way as Braithwaite and Carnap defended the inductive defense of induction against the circularity charge, viz. by claiming that it is only 'rule circular', not 'premise circular' (Ladyman 2002, p. 218, 2007, p. 343) .

However, IBE is debated in general as a rule of inference, where the following explication[5] is more or less presupposed, with a weak and a strong version.

> *IBE-E1:* If an explanation has so far proven to be the best one among the available empirically adequate theories, then conclude, for the time being, that
>
> - it is true, that is, true as a hypothesis about the domain, or even
> - it is the truth, that is, the strongest true hypothesis about the domain.

Elsewhere (e.g. Kuipers 2000, Chap. 11 of this book) I have put forward three general objections applying to both versions of this 'implicit explication':

(1) It is restricted to unfalsified theories, for empirically adequate theories have by definition only true observational consequences. Hence there is no possibility of conclusions to progress by false, but less false, theories.
(2) It is asymmetric in that it deals in the premises with the best of the available theories and in the conclusion with being 'true/the truth' simpliciter, that is, without reference to the available theories. In other words, the premises are comparative, whereas the conclusion is not.
(3) It lacks justification of the conclusion (3.1) when there is just one empirically the best theory available, let alone (3.2) when there are empirically equivalent (EE-) theories among empirically the best ones (i.e., in the case of strong underdetermination).

---

[5]In earlier publications (e.g. Kuipers 2000, 2004, revised version: Chap. 11 of this book) I have introduced a distinction between 'the best theory' and a stronger notion of 'the best explanation', viz., being the best theory which has not yet been (empirically) falsified. In this chapter I drop speaking of '(inference to) the best theory' and explicate '(inference to) the best explanation', leaving room for a falsified best explanation. Although it may sound a bit strange to talk about the best explanation even when it is known to be false, it makes the presentation much easier. Moreover, since I will here take non-empirical criteria into account for empirically equivalent theories it now meets the main criticism of Douven (2005, note 9, p. 291) of my previous version of 'inference to the best theory'.

Van Fraassen has raised three specific objections that may be seen as specifications of (3.2), hence dealing with EE-theories. They are known as:

(i) The Argument from Indifference
(ii) The Argument from the Best of a Bad Lot
(iii) The Argument from Bayesianism

and will soon be paraphrased.

These objections suggest the following extra condition of adequacy for the comparative realist explication tasks extended with IBE:

CA-IBE: An explication of IBE should also deal with the case that all available theories are false or even have been falsified (1), it should be symmetric between the formal nature of the premise(s) and the conclusion (2), and the (truth-related) conclusion should have some analytical justification and/or allow some empirical justification (3), when there is only one best theory available (3.1), and when there are EE-ones (3.2)

It is easy to check that the following revised explication of IBE satisfies at least CA-IBE1 and CA-IBE2.

IBE-E2: If an explanation has so far proven to be empirically the best one among the available theories, then conclude, for the time being, that it is the closest to the truth among the available theories.

As announced before, Sect. 16.4 will indicate to what extent CA1 and CA2 (see Sect. 16.2.1.1) can be satisfied and hence CA-IBE3.1, for that is essentially implied by them. However, IBE-E2 does not yet take EE-best theories into account, hence, it does not yet meet CA-IBE3.2.

There are some famous examples of EE-theories discussed by scientists themselves. Van Fraassen (1980) has also discussed some rather artificial examples. The best-known and by no means artificial case is of course provided by the various so-called interpretations of quantum mechanics. The first one being the Copenhagen interpretation (put together principally by Niels Bohr) followed by the hidden variable interpretation of David Bohm. Several other interpretations are discussed today. Another case is provided by the existence of at least four EE-versions of classical mechanics, which will be discussed in Sect. 16.4.1.3. Hence, there is a serious problem of how to choose between EE-theories. IBE has been put forward not least for this problem of choice. Let us look at Van Fraassen's objections, as formulated by Ladyman (2007, p. 344–5[6]), and evaluate them.

(i) The Argument from Indifference: "… since there are many ontologically incompatible yet empirically equivalent theories, we have no reason to choose among them and identify one of them as true. …"

[6]It is to be noted that the formulation of the Argument from Indifference in Ladyman (2002, p. 219) is a bit confusing, for it seems to coincide with the Argument from the Best of a Bad Lot (p. 220). The Argument from Bayesianism is not dealt with in Ladyman (2002).

This argument is to be rejected, for although there may be no compelling reasons, it disqualifies any role of non-empirical criteria as truth indicators.

(ii) The Argument from the Best of a Bad Lot: "… we are to think that the collection of hypotheses that we have under consideration will include the true theory. The best explanatory hypothesis we have may just be the best of a bad lot, all of which are false. …."

This is evidently a variant of our general asymmetry objection (2), but now specifically intended, though this remains here implicit, for EE-theories. However, also for EE-theories it is rather plausible. It is taken care of by CA-IBE2.

(iii) The argument from Bayesianism: "…any rule for the updating of belief that goes beyond the rules of Bayesian conditionalisation […] will lead to probabilistic incoherence."

This is also plausible and hence should be respected.

So let us reformulate CA-IBE3.2 in detail and in accordance with these evaluations, in particular, in contrast to (i).

CA3 (replacing CA-IBE3.2): As far as EE-explanations are concerned, an explication of IBE should only take those prevailing aesthetic criteria into account, *if any*, for which there are empirical or analytical reasons, however weak, to assume that they are truth-conducive in the field.

Note for the moment already that the prevailing aesthetic criteria in a certain field at a certain time constitute what McAllister (1996) has aptly called the *aesthetic canon*, arrived at by, and equally aptly called, *aesthetic induction*.

In accordance with (iii) we will require:

CA4: In a probabilistic version of an adequate explication of IBE 'EE-updating' of the relevant belief should remain within the confines of Bayesian conditionalization.

A merely probabilistic version will not do in general, for probabilistic updating makes only differential sense for unfalsified theories. However, probabilistic estimation of the distance from the truth (Niiniluoto 1987, 1999), using Bayesian updating, is generally applicable.

Of course, IBE-E2 trivially satisfies CA3 and CA4 when there are no EE-best theories. For the case that there are we extend it with an extra clause

IBE-E3:

(3.1) (=IBE-E2). If an explanation has so far proven to be empirically the best one among the available theories, then conclude, for the time being, that it is the closest to the truth among the available theories

(3.2) If some empirically equivalent explanations have so far proven to be empirically the best ones among the available theories, and if among them there is one that is according to the relevant aesthetic canon the best one, then conclude, for the time being, that it is the closest to the truth among the available theories

It is easy to check that (3.2) satisfies, like (3.1), CA-IBE1 and CA-IBE2. Section 16.4 will indicate to what extent it also satisfies CA-IBE3.2, that is, CA3 & CA4.

This concludes our survey of the main antirealist arguments and which conditions of adequacy comparative realist explications of the crucial notions have to satisfy in order to meet them. The resulting conditions of adequacy to be satisfied are on the one hand CA1 and CA2 and on the other hand CA3 and CA4, for CA-IBE was argued to be taken care of by IBE-E3 (meeting CA-IBE1 and CA-IBE2) and the resulting four conditions CA1-4 (meeting CA-IBE3).

16.3 Truth Approximation, Some Basics for Comparative Realism

In this short introduction to my favorite, qualitative theory of truth approximation, I will mainly restrict myself to its basic form, that is, without a (O-/T-) distinction between observational and theoretical terms and without other refinements. It is best represented within the structuralist theory of theory representation. Starting from a fixed vocabulary and a suitable similarity type of structures, let U indicate the set of structures of that type, also called the *conceptual possibilities* or *potential models* of the theory. Let the subset P of U indicate the set of *models* of theory $\underline{P}$, constituted by one or more postulates. Finally, assuming that our target is a fixed domain of physically or, more broadly, nomically possible constellations and events, let **T** indicate this domain 'as seen through U', hence a subset of U, and be called the set of (intended) *nomic possibilities* or the *domain of intended applications*. According to the *Nomic Principle*, we assume that such a unique fixed set exists, given U and an intended domain. Note that we do not yet suppose that we dispose of a general characterization of **T** as a subset of U. What we know is only that each intended application can be represented as a potential model. A general characterization of **T** is 'the great unknown' which theories, as represented by their models, are looking for. More formally, a theory is a triple of the form $\underline{P} = \langle U, P, \mathbf{T} \rangle$, together with the *exclusion claim* that **T** is a subset of P ($\mathbf{T} \subseteq P$), that is, not satisfying the postulates constituting P excludes being a nomic possibility. A theory is said to be true (false) when its (exclusion) claim is true (false). It is easy to check that (a general characterization of) **T** represents the strongest true claim, and hence may be called 'the truth' in this context, that is, the truth about the given domain in the given vocabulary.[7]

Now it is plausible to define what it means, for fixed $\langle U, \mathbf{T} \rangle$, that one theory $\underline{P^*}$ is closer (or more similar) to the truth than another $\underline{P}$, and hence what it means to say that $\underline{P^*}$ amounts to truth approximation relative to $\underline{P}$. Intuitively, when P* is moving from P in the direction of **T**. Formally, when $P^* - \mathbf{T} \subseteq P - \mathbf{T}$ ($\varnothing_1$-area empty in Fig. 16.1) and $\mathbf{T} - P^* \subseteq \mathbf{T} - P$ ($\varnothing_2$-area empty) and at least once it should be a proper subset relation ($\#_1$-area and/or $\#_2$-area non-empty). In terms of symmetric differences ($A \Delta B =_{def} (A-B) \cup (B-A)$), $P^* \Delta \mathbf{T}$ should be a proper subset of $P \Delta \mathbf{T}$. In

[7] Note that, in view of the supposed yes or no matter of being a member of **T**, the nomic truth is of a deterministic nature. This was assumed in the whole book.

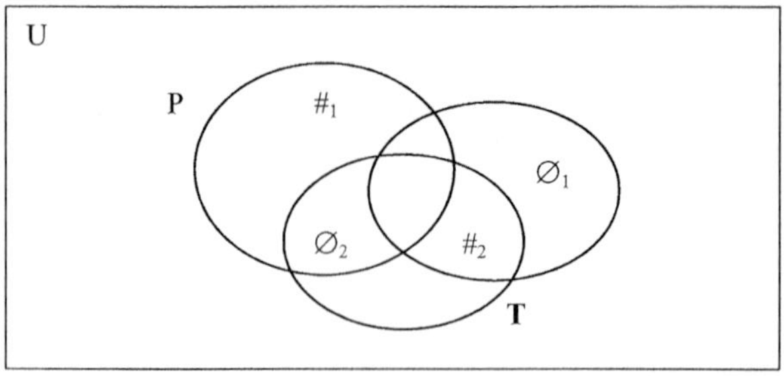

Fig. 16.1 Theory P* is closer to the truth **T** than theory P, where **T** represents the domain as seen through U

(Kuipers 2000) I have argued that, among other formulations, the two clauses of this definition amount to: (relative to **T**) P* has more true consequences than P ($\varnothing_1$ empty, $\#_1$ non-empty) and P* has more correct or desired models than P ($\varnothing_2$ empty, $\#_2$ non-empty), respectively. Of course, we get ‘at least as close to the truth’, when we replace the proper subset conditions by just subset conditions. We get an asymmetric (explanatory) form of ‘closer to’ by only requiring in addition that the first relation is a proper subset relation ($\#_1$- area non-empty).

Knowing what it means that one theory is closer to the truth than another is one thing, in view of the fact that we don’t know the truth, judging that this is in fact the case, in the light of our evidence, is quite another. Although there is, of course, no theorem guaranteeing that (empirical) more-successfulness entails closer-to-the-truth, there is a theorem, the *Success Theorem*, guaranteeing almost the reverse entailment, viz., at-least-as-close-to-the-truth entails at-least-as-successfulness.

Let R represent the set of realized (described) applications at a certain time, that is, the experimentally or otherwise realized nomic possibilities at that time. Let S indicate the strongest law induced on the basis of R. Of course, R will partly be the result of testing hypothetical laws. If no descriptive mistakes have been made in representing the realized possibilities, R is not only a subset of U, but even of **T**, for nomic *im*possibilities can’t be realized, by definition. Moreover, in the structuralist representation S is also a subset of U, such that it has been concluded, inductively, and hence at most provisionally, that conceptual possibilities outside S, that is, in U-S, are nomically impossible. In other words, we could speak of the (empirical) theory S. Finally, if our inductive jump from R to S is correct, S has to be a superset of **T**. To sum up: $R \subseteq \mathbf{T} \subseteq S \subseteq U$, provided no descriptive and inductive mistakes have been made.

Let us call R/S the data set. We define what it means that P* is, relative to the data set R/S, (empirically) more successful than P as follows: $R-P^* \subseteq R-P$ ($\equiv cP^*-cR \subseteq cP-cR$) and $P^*-S \subseteq P-S$ ($\equiv cP \cap cS \subseteq cP^* \cap cS$). The first condition amounts to ‘no extra established counterexamples’, to be called ‘instantially at least as successful’, and the second to ‘no loss of induced examples’ or, equivalently,[8] ‘no loss of

[8] Since the clause is also equivalent to $S \cup P^* \subseteq S \cup P$ it can also be paraphrased by “all induced laws entailed by P are entailed by P*, for any superset of $S \cup P$ is a superset of $S \cup P^*$.

explained established laws', for which reason to be called 'explanatorily at least as successful'.

Now the already announced *Success Theorem* states, assuming that R/S is correct, that is, $R \subseteq \mathbf{T} \subseteq S$, that $\underline{P}^*$ is (empirically) at least as successful relative to R/S as $\underline{P}$ if $\underline{P}^*$ is at least as close to the truth as $\underline{P}$. This theorem is such that *persistently* being-more-successful is functional or instrumental for truth approximation, though not guaranteeing it, in the sense that it is very difficult, and without the O–/T--distinction even impossible, for $\underline{P}$ to remain at least as successful as $\underline{P}^*$ when $\underline{P}^*$ is in fact closer to the truth than $\underline{P}$.

More in detail we may argue as follows. Let $\underline{P}^*$ be more successful than $\underline{P}$ relative to R/S at a certain time. This suggests the *Empirical Progress Hypothesis*, according to which it is hypothesized that this will remain the case, whatever experiments we design and perform. Testing of this hypothesis may result in the conclusion, at least for the time being, that this is in fact the case: $\underline{P}^*$ persistently remains more successful than $\underline{P}$. Note that this conclusion is a comparative inductive generalization, providing the paradigm situation for speaking of *empirical progress*, consisting of 'instantial' and 'explanatory' progress. That is, the paradigm case for applying the, essentially instrumentalist, 'rule of success', viz. replace, for the time being, $\underline{P}$ in favor of $\underline{P}^*$. Here $\underline{P}^*$ may well be known to be false in view of R/S, for which reason the rule is called instrumentalist rather than falsificationist. Assuming empirical progress in this sense, it is very hard to imagine that $\underline{P}^*$ is, despite appearances, not closer to the truth than $\underline{P}$. More specifically, it holds that (1) the 'closer to the truth hypothesis', the *TA-hypothesis*, predicts and explains the empirical progress, (2) it cannot be the reverse, that is, it is impossible that $\underline{P}$ is closer to the truth than $\underline{P}^*$, and (3) if $\underline{P}^*$ is in fact not closer to the truth than $\underline{P}$, there is a specific burden to explain the supposed empirical progress so far. In the last case, without theoretical terms, hence without O-/T-distinction, the apparent empirical progress must have resulted from an unhappy test history of the empirical progress hypothesis because the empirical progress conclusion can be broken by appropriate experiments. With O-/T-distinction, see below, the breaking of this conclusion is only excluded when $\underline{P}^*$, though presumably not closer to 'the theoretical truth' than $\underline{P}$, happens to be nevertheless at least as close to 'the observational truth' as $\underline{P}$. In sum, the TA-hypothesis provides the '*default explanation and prediction*' of empirical progress.[9]

So far I have presented the basic structuralist theory of truth approximation. It is basic in the sense that several idealizations have been made, requiring refinement or concretization. In Kuipers (2007) I have extensively illustrated the (philosophical)

[9] Surprisingly enough, the expression 'default explanation', let alone 'default prediction', does not seem to be in use in the philosophy of science. Here it is specifically used as an alternative to the expression 'functional for truth approximation'. As Gerhard Schurz has remarked, the default explanation 'closer to' is also a kind of *abductive* explanation, with the interesting restriction that the genuine abductive step in this explanation is only from 'at least as successful' to 'at least as close to', for 'not equally close' is a deductive consequence of 'more successful'.

method of concept explication by idealization and concretization by the example of truth approximation. Let us review the main concretizations as they have been elaborated or indicated in this volume for two-sided theories, with exclusion theories as one extreme type of theories.

One idealization was that we implicitly assumed that any established counterexample of a theory, that is, a realized possibility not being a model of the theory, is as bad for one theory as for any other. However, in Chap. 6 I have introduced the ternary relation of '(more) structurelikeness', that is, the idea that one structure may be more similar to another than a third one. In this way, the possibility arises that a counterexample is less dramatic for one theory than for another, because the former has a model that is more similar to the 'counter model' (i.e., the potential model representing the counterexample), than any model of the latter theory. Adapted definitions of more-successfulness, empirical progress, and closer-to-the-truth lead again to the conclusion that the TA-hypothesis provides the default explanation and prediction of empirical progress. As a matter of fact we only know of real life scientific examples of (potential) truth approximation, e.g. the Law of Van der Waals as successor of the ideal gas law, when this concretization is introduced. Without this, we only know of toy examples.

Another idealization was the assumption that there is just one domain. In (Kuipers 2000, Chap. 13) I have indicated that it is possible to introduce a domain vocabulary, as a subvocabulary of the (observational) vocabulary, such that it is possible to define the domain explicitly in that vocabulary, leaving the further behavior of the applications as the unknown to be specified. The advantage of this is that domains become comparable sets, that is, it enables us to say that an intended application of one theory is or is not an intended application of another. This opens the possibility of dealing with truth approximation by domain change, besides by theory change, as indicated above, and hence by the combination of domain and theory change, see Chap. 8.

Returning to theory change, it is evident that theories will seldom be comparable in terms of 'more successful' and 'closer to the truth'. These notions apply only in ideal cases. In practice it is important that in the case of 'divided success' between two theories, the crucial challenge is to improve both theories, that is, to design a third theory, a synthesis, that is more successful than both and, if so, plausibly closer to the truth than both.[10] This is called the 'principle of dialectics' in (Kuipers 2000, Chap. 6). In the theoretical context of the realism – antirealism debate, the move to the quantitative approach is more useful to deal with divided success. The basic form of being quantitatively closer to the truth is the condition, assuming finite U, $|P^{*}\Delta \mathbf{T}| < |P\Delta \mathbf{T}|$, see Chap. 5. For a refined quantitative approach the best proposal so far is of Ilkka Niiniluoto (1987, 1999), who has defined the so-called min-sum real-valued distance of a theory to the truth, which always enables comparison.

[10] At first sight, a more basic problem of comparability seems to arise from so-called incommensurable vocabularies. However, as long as there is the suspicion that two terms may have a different meaning it is plausible and possible to include both terms, with indices, in a fusion of vocabularies.

Since the only point of the coming exposition is the principled possibility of straightforward empirical progress and truth approximation I will not go here into details of this approach.

Finally, we did not make a distinction between (relatively) observational and theoretical terms, but such a O-/T-distinction is of course crucial for the realism-antirealism debate,. In Chap. 9 of (Kuipers 2000) I have spelled out how this distinction works out, with the main conclusion that empirical progress, of course, in observational terms, remains functional for truth approximation on the theoretical level, though with some greater risk of being wrong in this tentative conclusion.

For present purposes I will introduce the main features of this type of 'stratified (basic) truth approximation'. Let U_t now indicate the set of theoretical (cum observational) conceptual possibilities of the vocabulary. Let subset $\mathbf{T_t}$ of U_t indicate the unique subset of nomic theoretical possibilities, the theoretical truth. Let U_o indicate the set of partial or observational conceptual possibilities and $\mathbf{T_o}$ its unique subset of nomic observational possibilities, the observational truth. We assume that $\mathbf{T_t}$ and $\mathbf{T_o}$ can be characterized in some way. Let π project (subsets of) U_t onto (subsets of) U_o, stripping the (clauses involving) theoretical components. There are good reasons to assume that $\mathbf{T_o} = \pi(\mathbf{T_t})$ (see Chap. 7 of this book). Of course, theories $\underline{P}$ and $\underline{P^*}$ are empirically equivalent when $\pi(P) = \pi(P^*)$. The relations of 'empirically more successful', and hence of 'empirical progress', and of 'closer to the (observational) truth' between theories $\underline{P}$ and $\underline{P^*}$ now refer to these notions between $\pi(P)$ and $\pi(P^*)$ as defined above, but now on the level of U_o. On the other hand, '$\underline{P^*}$ is closer to the (theoretical) truth than $\underline{P}$' remains defined on the (theoretical cum observational) level of U_t, by the otherwise unchanged basic, that is, set-theoretic definition. Finally, it is possible to define 'the referential truth' as, roughly, the conjunction of positive or negative reference claims of theoretical terms that are correct according to $\mathbf{T_t}$, that is, those terms that do play or do not play a role in the shaping of $\mathbf{T_t}$ relative to $\mathbf{T_o}$, respectively. For each theory, a (total) reference claims is associated by a similar condition. In sum, the above assumptions enable definitions of 'closer to the observational, referential, and theoretical truth', and hence of 'observational, referential, and theoretical truth approximation'.

Besides the previously mentioned Success Theorem, now restricted to the observational level, some more TA-theorems and -conjectures become relevant, notably the Stratified Success Theorem (which is a kind of combination of the Success Theorem and the Projection Theorem in Chap. 7). In Sect. 16.4, I will give some more information. Together they support the main claim of comparative realism: (theoretical) truth approximation provides the (stratified) default explanation and prediction of empirical progress between non-empirically equivalent theories and of 'aesthetic progress' between empirically equivalent theories. Here 'aesthetic progress' will be defined in terms of the prevailing 'aesthetic canon' in the relevant field and period, that is, the prevailing non-empirical virtues of theories.

16.4 Evaluation of Comparative Realism

In Sect. 16.2 we have derived four desiderata for an adequate explication of a comparative realist stance that is claimed to be better defensible than other responses to the discussed antirealist charges. In this section the first task is the separate evaluation of comparative realism, that is, to show to what extent realism guided by the comparative perspective on success and on truth approximation, sketched in Sect. 16.3, fulfils these desiderata. The second task is its comparative evaluation, that is, to what extent is it superior to the other responses to antirealism.

16.4.1 In Defense of the Main Claim of Comparative Realism

Recall that we have summarized the way in which comparative realism is claimed to fulfill the desiderata a couple of times by its 'main claim': truth approximation provides the default explanation and prediction of empirical and aesthetic progress. This claim is supposed to be realized by the stratified (basic) explication of the notions of 'more successful' and 'closer to the truth', and the resulting mutual relations. Let us start with noting a number of terminological equivalencies that will be presupposed in the defense of the claim.

Empirical progress combines instantial progress and explanatory progress, and a new theory amounts to empirical progress relative to its predecessor when it is concluded on the basis of 'sufficient' comparative testing that the new theory is persistently empirically more successful than the old one. Empirical success preservation provides weak versions of these kinds of progress.

Aesthetic progress of a new theory relative to an old one, according to the prevailing aesthetic canon and in particular among empirically equivalent (EE-)theories amounts to an increase of having desirable-aesthetic properties and a decrease of having undesirable ones. Below I will indicate that empirical progress and aesthetic progress have in some formal sense the same structure. Aesthetic success preservation is again a weak version.

In the expression 'closer to the truth', 'the truth' always refers to the strongest true theory about a given domain in a given vocabulary and 'truth approximation' (TA) is short for 'getting closer to the truth'. In the stratified context we get the equivalences: getting closer to the theoretical/observational/referential truth = theoretical/observational/referential truth approximation. In this context, 'truth approximation' and TA without qualification are supposed to refer to 'theoretical truth approximation'. The weak version of 'theoretical truth approximation' is called 'theoretical truth preservation'. Similarly, the weak version of 'referential truth approximation' is 'referential truth preservation'.[11]

[11] Note that we do not call 'theoretical, referential, observational truth approximation' progress of that kind, for we like to reserve the term 'progress' for kinds of progress that do not presuppose to know the relevant truth, such as empirical and aesthetic progress.

We will first deal with CA1 and CA2, in which EE-theories and aesthetic progress are not yet at stake, and later with CA3 and CA4. In both cases the claim is of course that the crucial notions can be explicated such that (it is reasonable to assume that) these conditions of adequacy are satisfied. In the first case the justification is primarily analytical, allowing empirical illustrations. In the second case the justification will be a combination of empirical and analytical arguments. The analytical justification will be based on a number of 'TA-theorems and -conjectures'. The empirical justification will be twofold. Moreover, I will first restrict the attention to the cautious, very demanding, explications of the crucial notions, that is, the stratified (basic) explications.

16.4.1.1 Empirical Progress

Let us first focus on matters of empirical progress in relation to theoretical truth approximation, and address matters of referential truth preservation later. For convenience, I repeat the first two conditions of adequacy restricted in this way:

CA1-EP: The explications of the crucial phrases should leave room for the possibility that false theories are not only (persistently) empirically more successful but even closer to the (theoretical) truth.

CA2-EP: The explications should be such that empirical progress is, *as a rule*, due to theoretical truth approximation, which on its turn entails, as a rule, novel predictive success.

In Sect. 16.3 I have already specified the stratified basic definitions of 'empirically more successful', enabling the definition of 'empirical progress', and of 'closer to the theoretical truth'. From these definitions it is easy to see that CA1-EP is satisfied, for the definitions do not make a crucial difference between theories that are true or false, in the (weak) sense, that is, theories that don't exclude nomic theoretical possibilities and theories that do, respectively.

Regarding CA2-EP, recall, the (empirical) Success Theorem, which can be formulated for the stratified case as follows?

Stratified Empirical Success Theorem:

- Theoretical truth *preservation* of one theory relative to another entails explanatory success preservation and, as a rule, instantial success preservation. Exceptions to instantial success preservation concern extra pseudo instantial successes, to be explained below.[12]
- Theoretical truth *approximation* entails in adodition, assuming empirically non-equivalent theories, that at least some extra (empirical) success can be obtained, and hence the prospect of empirical progress.

[12] In brief: an extra pseudo instantial success of the old theory relative to the new one concerns a realized, hence, observational, nomic possibility that is a counterexample of the new theory but that is not excluded by the old theory thanks to a non-excluded nomic *im*possibilities on the theoretical level, in Chap. 7 called 'extra lucky miss'.

The proof of the explanatory part in the first clause is straightforward, as in the unstratified case. However, the instantial part is now not unconditional, for in the stratified situation there may be 'instantial pseudo successes' of P relative to P*. Such a success amounts to an observational nomic possibility that is allowed by P but excluded by P*, and hence a counterexample of P* (and therefore breaking observational truth approximation of P* relative to P), which, however, on the theoretical level, assuming truth approximation on that level, must be due to the allowance by P of a (theoretical) nomic *im*possibility, which is rightly excluded by P*. Of course, any instantial success of P or P* may be a pseudo success, which is something which we cannot observationally distinguish from a genuine instantial success. However, this is only obscuring comparative theory evaluation as far as *extra* instantial successes are concerned. By the way, the epistemic asymmetry between explanatory and instantial successes may well explain why explanatory successes of proper theories (theories with theoretical terms) seem more impressive than counterexamples.

From this theorem we may conclude that theoretical truth approximation provides the realist default explanation and prediction of empirical progress. For, assuming at least some initial extra explanatory success, the hypothesis of theoretical truth approximation, the TA-hypothesis, enables the explanation and prediction of *explanatory* progress, including some (differential) novel predictive success, as we will see. Moreover, the hypothesis enables the conditional explanation and prediction of *instantial* progress, including some (differential) novel predictive successes, as we also will see. The condition being that the old theory does not have or get extra pseudo successes. If the TA-hypothesis is in fact false, new experiments will break the empirical progress conclusion in the long run. However, as far as breaking the instantial progress conclusion is concerned this may be due to pseudo successes.

In some more detail, the situation is as follows, assuming empirical progress, that is, a well confirmed empirical progress (EP-)hypothesis of some theory P*, relative to P, on the basis of evidence R/S. If P* is in fact not closer to the theoretical truth than P, then there is some theoretical nomic possibility, say x, excluded by P* but not by P and/or some theoretical nomic *im*possibility, say y, allowed by P*, but not by P. In the first case there is room for an experimental counterexample of P* which is not a counterexample of P, viz. when the projection of x is also excluded by P*, and hence is no pseudo success of P*. In the second case an observational law, according to which at least the projection of y is impossible, can be experimentally established that is an explanatory success of P, but not of P*, provided the projection of y is also excluded by (the projection of) P.

Regarding the claim of some (differential) novel predictive success the situation is as follows. It is rather difficult to imagine that, at the start of testing an EP-hypothesis, see Sect. 16.3, where the success superiority of P* relative to P is based on previous findings (R/S), these findings exhaust every kind of thing there is to know of the observable differences between P and P*. In other words, except in finite cases, and perhaps some very simple infinite cases, at least some room for differential 'novel predictive success' is plausible, in whatever sense 'novel' precisely is meant.

The foregoing not only shows that the TA-hypothesis provides the default explanation and prediction of empirical progress when the crucial notions are explicated in the stratified basic way, but also that CA2-EP then is satisfied.

16.4.1.2 Referential Truth Preservation

The first two conditions of adequacy also deal with reference claims, specifically:

CA1-Ref: The explications of the crucial phrases should leave room for the possibility that false theories, even with some non-referring theoretical terms, are closer to the (theoretical) truth and refer at least as well as other theories.

CA2-Ref: The explications should be such that theoretical truth approximation entails, as a rule, referential truth preservation.

In Sect. 16.3 I have already indicated the definition of 'reference' and of 'closer to the referential truth'. Here I will elaborate these indications a bit further, partly because of some weak spots in earlier formulations, but the main ideas remain the same.

In the definition of a subset P of U_t, constituting theory $\underline{P}$, with the exclusion claim "$\mathbf{T_t} \subseteq P$", a theoretical term may or may not play a substantial role in the sense that if we project P on the conceptual possibilities generated by the U_t-vocabulary minus that term, and then recombine these possibilities again with that term in all possible ways, the resulting subset of U_t may or may not be a proper superset of P. If it is, $\underline{P}$ is said to claim that the term does refer, for it makes a difference. If it is not, $\underline{P}$ is said not to refer, for it does not make a difference.

The definition still needs some qualification. It seems adequate primarily for theoretical terms of which $\underline{P}$ claims that they refer to attributes (properties, relations, and functions); attribute terms for short. For theoretical terms claimed by $\underline{P}$ to refer to classes of entities, used as domain-sets for attribute terms – entity terms for short – this definition does not seem to work. However, it is plausible to say that a (theoretical) entity term refers according to $\underline{P}$ if and only if there is at least one (theoretical) attribute term that is claimed to refer and that uses the entity term as (one of) its domain-set(s). The consequence is that an entity term does not refer according to $\underline{P}$ if there are no attribute terms using it as domain-set. This is plausible, for in this case it is difficult to see how that theoretical domain could play a substantial role in the definition of $\underline{P}$, which is precisely the reason for the detour via attribute terms. If this detour is not possible, the theoretical entities hang in the air according to $\underline{P}$ as unconstrained entities, not distinguishable from genuine fictitious entities, at least not with the means provided by the vocabulary. For example, the Higgs-particles, postulated for theoretical reasons by the so-called Standard Model in elementary particle physics, have theoretical properties with observational consequences. In 2012 the latter were sufficiently confirmed by the experiments with the

Large Hadron Collider (LHC) at CERN (near Geneva), which started in 2008, to claim that the existence of Higgs-particles, that is, that they were not fictitious.

Of course, assuming that **Tt** can be defined with the means of the vocabulary behind Ut, **Tt** also claims, according to this definition of reference claims, of certain theoretical terms that they refer. Of the remaining theoretical terms it does not only not claim that they refer, but being the strongest true theory, it claims in fact that they do not refer. Of course, all (non-) reference claims of **Tt** itself may be supposed to be correct. Note that this definition of (non-)reference amounts to a direct, though highly theoretical, reference definition for attribute terms and an indirect one for entity terms. Note in passing that the very possibility to define reference of entity terms on the basis of the definition of reference of attribute terms is a good reason to extend the idea of 'entity realism' to 'referential realism'. Note finally that the given definition of reference is explicitly $\mathbf{T}_t$-relative.

According to the *Nomic Principle*, applied to Uo as well as Ut, we have assumed that unique fixed sets $\mathbf{T}_o$ and $\mathbf{T}_t$ exist, given an intended domain. Moreover, we have implicitly assumed that both can be characterized in some way by the available vocabularies. Regarding $\mathbf{T}_t$, however, the Nomic Principle is somewhat dubious, for it seems to presuppose that the reference of the theoretical terms is given beforehand. More cautiously than we have done before, we have to assume that, if we would have some kind of superhuman observational access to U_0-relevant theoretical entities and attributes it is possible to couple none, some or all theoretical terms constituting U_t to them such that $\mathbf{T}_t$ arises. Of course, it is not at all evident that this can be done only in one way. But we may assume that the number of terms that can be meaningfully coupled at once has a maximum and that this maximal coupling is unique and gives rise to $\mathbf{T}_t$. In other words, we assume that there is unique maximal referential use of the theoretical vocabulary and that this is which we are looking for.

For the remaining explications, let us first assume that our theories are maximal, i.e., claim that e.g. P = $\mathbf{T}_t$, in which case not claiming reference by $\underline{P}$ amounts to a non-reference or negative reference claim. The conjunction of the (non-)reference claims of $\mathbf{T}_t$ is called 'the referential truth' and the conjunction of all (non-)reference claims of theory $\underline{P}$ is called its (total) referential claim. Finally, theory $\underline{P*}$ is closer to the referential truth than theory $\underline{P}$ if $\underline{P*}$'s referential claim agrees on more terms with the referential truth than that of $\underline{P}$. Referential truth preservation is defined as: theory $\underline{P*}$ is at least as close to the referential truth as theory $\underline{P}$ if $\underline{P*}$'s referential claim agrees on at least the same terms with the referential truth as that of $\underline{P}$.

Let us call two (maximal) theories referentially equivalent when they have the same (total) referential claim, and let us call them referentially semi-equivalent when they have the same number of positive reference claims. The uniqueness assumption about $\mathbf{T}_t$ now amounts to the assumption that a (maximal) referentially semi-equivalent theory to $\mathbf{T}_t$ is referentially equivalent to $\mathbf{T}_t$. Of course, referential (semi-) equivalence does not imply nor exclude empirical equivalence.

Returning to (non-maximal) exclusion theories, we remain calling the conjunction of the (non-)reference claims of $\mathbf{T}_t$ 'the referential truth'. The conjunction of all (positive) reference claims of theory P is called its positive referential claim. Finally, theory P* is closer to the referential truth than theory P if P*'s positive referential claim agrees on more terms with the referential truth than that of P. Referential truth preservation is defined as: theory P* is at least as close to the referential truth as theory P if P*'s positive referential claim agrees on at least the same terms with the referential truth as that of P.

From the above it is clear that whether or not (the claim of) a theory is true or false, even with some non-referring theoretical terms, does not play some crucial role in the definitions of theoretical truth approximation and referential truth preservation. Hence, CA1-Ref is satisfied.

Regarding CA2-Ref, we submit the following (diagnostic) conjecture:

Referential Truth Conjecture:
Theoretical truth approximation entails, as a rule, referential truth preservation.

If not, the old theory has *extra lucky* reference claims.

Note first that we do not add a second clause to the conjecture, similar to that of the Stratified Empirical Success Theorem, which would now amount to: theoretical truth approximation entails in addition (to theoretical truth preservation), as a rule, referential truth *approximation*. In the present case, it is difficult to argue for it, and plausibly so, because this would (almost) exclude the possibility of improving theories without improving its reference.

Let me explain what lucky reference claims, mentioned in the conjecture, are. One might think that the assumption that P* is closer to $\mathbf{T_t}$ than P does entail the claim that P* is referentially at least as close to the truth as P. A specific reason to expect this is the fact that the theoretical claim of a theory, e.g. "$\mathbf{T_t} \subseteq P$", entails its referential claim "$V_r(P) \subseteq Vr(\mathbf{T_t}) =_{def} V_r$", where $V_r(P)$ indicates the set of theoretical terms that refer according to P, and V_r is the one corresponding to $\mathbf{T_t}$. However, a proof for the suggested unconditional conjecture is not possible, for interesting reasons. Suppose that P* is closer to $\mathbf{T_t}$ than P. What we would like to prove amounts to the following: all τ in $V_r(P^*)$-$V_r(P)$ belong to V_r. Let us suppose that τ in $V_r(P^*)$-$V_r(P)$ does not belong to V_r. Hence, P* wrongly claims that it does, whereas P rightly claims that it does not. However, P's claim may be based on a wrong aspect of its theoretical claim "$P \subseteq \mathbf{T_t}$", whereas P* may not yet be so good that its theoretical claim implies the reference claim for the right reasons. It is important to note that the suggested proof already fails in the case of just one theoretical term.

In sum, we have to admit the possibility that P has such lucky reference claims relative to P*. This does not mean, of course, that 'to be as close to the theoretical truth' does not provide good arguments for believing 'to be at least as close to the referential truth'. As can be learned from the suggested proof attempt, the suggested entailment is only violated when a theory bases (part of) its referential claim on wrong reasons in precisely the right direction, which will be rather exceptional because it is rather artificial.

So far, for the analytical justification of the claim that the stratified (basic) definition of truth approximation satisfies to a considerable extent the conditions of adequacy regarding empirical progress and referential truth preservation, viz. CA1 and CA2. After dealing with aesthetic progress, I will come back on the possibility of empirical support of this claim.

16.4.1.3 Aesthetic Progress

For the moment we will focus on CA3, for CA4, dealing with probabilistic versions of IBE, evidently exceeds the boundaries of the qualitative approach.

CA3: As far as EE-explanations are concerned, an explication of IBE should only take those prevailing aesthetic criteria into account, *if any*, for which there are empirical or analytical reasons, however weak, to assume that they are truth-conducive in the field.

As we already noted, the prevailing aesthetic criteria in a certain field at a certain time constitute what James McAllister (1996) has called the *aesthetic canon*, arrived at by *aesthetic induction*. In contrast to observational criteria, where we may assume, and have argued in some detail in the first subsection, that they are, as a rule, truth-conducive in the comparative sense, empirical information will have to support that certain aesthetic criteria are truth-conducive. However, before we enter the empirical justification, we will first show that there may well be truth-conducive non-empirical criteria at all. We will show this by a theorem that is a general version of (the explanatory part of) the success theorem. Of course, 'the truth' has not only observational consequences, but also theoretical (and mixed) consequences. A (purely) theoretical consequence of $\mathbf{T_t}$ amounts to a superset C of $\mathbf{T_t}$, of which the projection equals Uo, it does not exclude something observational. Note that such a consequence can also be interpreted as a property or feature of all nomic possibilities, to be called a *nomic* property.[13]

Nomic Theorem
Theoretical truth preservation (whether or not among EE-theories) entails being at least as successful regarding nomic properties.

A generally interesting question is which non-empirical properties are nomic, hence truth-conducive, properties. However, since observational differences will have priority in truth approximation evaluations, the use of this theorem naturally focuses on EE-theories. Hence, the main question becomes which non-empirical properties may be or have been supposed to be truth-conducive in the indicated sense.

We have already suggested, following McAllister and, for example, Weinberg, calling the non-empirical properties of interest *aesthetic* criteria, for such properties

[13] Similarly, we could introduce 'strongly non-nomic' properties, properties corresponding to consequences of the negation of $\mathbf{T_t}$, i.e., supersets of the complement of $\mathbf{T_t}$, enabling a kind of mirror theorem to the nomic theorem. See (Kuipers 2001, Chap. 10) for a general exposition of truth approximation in terms of two kinds of features.

are as a rule also aesthetically appreciated. The empirical justification of claims to the effect that certain aesthetic criteria are truth-conducive in the field may be given along two different lines, and their combination. In the first line, due to Igor Douven, the claim is tested with hindsight, here called 'aesthetic testing', in the second line, the claim is based on 'aesthetic induction' as introduced by McAllister.

Douven (1996, 2002, 2005) has described a way to (bootstrap) test applications of IBE on EE-theories with hindsight, which amounts briefly to the following. Although at the time of the application of IBE the theories were empirically equivalent, this may change in the course of time. New developments, e.g. the development of a sophisticated type of microscopes or the building of higher energy particle colliders, may lead to "a shift in the boundary between what can and cannot be observed" (2005, p. 297), which he aptly called an OUD-shift in his (1996), where OUD is short for 'Observable-Unobservable Distinction'. In our approach this amounts to enriching U_0 to some U_0'. Assuming that the new microscope or collider is veridical, which is the bootstrap condition of the testing, the interesting question becomes whether the previously best EE-theory, based on certain non-empirical properties, turns out to be empirically the most successful theory according to the new observational evidence acquired with the new means. If so, this confirms the claim that the (still) non-empirical properties were in fact truth-conducive.

The second line of empirical justification is provided by the 'naturalistic-cum-formal' analysis of the relation between aesthetic criteria, empirical success and truth (Kuipers 2002; Chap. 9 of this book). It starts with the notion of 'aesthetic induction'. As McAllister has shown in his very inspiring book *Beauty and Revolution in Science* (1996), our aesthetic judgments are subject to change. We are not only inclined to find the heliocentric worldview of Copernicus more beautiful, because it is simpler, than the geocentric view of Ptolemy, but we are also inclined to find Kepler's elliptic planetary orbits at least as beautiful as Copernicus' circular orbits. However, ellipses are undoubtedly more complicated than circles, and this is precisely the reason why they were found less beautiful, if not problematically ugly, at the time. Moreover, I would like to add, aesthetic criteria not only change with time within a discipline, but may also differ greatly between disciplines. For example, after expressing in an interview series (Kayzer 2000) his agreement with (physicist) Steven Weinberg about the importance of beauty considerations, (biologist) Stephen Gould hastens to stress that his criteria for beauty totally differ from those of Weinberg. Whereas Weinberg mentions 'inevitability' of desired consequences as his dominant aesthetic criterion – as exemplified by Einstein's theory, which, in contrast to Newton's theory, made the inverse square in the law of gravitation inevitable – Gould stresses that, besides diversity, unrepeatable contingencies and irregularities are the sources of his ultimate aesthetic satisfaction.[14]

Having observed the variation of aesthetic criteria, McAllister's main claims are as follows. First, scientists normally use aesthetic criteria in addition to empirical

[14] This is not to suggest that standard examples of aesthetic features mentioned by physicists do not play a role in biology. For example, Gould mentions order a number of times and Sober (2002) points out that simplicity in the form of parsimony plays a considerable role in taxonomy.

criteria for theory evaluation. Second, and most importantly, the aesthetic criteria of the time, the 'aesthetic canon', is based on 'aesthetic induction' regarding non-empirical features of paradigms of empirically successful theories which scientists have come to appreciate as beautiful. Third, aesthetic criteria can play a crucial, schismatic role in scientific revolutions. Since they may well be wrong in a particular case, they may, in the hands of aesthetic conservatives, retard empirical progress, but this does not occur in the hands of aesthetically flexible, 'revolutionary' scientists.

My earlier analysis (Kuipers 2002; Chap. 9 of this book) of the relation between the empirical success and the beauty of scientific theories elaborates and supports McAllister's claims. Like McAllister, I focus on non-empirical aesthetic features, that is, features with aesthetic value but without empirical content, although empirical features may also be aesthetically valued. In the spirit of naturalized epistemology, one may first argue that the phenomenon of aesthetic induction may be a variant of the so-called 'mere-exposure effect' (Zajonc 1968, 2001), viz. non-conscious affective priming (Zajonc 1994). In this perspective one may decompose the notion of aesthetic induction into aesthetic induction proper, or affective induction, and a related cognitive (meta-)induction. Together they lead to correlations between non-empirical features that are found beautiful on the one hand and empirically successful theories on the other. Such correlations may be called 'beauty-success correlations'. The corresponding received or 'canonical' aesthetic features are non-empirical features that have acquired (positive) aesthetic value and (empirical success related) meta-inductive support. This makes it plausible to explicate the notion of an 'aesthetic feature' more precisely as an aesthetically (positively) valued non-empirical feature. Now it is possible to argue that the co-production of the two types of induction (aesthetic induction proper and cognitive meta-induction) is functional for empirical progress as far as the cognitive meta-induction is reliable. That is, aesthetic features may rightly have become indicative of empirical success.

In view of the nomic theorem and our formal analysis of the relation between (more) empirical success and truth approximation, indicated in the first subsection, such features become indirectly indicative of truth approximation, that is, truth-conducive, provided they are of a certain formal nature. Since aesthetic features correlate with empirical success and empirical success with truth approximation, it has some plausibility to submit that such features of EE-theories are nomic properties, in which case they strongly correlate with truth approximation. However, this formal account only holds for aesthetic features of a certain formal kind, called distributed features. In order to possibly be a nomic property, such a feature should apply to all models of the relevant theory, in which case they can also be represented as a consequence of the theory. To be sure, the link between distributed aesthetic features and truth approximation is weak, but it makes some sense. It essentially consists in the combination of the stratified empirical success theorem and the nomic theorem. Below we will briefly discuss which aesthetic features are or may be distributed.

Of course, aesthetic induction will cover (positive) results of previous aesthetic testing, in the sense of Douven, whether or not such testing has been undertaken explicitly for this purpose. Hence, like aesthetic induction, aesthetic testing contributes to the relevant 'aesthetic canon'. This strengthens the conclusion that there are good, though not strong, reasons to assume that the aesthetic canon, as far as consisting of distributed features, may rightly hypothesize what are truth-conducive theoretical nomic properties, and hence for the following conjecture:

Aesthetic Success Conjecture:

Theoretical truth *preservation* among EE-theories entails (increase of theoretical nomic properties), and hence, in view of the nature of aesthetic testing and aesthetic induction, as a *weak* rule, aesthetic success preservation according to the aesthetic canon as far as distributed. If not, which may appear by new aesthetic testing or a major breaking of aesthetic induction, previous aesthetic testing and induction prove to have their limits.

Theoretical truth *approximation* among EE-theories entails (increase of theoretical nomic properties), and hence, as (an at least as weak) a rule, aesthetic progress according to the aesthetic canon as far as distributed.

The important remaining question is which aesthetic features are, or can have a distributed form. Let us review the features that have been mentioned earlier, now grouped in five categories. Each of them may have other examples.

- inner beauty, elegance,
- simplicity, symmetry, visualizability, inevitability or rigidity, coherence with metaphysical suppositions
- explanatory power, novel predictive power, fertility,
- coherence with background theories,
- non-adhocness, parsimony.

I will deal with these categories in the listed ordering. The *first* category lists features that ask for some kind of objective specification, such as those of the other categories.

The *second* category typically represents features that are non-empirical and admit an objective specification. Moreover, although they can apply to theories as wholes they can also apply, or not, to separate conceptual possibilities, and hence to all models of a theory, or not, that is, they can distributively apply to a theory, or not. Moreover, their precise nature and their aesthetic value may typically vary historically and between disciplines. In sum, they typically form the kind of features to which the aesthetic success conjecture is oriented.

The *third* category typically contains features that first of all promise to be accounted for in empirical success: 'explanatory power' in explanatory success, 'novel predictive power' in novel predictive success, instantial or explanatory, and finally 'fertility' frequently is used to express the two previous kinds of power. In all three cases, it may well be that the aesthetic value they have, derives from these promises of empirical virtues. Although such promises may turn out to be realized, it is important to note that such empirical pay-offs can only arise when there are such promises. All this holds, whether or not the features may apply in a distributed form, which seems to be possible for 'novel predictive success', and hence for a certain kind of fertility.

The *fourth* category is a kind of mirror category of the previous one. Incoherence with background theories is a 'promise' of empirical problems, the stronger the more established the background theories are. Hence, this type of incoherence may well have become negatively valued aesthetically as a predictor of empirical problems. However, it may also be a (very weak) predictor of a scientific revolution.

Finally, 'non-adhocness' in the *fifth* category, is generally valued positively because adhocness is disvalued, probably because of similar reasons as the previous type of incoherence. The saving of a theory by an ad hoc repair frequently turns out to have its empirical price. However, in this case it is important to note that it may also turn out to be a smooth kind of truth approximation. This applies *mutatis mutandis* to parsimony, for again non-parsimoniousness may show to have its empirical, notably explanatory, price, but it may also be a step in the good direction, the truth.

Let us finally confront the proposed way of dealing with EE-theories in accordance with CA3 with the provoking paper of Roger Jones (1991), entitled "Realism about what?" (See also, (Ladyman 2002, 252–7), from which I will borrow crucial formulations). In this paper, Jones reminds us of four alternative formulations of classical mechanics, as for example applied to planetary motion. Though empirically equivalent, they suggest nevertheless a quite different ontology and metaphysics. Newton's own approach, based on the three laws of motion and force laws, suggests an ontology of point particles and forces acting at a distance. The gravitational field approach assumes an ontology of action by contact in conformity with a principle of local causality, but postulates a new type of entity, the field. In the curved space approach, in which the gravitational field is absorbed into the structure of space (analogous to the way in which this field is absorbed into the structure of spacetime in general relativity). In this approach the field is represented by the local degree of curvature and space itself is made part of the fundamental ontology of the theory having a causal efficacy of its own. Finally, the analytical mechanics approach, in which the force laws and the laws of motion can be derived from so-called minimum principles, seems to be independent of causal thinking, but requires instead a kind of teleology for particle behavior according to which it is the properties of complete paths between points of space that determine the motion of a body.

The question is of course which ontology (cum metaphysics) is suggested by a realist attitude to these EE-versions of classical mechanics. The suggestion of our analysis is that this depends on which formulations score the best according to the prevailing relevant aesthetic canon, dealing with ontological, epistemological, methodological and pragmatic aspects. However, apart from the difficult question as to which canon is prevailing and relevant, given so long a period of time in developing these formulations, the preliminary question is whether the suggested ontologies are really as different as they seem at first sight. The first approach (with forces acting at a distance) and the field approach may well be seen as compatible ontologies, provided the forces are not assumed to act at a distance instantaneously. With this (non-Newtonian) assumption they are related as the manifest global or macro-ontology to the theoretical local or micro-ontology as seems, for example, the standard view in response to Eddington's non-existence claim of his famous table. The relation between the field approach and the (non-relativistic) curved space approach does not seem to be a matter of compatible ontologies. However, looking back from general relativity theory, and its empirical success, the aesthetic appreciation of a curved spacetime and hence presumably of a curved space (both of which can be distributed properties of a theory), seem to have become (much) higher than that of the (relevant) field approach. Hence, the curved space version of the classical theory of gravitation seems to deliver the most plausible realist ontology, and hence would seem so far to be the best theory in accordance with CA3, and hence for inferring, by IBE-E3.2, that it is the closest to the theoretical as well as to the referential truth. This judgment will not be changed by the last, minimum principles formulation of mechanics, for that delivers from a realist point of view at most an 'as if metaphysics'. However, as such it does no theoretical work, let alone observational work, for in the present deterministic setting this kind of 'as if metaphysics' is purely a matter of logico-mathematical equivalences.

This concludes our articulation and defense of the claim that CA3 can be satisfied, at least to some extent.

As we stated already, CA1 and CA2 and our main claim have primarily analytical justification. Of course, this does not exclude empirical support. On the contrary, their justification asks for real life illustrations. In particular for CA2 one may think of the kind of (implicit or explicit) testing proposed by Douven on the basis of OUD-shifts: do previous 'empirical progress' and 'referential truth preservation' conclusions survive after the shift or even, when *all* theoretical terms have become observable, can we now conclude to 'referential truth approximation'?[15] Unfortunately, the demands for stratified basic truth approximation are very high, so that we will have to extend the analysis to 'refined qualitative' or 'quantitative versions'. This brings us to CA4.

Regarding CA4, the situation is not difficult. Recall:

CA4: In a probabilistic version of an adequate explication of IBE 'EE-updating' of the relevant belief should remain within the confines of Bayesian conditionalization.

The strictures of the 'argument from Bayesianism', that is, resulting from non-Bayesian conditionalization, can be avoided in at least one way, viz. by taking aesthetic features only into account in the (expert) prior distribution. As far as I know there is no other way.[16] As I already remarked in Sect. 16.2.2, a merely probabilistic version will not do in general, for probabilistic updating makes only differential sense for unfalsified theories. However, if we leave behind the strict conditions of

[15] Of course, OUD-shifts will also lead to reconsideration of previous aesthetic inductions: do they survive or has the canon (to be) changed. According to McAllister, a change of the canon is a characteristic of scientific revolutions.

[16] However, Niiniluoto reminded me of the fact that Hintikka's α-γ-system of inductive probabilities converges to the simplest generalization that is still compatible with the evidence. Hence, it need not be necessary to have to take explicitly care of aesthetic considerations in order to be respected.

qualitative stratified basic truth approximation, probabilistic estimation of the distance from the truth (Niiniluoto 1987, 1999), using Bayesian updating, is generally applicable.

16.4.2 Review and Extension of the Comparison with Other Realist Responses to Antirealism

Some preliminary remarks are in order before we can take up the comparison of the comparative approach with other realist responses. First, there are, of course, no *compelling* reasons neither for comparative realism nor for antirealism, whether the latter concerns theoretical, inductive or even experiential skepticism. However, Van Fraassen's adagium "what is rational to believe includes anything that one is not rationally compelled to disbelieve" (1989, p. 171–2) is a license for all kinds of wishful thinking, blocking a 'good reasons' debate. Although there is not much hope to convert Van Fraassen, a paradigm believer in antirealism, more specifically, constructive empiricism, into a believer in comparative realism, I have some hope that analytical philosophers with realist inclinations but doubts based on the antirealist charges, may become comparative realists instead of remaining adherent of one of the other retreats from full blown realism.

Let me first review the realist responses already dealt with so far. The retreat to claims of 'approximately true theories' instead of 'true theories' was put aside at the start as of no help, for that remains basically non-comparative. Although we certainly prefer giant leaps to the truth, any step in the direction of the truth, from whatever to whatever distance to the truth should be welcomed by the realist as a kind of progress. Moreover, the retreat to 'approximate truth' requires a necessarily arbitrary threshold. Further we came across restricting realism to 'mature' theories, or to theories with 'novel' predictive success or to the 'essential' parts and aspects of theories. The general tenet of our criticism of these moves was a kind of redundancy. From the comparative perspective all theories may remain in the game, no ban of certain kinds or aspects of theories are required. This is not to say that some of the suggested distinctions may not be useful for other purposes, related or not. The same holds for the one opposite kind of response we have met, viz. stretching realism by stretching the (causal) theory of reference underlying many conceptions of realism: it is not necessary for defending realism. As we have seen, such a defense is possible by using a non-standard criterion of reference, which may well be compatible with the causal theory of reference, or some stretched refined version.

Let me now turn to the extension of the comparison, in particular, with other realist responses by restricting realism. To begin with, 'entity realism', that is, realism with respect to the referential claim attached to theoretical entity terms, has been motivated, at least partially, by the idealizational, hence truth-abusing, nature of scientific practice, in particular model construction. From the comparative perspective this motive for restricting realism in order to save the important idealization practice is not relevant, for the question is whether of two successive theories, the second, being a concretization of the first, idealized theory, is closer to the truth than the first. In (Kuipers 2000, Chap. 10; Chap. 6 of this book), I have shown and illustrated for theories typically dealing with model construction, viz. inclusion theories, that the refined version of the truth approximation theory leaves perfect room for a positive answer to this question, roughly due to the fact that the second theory has models that are more similar to correct models than the models of the first. A mirror version of this argument for exclusion theories is well conceivable. Of course, the same argument can be applied to the less restrictive version of entity realism, called 'referential realism', that is, realism with respect to the referential claim of all theoretical terms, entity as well as attribute terms. Recall that I have remarked in passing that the transition from entity to referential realism is particularly plausible in view of the non-standard criterion of reference.

Finally, another retreat, which has become much debated in the last decade, is so-called structural realism, according to which typical realist features of successful theories consists of their structural features as expressed by the formalism specifying the formal relations. This retreat is mainly motivated by frequent referential failures, as highlighted by the pessimistic meta-induction. Again, it is an unnecessary retreat from the comparative perspective, now in view of the possibility of referential truth approximation.

16.5 Concluding Remarks

I would like to conclude this chapter with a general remark and a survey of epistemological positions.

All realist responses to the antirealist charges are retreats of realism of a non-comparative nature. As already suggested by my puzzling question in the introduction, it seems that all responses are non-comparative because of the lack of awareness that in the present debate relations, in particular, comparative ones, may be crucial and should be taken into account. Of course, comparative realism is also a kind of retreat. However, since it is a kind of piecemeal realism, it fits perfectly well into what may be seen as the core idea of realism, namely that our theories succeed more and more in giving 'a literally true story of what the world is like', to use Van Fraassen's well-known characterization of realism (Van Fraassen 1980, p. 8).

In my *From Instrumentalism to Constructive Realism* (2000), I have presented a hierarchical survey of epistemological positions. In the course of time I have refined and revised it at some places, but the core is still the same. I would like to present here the latest version, for it captures the main lines of thought in this chapter rather well. It is structured by a number of epistemological questions, however starting with an ontological one (Scheme 16.1).

A non-standard kind of summary of the main lines in this chapter, some of it implicit, is as follows, in the scheme indicated by the starred arrows. There is a human-independent natural world, about which true claims may be (non-

Q0:	independent natural world?	No $\Rightarrow$	ontological idealism
	$\Downarrow$* Yes: ontological realism		
Q1:	true claims about it possible?	No $\Rightarrow$	epistemological skepticism
			- experiential skepticism
	$\Downarrow$* Yes: epistemological realism		- inductive skepticism
Q2:	beyond the observable?	No $\Rightarrow$	theoretical skepticism
			- instrumentalism
	$\Downarrow$* Yes: scientific realism		- constructive empiricism
Q3.1:	beyond reference?	No $\Rightarrow$	referential realism
			$\Rightarrow$ entity realism
Q3.2:	beyond structures?	No $\Rightarrow$	structural realism
	$\Downarrow$* 2 x Yes: theory realism		
Q4:	ideal conceptualization?	No $\Rightarrow$*	constructive realism
	$\Downarrow$ Yes: essentialistic realism		

Scheme 16.1 Hierarchy of epistemological positions

compellingly) justified, not merely restricted to what is observable, but also with respect to theoretical terms and statements, provided we take the comparative piecemeal perspective, being a core feature of comparative 'theory realism'. However, as to the question whether there is some ideal vocabulary fitting the natural world, and hence leading to a kind of essentialistic realism, my answer is negative, for which reason I speak of 'constructive' realism, e.g. in the title of my 2000-book. The main message of this chapter is that constructive realism can best be defended against antirealist charges in its comparative form. Finally, I would like to stress that comparative constructive realism is a kind of nomic realism. As became clear in the present chapter, according to my view, theory formation and revision is not directed at truth approximation with respect to the actual world but at the nomic world, that is, the realm of what is nomically, e.g. physically, possible. In sum, comparative constructive nomic realism is the full name of the position I have defended.

References

Douven, I. (1996). *Defence of scientific realism*. PhD-thesis. Leuven: University of Leuven.

Douven, I. (2002). Testing inference to the best explanation. *Synthese, 130*, 355–377.

Douven, I. (2005). Empirical equivalence, explanatory force, and the inference to the best theory. In R. Festa, A. Aliseda, & J. Peijnenburg (Eds.), *Confirmation, empirical progress, and truth approximation* (Essays in debate with Theo Kuipers, Vol. 1, pp. 281–309). Amsterdam: Rodopi. (Reply by T. Kuipers, 310–313).

Jones, R. (1991). Realism about what? *Philosophy of Science, 58*, 185–202.

Kayzer, W. (2000). *Het boek over de schoonheid en de troost*. Amsterdam: Contact.
Kuipers, T. (2000). *From instrumentalism to constructive realism. On some relations between confirmation, empirical progress, and truth approximation* (Synthese library, Vol. 287). Dordrecht: Kluwer Academic Publishers.
Kuipers, T. (2001). *Structures in science. Heuristic patterns based on cognitive structures. An advanced textbook in neo-classical philosophy of science* (Synthese library, Vol. 301). Dordrecht: Kluwer Academic Publishers.
Kuipers, T. (2002). Beauty, a road to the truth. *Synthese, 131*(3), 291–328.
Kuipers, T. (2004). Inference to the best theory, rather than inference to the best explanation. Kinds of abduction and induction. In F. Stadler (Ed.), *Induction and deduction in the sciences*. Proceedings of the ESF-workshop induction and deduction in the sciences, Vienna, July, 2002, pp. 25–51. Dordrecht: Kluwer Academic Publishers.
Kuipers, T. (2007). Empirical and conceptual idealization and concretization. The case of truth approximation. In J. Brzezinski, et al. (Eds.), *The courage of doing philosophy* (Essays presented to Leszek Nowak, pp. 75–101). Amsterdam: Rodopi.
Ladyman, J. (2002). *Understanding philosophy of science*. London: Routledge.
Ladyman, J. (2007). Ontological, epistemological, and methodological positions. In T. Kuipers (Ed.), *General philosophy of science: Focal issues* (Handbook of the philosophy of science, Vol. 1, pp 303–376). Amsterdam: Elsevier.
McAllister, J. (1996). *Beauty and revolution in science*. Ithaca: Cornell University Press.
Niiniluoto, I. (1987). *Truthlikeness* (Synthese library, Vol. 185). Dordrecht: Reidel.
Niiniluoto, I. (1999). *Critical scientific realism*. Oxford: Oxford University Press.
Niiniluoto, I. (2007). Evaluation of theories. In T. Kuipers (Ed.), *General philosophy of science: Focal issues* (Handbook of the philosophy of science, Vol. 1, pp. 175–217). Amsterdam: Elsevier.
Panofsky, W., & Phillips, M. (1962). *Classical electricity and magnetism*. London: Addison-Wesley.
Psillos, S. (1999). *Scientific realism: How science tracks truth*. London: Routledge.
Sober, E. (2002). *Philosophy of biology*. Oxford: Oxford University Press.
van Fraassen, B. (1980). *The scientific image*. Oxford: Clarendon Press.
van Fraassen, B. (1989). *Laws and symmetry*. Oxford: Clarendon Press.
Weinberg, S. (1993). *Dreams of a final theory*. London: Vintage.
Zajonc, R. (1968). *Attitudinal effects of mere exposure*. Monograph supplement 9 of The Journal of Personality and Social Psychology.
Zajonc, R. (1994). Evidence for nonconscious emotions. In P. Ekman & R. J. Davidson (Eds.), *The nature of emotion* (pp. 293–297). Oxford: Oxford University Press.
Zajonc, R. (2001). Mere exposure: A gateway to the subliminal. *Current Directions in Psychological Science, 10*(6), 224–228.

Acknowledgements

First of all I like to thank all those who have commented on drafts of one or more papers that underlie most of the chapters of this book; the chapter numbers are indicated. Atocha Aliseda (10), David Atkinson (4, 8, 9, 13 16), Sander Bais (9), Johan van Benthem (10), Alexander van den Bosch (10), Gustavo Cevolani (2, 4, 14, 15), Luis-Alberto Cordero-Lecca (16), Richard Dawid (16), Gordana Dodig-Crnkovic (8), Joop Doorman (9), Igor Douven (11, 13), Job van Eck (9), Roberto Festa (2, 4, 9, 14, 15), Adam Grobler (11), Dale Jacquette (13), Lars-Göran Johansson (16), Hidde de Jong (10), James Joyce (16), Barteld Kooi (16), Jeffrey Koperski (9), Erik Krabbe (4, 9), Nico Krijn (8), Ladislav Kvasz (11), Jan- Alber van Laar (9), Anne Ruth Mackor (9), Jacek Malinowski (13), James McAllister (9), Joke Meheus (11), David Miller (9), Ulianov Montano (16), Fred Muller (4, 8), Ilkka Niiniluoto (6, 11, 13, 16), Robert Nola (16), Jeanne Peijnenburg (9, 13, 16), Isabelle Peschard (16), Dick Povel (9), Wlodek Rabinowicz (14), Henk de Regt (9), Andrés Rivadulla (8), Jan-Willem Romeijn (8, 13,16), Gerhard Schurz (6, 14, 16), Ann Simpson (14), Elliott Sober (9, 13), Wolfgang Spohn (11), Jan Sprenger (2), Allard Tamminga (16), Rein Vos (8), André de Vries (9) en Sjoerd Zwart (16). Besides them, I like to thank all anonymous referees that provided as a rule also very useful comments on the papers and one on the whole book manuscript.

Second, several papers have been presented at various conferences and other occasions, usually leading to a number of clarifying questions and useful suggestions. To begin with, the presentations in the local research group Promotion Club Cognitive Patterns (PCCP) in Groningen provided frequently an ideal point of departure for rethinking matters. I like to specify a selection of the other occasions in relation to the relevant chapter. The titles of the invited presentations or submitted papers may have been a bit different.

Chap. 2. Nomic Truth Approximation by Empirical Progress Revisited

Conference *The Progress of Science*, *5th Sydney – Tilburg conference*, Tilburg, 25–27 April, 2012; Guestlecture University of Düsseldorf, April 23–26, 2013;

T. A. F. Kuipers, *Nomic Truth Approximation Revisited*, Synthese Library 399,
https://doi.org/10.1007/978-3-319-98388-2

Conference *Philosophy of Science in a Forest* (PSF), NVWF, Leusden, May 23–25, 2013.

Chap. 3. Perspectives and guidelines for theories of (nomic) truth approximation by empirical progress

Chap. 4. Models, postulates, and generalized nomic truth approximation

PCCP, Groningen, October 3, 2013; EIPE, Rotterdam, November 25, 2013; CamPos Seminar meeting, Cambridge, March 12, 2014; Department of Philosophy, University of Bayreuth, June 24, 2014; CNRS, Paris, December 5, 2014.

Chap. 5. Quantitative nomic truth approximation

Chap. 6. Refined nomic truth approximation

Chap. 7. Stratified nomic truth approximation

Under the title "Concretizations of two-sided nomic truth approximation: quantification, refinement, and stratification": VAF-conference, Groningen, January 29–31, 2014; EPSA2015, Düsseldorf, September 23–26, 2015; *Logic and Interactive Rationality* (LIRa-) seminar, Amsterdam, January 21, 2016; Conference *Philosophy of Science in the Forest* (PSF), NvWF, Doorn, May 21, 2016.

Chap. 8. Nomic truth approximation by domain revision

Workshop *Structures and Dynamics of Science*, École Normale Supérieure (ENS), Paris, April 23–24, 2004; Conference *Model-Based Reasoning in Science and Engineering*, Pavia, December 16–18, 2004; Colloquium in Honor of Gerhard Schurz 60th Birthday, Düsseldorf, December 1–3, 2016.

Chap. 9. Beauty, a road to the truth?

Department of Philosophy, University of Triëst, June 3, 1999; Forum voor Europese Cultuur, Amsterdam, April 14, 2000; Department of Philosophy, Catholic University of Lublin, Lublin, June 15, 2000; Congress British Society for the Philosophy of Science (commentary by David Miller), Sheffield, July 7, 2000; 4th International Congress of the Gesellschaft für Analytische Philosophie, Bielefeld, September 26–29, 2000; Centrum voor Epistemologie en Cognitie (co-referaten van James McAllister (Leiden) en Sander Bais (UvAmsterdam), VU-Amsterdam, Frebruari 16, 2001; RUG-voorjaarsdiner, Statenzaal Provinciehuis, Groningen, April 19, 2001; Vakgroep Anatomie en Embryologie, Groningen, Mei 31, 2001; Farewell symposium Kees Doets, Amsterdam, February 13, 2002; NIAS-lecture, NIAS, Wassenaar, October 15, 2009.

Chap. 10. Abduction aiming at empirical progress or even truth approximation

International Congress on *Discovery and Creativity*, Gent, 14–16 May 1998.

Chap. 11. Inference to the best theory, rather than inference to the best explanation

Fourth Dutch-German Workshop on *Nonmonotonic Reasoning*, Amsterdam, March 25–27, 1999; ESF-workshop *Induction and Deduction in the Sciences* (commentary by Adam Grobler), Vienna, July 7–9, 2002; CEPON-colloquium on realism, Erasmus University Rotterdam, January 31, 2003; Guestlecture University of Düsseldorf, February 5, 2003.

Chap. 12. Pragmatic aspects of truth approximation

Wittgenstein conference, Kirchberg am Wechsel, August 10–16, 1997

Chap. 13. Empirical progress and nomic truth approximation by the 'Hypothetico-*Probabilistic* Method'

GAP-workshop *Philosophy and Probability*, Bielefeld, September 20–22, 2003; CEPON-colloquium on scientific methodology, Rotterdam, October 24, 2003; "Degrees of Belief" conference, Konstanz, 22–24 July, 2004; One-day workshop on the occasion of the appearance (end 2005) of 2 Volumes in Debate, Trieste, January 27, 2007; VI-th Foundations of Formal Sciences Conference, May 2–5, ILLC, Amsterdam, May 2–5, 2007; 4e Congres voor Algemene en Speciale Wetenschapsfilosofie, ISVW, Leusden, May 18–20, 2007.

Chap. 14. Basic and refined nomic truth approximation by evidence-guided belief set revision

Journée d'études: Revision des croyances et decision, Sorbonne IV, Paris, December 7, 2001; EPSA2007, Madrid, November 14–17, 2007; International Workshop *Probability, Confirmation, and Verisimilitude*, September 12, Trieste, 2009.

Chap. 15. Dovetailing belief base revision with truth approximation

CIRSFIS-workshop Science, Truth, and Ontology on the SIFA-conference in Padua, September 23–25, 2010; *Logic, Reasoning & Rationality* congress, Gent, September 20–22, 2010; LMPS14, Nancy, July 19–26, 2011; International Conference Philosophy of Science in Europe – *European Philosophy of Science and the Viennese Heritage*, on the occasion if the 20th anniversary of the Institute Vienna Circle, December 5–7, 2011; Guestlecture University of München, January 25, 2012; Guestlecture University of Düsseldorf, April 23–26, 2013.

Chap. 16. Comparative realism as the best response to antirealism

LMPS13, Beijing, August 9–15, 2007; NVWF-ANTW symposium, Amsterdam, January 12, 2008; Farewell lecture, University of Groningen, March 1, 2010; Guestlecture University of Düsseldorf, April 23–26, 2013.

Finally, I like to thank the Netherlands Institute of Advanced Study (NIAS) in Wassenaar (since 2016 in Amsterdam) for providing several short periods of paradisiacal conditions for research and writing.

March 2018
Bloemendaal, NL

Curriculum Vitae

Theo A. F. Kuipers (1947) studied mathematics and philosophy in Eindhoven and Amsterdam. He got his PhD in 1978 in Groningen, based on *Studies in Inductive Probability and Rational Expectation* (Synthese Library 123). From 1988 till 2010 he was full professor of philosophy of science at the University of Groningen. A synthesis of his work on confirmation, empirical progress, and truth approximation, entitled *From Instrumentalism to Constructive Realism* appeared in 2000 as Volume 287 in the Synthese Library of Kluwer AP. A twin synthesis of his work on the structure of theories, research programs, explanation, reduction, and computational discovery and evaluation, entitled *Structures in Science*, appeared in 2001 as Volume 301 in the Synthese Library.

He was for 2 years (1982/3, 1996/7) a regular fellow of the Netherlands Institute of Advanced Study (NIAS) in Wassenaar (since 2016 in Amsterdam). Several short stays followed. In fall 2009 he was Guest-of-the-Rector.

In December 2005 there appeared two volumes of Essays in Debate with Theo Kuipers, with in total 34 essays related to the books of 2000 and 2001, respectively. Both volumes start with a synopsis of the corresponding book and each essay is followed by a reply of Kuipers:

- Roberto Festa, Atocha Aliseda and Jeanne Peijnenburg (eds.), *Confirmation,Empirical Progress, and Truth Approximation, Essays in Debate with Theo Kuipers*. Vol. 1 (Poznan Studies in the Philosophy of the Sciences and the Humanities, vol. 83). Amsterdam/New York: Rodopi, 2005.
- Roberto Festa, Atocha Aliseda and Jeanne Peijnenburg (eds.), *Cognitive Structures in Scientific Inquiry, Essays in Debate with Theo Kuipers*. Vol. 2 (Poznan Studies in the Philosophy of the Sciences and the Humanities, vol. 84). Amsterdam/New York: Rodopi, 2005.

In 2007 appeared *General Philosophy of Science: Focal Issues, Handbook of the Philosophy of Science,* Vol 1, edited by Theo Kuipers. Series editors Dov Gabbay, Paul Thagard and John Woods. Amsterdam: Elsevier.

T. A. F. Kuipers, *Nomic Truth Approximation Revisited*, Synthese Library 399,
https://doi.org/10.1007/978-3-319-98388-2

In 2009 he became Ridder in de Orde van de Nederlandse Leeuw (Knight in the Order of the Dutch Lion) and in 2010 he became a member of the Koninklijke Hollandse Maatschappij der Wetenschappen (Royal Holland Society of Sciences and Humanities). He retired in 2010 with the farewell lecture "Waarheidsbenadering in theorie en praktijk" ("Truth approximation in theory and practice").

Email: T.A.F.Kuipers@rug.nl
Webpage: http://www.rug.nl/staff/t.a.f.kuipers

Author Index

T. A. F. Kuipers, *Nomic Truth Approximation Revisited*, Synthese Library 399,
https://doi.org/10.1007/978-3-319-98388-2

Subject Index

T. A. F. Kuipers, *Nomic Truth Approximation Revisited*, Synthese Library 399,
https://doi.org/10.1007/978-3-319-98388-2

Printed by Printforce, the Netherlands